Good or bad? New or old? The rich connotations of the word 'biotechnology' reflect a history that surprisingly stretches back more than seventy years. To some, the concept describes the evolving crafts of industrial production using microorganisms. To others, biotechnology is a product of recombinant DNA techniques only recently developed by molecular biologists. It has been seen simply as a means of wealth production and as a new kind of technology – sometimes as distinctively benevolent and, at other times, as particularly dangerous.

Robert Bud shows how the hopes and fears for the combination of biology with engineering have been an integral part of the history of the twentieth century, including the Great Depression of the 1930s, the two world wars, and the more recent anxieties over genetics and entrepreneurial industry. The problems and opportunities of agricultural surpluses provide an enduring theme. Skillfully, the author relates biotechnology's origins in the chemistry and microbiology of the nineteenth century. Personalities with influential roles in its subsequent development – the future first president of Israel, Chaim Weizmann; a pioneer of industrialized agriculture and Hungarian pig farmer, Karl Ereky; the British biologist and town planner Patrick Geddes; his friend the writer Lewis Mumford; the Nobel Prize–winning American geneticist Joshua Lederberg; the sceptical campaigner Jeremy Rifkin; among many others – are discussed. Analysis of the changing roles and hopes for biotechnology in government and society takes the book to the end of the 1980s, when recombinant DNA techniques had become the dominant driving force behind what today we think of as biotechnology.

This first history of biotechnology provides a readable and challenging account for anyone interested in the development of this key component of modern industry, not just for biologists, chemists, engineers, and historians of science and technology.

The uses of life

The uses of life
A history of biotechnology

ROBERT BUD
The Science Museum, London

CAMBRIDGE
UNIVERSITY PRESS

Published by the Press Syndicate of the University of Cambridge
The Pitt Building, Trumpington Street, Cambridge CB2 1RP
40 West 20th Street, New York, NY 10011-4211, USA
10 Stamford Road, Oakleigh, Melbourne 3166, Australia

First published 1993
First paperback edition 1994

Printed in the United States of America

Library of Congress Cataloging-in-Publication Data

Bud, Robert.
The uses of life : a history of biotechnology / Robert Bud.
p. cm.
Includes bibliographical references and index.
ISBN 0-521-38240-8
1. Biotechnology – History. I. Title.
TP248.18.B83 1993
660'.6'09 – dc20 92-19513
CIP

A catalog record for this book is available from the British Library.

ISBN 0-521-38240-8 hardback
ISBN 0-521-47699-2 paperback

**Transferred to
Digital Reprinting 1999**

**Printed in the
United States of America**

For Alexander
In memory of Konrad and Martha Bud

Contents

Illustrations

Foreword

There are too many books about biotechnology, so any addition to that number must justify its existence by an introductory apologia. This one has the unusual merit of being a history book – a work of scholarship – rather than a textbook, a national plan, or a research proposal – these last being the commoner species of the genus *Liber biotechnologicus*.

Biotechnology has a long history, and due reference to the brewers of Babylon or ancient Egypt is all but obligatory in many a conference presentation. Robert Bud has invested the time and effort, in libraries and in conversations with the living, young and old, to go beyond the token references: to demonstrate the historical depth, the cultural, linguistic and multidisciplinary breadth; the diversity and complexity, but yet the cladistic continuity, of biotechnology's long history of punctuated equilibria, from zymotechnology to the human genome. Those punctuation marks, the pauses and the periods of turbulent evolution, are here presented in fresh detail, brought together on a single canvas. The key individuals, discoveries, transfers, re-discoveries and adaptations, are recorded in a coherent whole.

Life (like language) is older than recorded history; and organisms carry within themselves not only the blueprints or software for their posterity, but the messages and memories of their ancestry. The 'uses of life' are therefore a natural topic for a historian; and this is a history of understanding and application, of discoveries and technologies, and above all, of ideas – and of the words developed to convey these ideas.

The author is rightly attentive to the use of words: as linguists and geneticists interact in anthropology, we see that etymology may become a branch of biology. Since the dawn of culture, *Homo sapiens* has been shifting his involuntary allegiance from Darwin to Lamarck, transmitting across generations, centuries, even a few millennia, tools, technologies and words, developed and honed in the hard school of experience. Since the Renaissance, or the Enlightenment, over five centuries or especially the last two or three, we have seen an acceleration of those breakthroughs in understanding and application which we call 'science and technology'. The cross-fertilization from physics and chemistry stimulated particularly sharp acceleration in the twentieth century, and the coming of the global electronic village in recent decades has ensured biotechnology's worldwide diffusion.

In this exhaustive history of the progress of biotechnology, mainly over the past century, Robert Bud has achieved more than a coherent record of deeply and multiply interconnected events. He has captured the excitement and the Angst, the deep anxieties and disquiet, provoked by the juxtaposition of a seemingly omnipotent and uncontrollable science and technology with the familiar, value-laden sacraments of birth, reproduction, and death.

There is something slightly incestuous about contributing to this introspection and history-writing from the field of ongoing policy battles; but one has to respect the courage of Britain's Science Museum for its determination to demonstrate that history is not something that ended a generation ago. It is a continuum, and we are all parts of it.

The ongoing history of biotechnology now embraces more than science and technology; more than enterprise and public policy: it includes a growing element of introspection, of which indeed this book of history is both record and symptom. Self-consciously introspective, our continued involvement with the 'uses of life' is accompanied by profounder questions of ethics and philosophy.

Self-consciousness has been with us since Adam; molecular biology, molecular genetics, force us to new depths of self-awareness, to new perceptions of the nature and origins of man. These unsought insights, casual offspring of curiosity-driven research, are disquieting in their implications, subversive of established custom, belief and authority. Public anxieties and the public authorities which mediate them do not (in most countries) stop the research, but they are directing a growing proportion of the funds from the excitement of the pursuit of research, to deliberation, forecasting and assessment about where its pursuit may lead us.

Yet the growth of knowledge is cumulative; the global knowledge mill keeps turning, and even if we do not try to observe and understand its operations, we shall nonetheless be obliged or condemned to live with its consequences. With intelligent introspection, such as this book offers, we may better ensure that those consequences remain benign.

M.F. Cantley
Head of Unit
Concertation Unit for Biotechnology in Europe (CUBE)
Directorate General for Science, Research and Development
Commission of the European Communities

Acknowledgements

During this book's long gestation, it has often seemed that biotechnology was moving faster than the writing of its history. Any success in catching up has been primarily due to the enormous support provided by a network of people and institutions around the world. This is not primarily an 'oral history'; however, I am indebted to those who took the trouble to provide me with their view of such recent events as are explored here.

I am therefore particularly appreciative of the advice given by Dr Klaus Buchholz, Dr Mark Cantley, Mr Gradon Carter, Mr Keith Copeland, Professor E.M. Crook, Dr Edgar DaSilva, Dr G.A. Dummett, Professor Derek Ellwood, Mr Charles Evans, Professor Robert Finn, Professor Elmer Gaden, Professor Walter Gilbert, Dr Zsolt Harsanyi, Professor Carl-Gören Hedén, Dr Leo Hepner, Professor O.B. Jørgensen, Professor Joshua Lederberg, Mr Gregory Ljunberg, Professor Everett Mendelsohn, Dr Dreux de Nettancourt, Mr Keith Norris, Mr N.W. Pirie, Professor John Postgate, Dr J.R. Ravetz, Dr Peter Rogers, Mr Ken Sargeant, Mr Nelson Schneider, Dr Gerald Solomons, Professor Myron Tribus, Dr Salomon Wald, and Professor Lord Zuckerman.

For the accumulation and hospitable provision of the printed materials on which this book is based, I am indebted to the work of many archives and libraries. A primary debt is owed to the magnificent Science Museum Library and to the British Library. In addition to the assistance of many other academic libraries, I do owe a special obligation to those holding private collections and giving me access, CUBEDOC in Brussels, IVA in Stockholm, the Institute of Brewing, London, and the Siebel Institute of Technology, Chicago. General Sir Christopher Hartley graciously gave permission to examine the papers of his father, Sir Harold Hartley, at Churchill College, Cambridge, and the late Lady Chain gave access to the papers of Sir Ernst Chain at the Wellcome Institute, London. In the history of such a young subject, many of the key papers are not yet in any collections, and I am therefore particularly indebted to those individuals who provided access to documents in their possession: Professor D. Behrens, Professor H. Dellweg, Professor E. Fiechter, Professor R. Finn, Mr N.W. Pirie, Professor John Postgate, Dr Brian Richards, Professor Allen Rosenstein, Mr W. Siebel, and Professor George Sines. With material drawn from many nations, I am obliged to those who have helped me with translation, including Mrs Patricia Crampton, who translated from Swedish, and Ms Judit Brody, who penetrated Hungarian on my behalf.

Drafts of the entire manuscript were most carefully read by Dr Peter Morris, Dr J.R. Ravetz, and Dr Tony Travis, and for their merciless criticism and helpful improvements I am really very grateful. Early drafts of individual sections were read, criticized, and improved by Dr Bernadette Bensaude, Dr Klaus Buchholz, Dr Mark Cantley, Mr Gradon Carter, Dr Edgar DaSilva, Dr Alastair Duncan, Professor John Durant, Professor Derek Ellwood, Professor Robert Finn, Professor Carl-Gören Hedén, Professor Koki Horikoshi, Professor Robert Kohler, Dr Robert Olby, Professor John Postgate, Dr Peter Rogers, Mr Ken Sargeant, Mrs Margaret Sharp, Dr Gerald Solomons, Dr Myron Tribus, and Dr Paul Weindling. For their efforts I am indebted, and for any errors that remain only I can be responsible.

Some of the material presented here has been tested at a variety of conferences, and to the audiences I am thankful for their suggestions. This book also draws on material I have previously published. I must thank the British Society for the History of Science for permission to reprint material which previously appeared in the *British Journal for the History of Science,* Dr Rod Greenshields, Editor of *The Genetic Engineer & Biotechnologist,* for permission to reprint material that appeared in *International Industrial Biotechnology,* and Sage Publications for permission to reprint material that appeared in *Social Studies of Science.* Referees of those journals also helped me correct lapses in the argument. For permission to reproduce manuscript material quoted in the text, I am obliged to Dr Norman Carey, Professor Elmer Gaden, Professor John Postgate, Dr Brian Richards, and to MIT Libraries, G.D. Searle & Co., and Shell Research Ltd.

The Science Museum has proved a most sustaining environment in providing the resources of time, the financial support of travel and other expenses, its peerless library, and supportive colleagues who have given expert advice and facilitated the complex arrangements occasionally required. The moral support of Dr Derek Robinson and

Dr Tom Wright over the years made this book possible. For their assistance, I should like to thank also Miss Marjorie Castle, Miss Sarah Marshall, Miss Suzanne Tagg, and Mr Peter Tajasque.

Finally, authors too guiltily know the amount they exploit the goodwill of those around them. I gratefully acknowledge the cheerful tolerance of my wife, Lisa, and my son, Alexander, who bore the brunt of this work for so many years, and hope that they will think it was worthwhile.

Introduction

What other single word is itself the subject of worldwide polling? In the Netherlands, 1,700 adult consumers were telephoned during 1988 and asked about their knowledge of 'biotechnology'. More than half the respondents were familiar with the term, even if a third of those had no clear idea of its meaning.[1] Such findings have been repeated around the globe. Even among experts, despite a wide agreement on its importance, there has been little accord on the meaning of biotechnology. To some, the concept describes the evolving crafts of industrial production using microorganisms. To others, it is the product of the recombinant DNA techniques recently perfected by curious geneticists. Opinion is divided as to whether it is simply a means of wealth production or whether it threatens the roots of our ethical system. Such differences have been dismissed as the results of cynicism, immaturity, or plain disagreement, but they have been a key characteristic of a subject whose identity has been suspended between craft and science. Curiously, although this is one of the handful of technologies – together with computing and materials – whose rapid advance is widely seen to be potentially revolutionary, there is no established historical literature to provide a context. When the past is mentioned, ancient processes – baking and brewing – have come first to mind, and between the hypermodern and the ancient, there has been little space for history.

Although often described as new, the term 'biotechnology' itself was coined as long ago as the year of the Russian revolution, 1917. Today, the best-known definition is perhaps that spelled out by the Organization for Economic Co-operation and Development (OECD): 'Biotechnology is the application of scientific and engineering principles to the processing of materials by biological agents to provide goods and services'. This may be all-encompassing, yet despite many attempts at refinement, all such short expressions seem inadequate. Ideas and interests have not been painlessly replaced; rather they have been overlaid one upon another, and the word has acquired many connotations over the past century. Whereas some may have been reworked and, in places, others largely forgotten or consciously rejected, in such a cosmopolitan and worldwide science, they have not been lost. Through historical study, those still-active, if scarcely recognized, associations may be recovered and shared. Thus, not only should a historical perspective revive a neglected aspect of our heritage, it may also make better sense of the various connotations the subject carries in different nations, from Germany to Japan, and for different social sectors: laypeople, microbiologists, theologians, engineers, accountants, and politicians. It may help us answer the question of how apparently disparate concepts have come to be related under the single label of 'biotechnology'.

The emphasis, here, is on the past hundred years, despite the ancient lineage. An Egyptian stela dating from 2300 B.C. does show stages in the brewing process that today we recognize as using microorganisms, but it would be wrong to grant equal antiquity to the concept of 'biotechnology' itself. Those Egyptian craftsmen thought of their work in a way quite different from the modern technologist.[2] Our term 'biology' was invented no earlier than the end of the eighteenth century. Courses in the science, seen as a whole and separated from medicine, were first given from about 1860, and this was the time that hopes for 'biologically'-based technologies began to emerge.[3] In the twentieth century, with growing bonds between biology and engineering, the word 'biotechnology' has repeatedly connoted aspirations towards a new industrial revolution, based on the science of biology, just as earlier industrial revolutions had been based on physics and chemistry.

Although now most often associated with the development of remarkable drugs, historically, biotechnology has been principally associated with food, confronting such issues as the challenge of agricultural surpluses and the apparent opposite, dread of malnutrition and famine. Its detailed development has been so interdependent with its interpretations of its meaning within society as a whole (and even fantasy about it) that the development of the two have to be taken together. Dreams of a new technology have been based on the exuberance over the progressive power of science, and on distaste for the dehumanizing effects of technology based on coal and iron, particularly in war, whether symbolized by the tanks of Cambrai or the B52. Moreover, in its very structure, the word uniting 'bio' and 'technology' reflects persistent concerns of society right through the twentieth century. 'Bio' suggests natural; it connotes all those living things whose lives, it often seems, would be better but for the human species. By contrast 'technology' evokes human control over nature. The combination of the two has often seemed deeply disturbing, even monstrous, as amalgams of people and machines have been described. The menacing word 'cyborgs' is used in the language of science fiction.[4] In the history of biotechnology, therefore, lies the story of twentieth-century wrestling with the concept of life. There are obvious parallels with the history of atomic power, recounted by Spencer Weart, in which, over roughly the same span, the public has wrestled with concepts of the power of the atom, seen as both a potential blessing and a threat.[5]

For biology and engineering as professions, guardians respectively of 'bio' and 'technology', biotechnology has acted as a common frontier. Typically, border objects tend to have shifting meanings, depending on which neighbour is dominant. With such a powerful word at stake, there have been enduring debates over ownership of the 'true' meaning of the term. Though its constituent traditions may appear to have had little in common, we can come to understand the processes by which they interrelated through the variety of the uses of the term, with its evolving coverage and connotations. It is startling to see, in bureaucratic tangles, apparently scholastic disputes over definition far from their more likely source in discipline diplomacy. The U.S. Office of Technology Assessment, despairing of finding a meaning of the word common across the federal government, recently printed each agency's own definition as an introduction to policy descriptions.[6]

Of course, interpretations have altered in the interplay between science, individuals, government, and industry. This process has been worldwide, which is one reason why the periodic enthusiasm has so often been overlooked by historians focusing on individual places. Here we examine developments in the United States and Japan and across Europe: in Hungary, Czechoslovakia, Germany, Denmark, Sweden, France, and Britain. Many of the key players have lived in the smaller and agricultural countries that do not normally figure quite so prominently in the history of technology. Even in the United States, the story, until recently, did not particularly involve the great East Coast universities, but rather state universities and obscure consultancies in Chicago. Though the development of the concept of biotechnology has been part of the great dramas of the twentieth century, it has, generally, been in the wings.

Recently, however, the use of visions has become very practical and material. In the early 1980s when prophesies of genetic engineering were articulated and employed to win resources, the seers foretold a world in which wealth would relate to the ability to manipulate the new science. After numerous reports extolled its importance and power, every major industrial power invested heavily. For instance, from 1980, the European Commission's Programme of Forecasting and Assessment in Science and Technology (FAST) sponsored research into the 'Biosociety' whose impact would follow the more immediate 'Information Society'. The inhouse team that sponsored the report then became the nucleus for the commission's permanent Concertation Unit on Biotechnology in Europe (CUBE), and prophetic vision was harnessed to bureaucratic authority. The funding chariots were greatly accelerated by research findings, but their momentum cannot simply be attributed to the influence of molecular biology alone. The first of the government-inspired reports, *'Biotechnologie'*, published by the German Ministry of Research and Technology in 1974, hardly mentioned that science.

Negotiations over the control of the term 'biotechnology' have indicated the balance of key intellectual, social, and bureaucratic forces effecting the subject more widely. Furthermore, those interconnecting and evolving networks of scholars and policy makers which brought together disparate meanings themselves played a key role in sustaining and changing the significances of biotechnology. Earlier in the century, the links between members were informal and personal. Even when, later, they became bureaucratic and public, networks were maintained by passionate individuals such as Patrick Geddes, Carl-Göran Hedén, Joshua Lederberg, and Mark Cantley, whose key places in biotechnology's development this book highlights.

Thus, biotechnology's history has interwoven the development of a wide range of novel techniques, particularly in microbiology, with prophecies of their power and of its dangers. The consequent connections between practice, philosophy, and public perception are at the core of this book. Such a definition of the general problem raises a host of questions: What was the relationship between brewing technology and biotechnology? How were early formulations of biotechnology connected to modern conceptions? Why is there still uncertainty in our definition? In answer, three aspects of the relationship between biology and engineering are highlighted: first, the various crystallizations of industrial microbiology, particularly in relation to chemistry; second, conflicting conceptions of a technology of human development; and third, the interrelations between two such apparently disparate traditions.

Even in these terms, the treatment cannot be comprehensive, and it certainly could not, at the same time, be the history of industrial microbiology or of the application of recombinant DNA techniques, however necessary these are. Key aspects which are well known, such as the work of Pasteur, the discovery of penicillin, and the discovery of DNA's structure, are cited rather than explored in detail. Less familiar aspects concerning the engineering context of biology, such as the industrial roots of Weizmann's work, and such developments as continuous fermentation are discussed in depth to illustrate more general themes.

The emergence of biotechnology from zymotechnology in research institutes, consultancies, and industry in Germany, Denmark, the United States, and Britain is the theme of the first two chapters: the first dealing with long-running trends in the nineteenth century, the second with a critical constellation of developments around World War I. Biotechnology's other roots (principally Austrian and British) in the philosophies of a vitalistic technology are traced in Chapter 3. The integration of these two traditions and their translation into engineering conceptions of a biological technology within enduring institutional forms in the United States and Sweden is explored in Chapter 4. In the wake of World War II, one aspect of the technologies these encompassed, industrial microbiology, seemed to promise immense growth, and Chapter 5 thus picks up the issues introduced at the book's beginning. The new techniques appeared to be peculiarly environmentally benign and resource efficient and, therefore, specially appropriate to developing countries, as explored in Chapter 6. The apparent potential of such new products as single-cell protein and the maturing of older industries, during the 1960s and 1970s, moved biotechnology towards the centre of public policy in Germany, Japan, the United States, and Britain, as we see in Chapter 7. Against that background, Chapter 8 explores 1960s and 1970s hopes for the new molecular biology, the attempt to counter the perceived threats of genetic engineering by its technological potential and its translation into biotechnology. Chapter 9 closes this examination of biotechnology in the twentieth century. It focuses on 1980s reinterpretations in the light of a dispersal throughout diverse industries, and the still persistent cultural identity reinforced by debate between promoters, opponents, and the regulators called on to decree what biotechnology now really was.

Too easily, such an account, with its familiar words, could appear a march towards the present. If it does, it might sometimes appear to have been a tragic tale of gigantic hopes deferred. We are, however, constantly reminded of the strangeness of the past, even when it is so recent that it seems to have been just yesterday.

1

The origins of zymotechnology

> Not unlike the agricultural experimental stations which also flourish in the United States, foreign governments do richly subsidize similar establishments, the object of which is the furtherance of various agricultural industries, among which those relating to fermentation are considered the most important, by scientific investigation and research.
>
> (*The Zymotechnic Institute*, 1891, p. 1)[1]

Introduction

In addition to the familiar four elements of the ancient world, earth, air, fire and water, the Greek philosopher Aristotle postulated a fifth from which the heavenly spheres were made. Later, that 'quintessence' came to be sought by the alchemists, and the fourteenth-century scholar John of Rupescissa, learning of the powers of the recently introduced distilled alcohol, believed it to have been found.[2] Thus, even in the Middle Ages, the manufacture of alcohol evoked the combination of cosmic theory with practical skill. Five hundred years later, the crafts of alcohol production would be integrated with science to drive the development of modern biotechnology. That process began with the formation of the dynamic subject called, in the nineteenth century, 'zymotechnology'.

Derived from the Greek root 'zymē', meaning leaven, zymotechnology could connote all types of industrial fermentation and not just brewing, its dominant interpretation. Because of the new ability to control a wide range of applications, from the curing of leather to the manufacture of citric acid, it was widely commended as a future superstar of the economic stage by 1900. 'In this entire field a new era has now commenced' proclaimed a pioneer, the enthusiastic Dane Emil Christian Hansen.[3] Berlin's Institut für Gärungsgewerbe came to boast a magnificent building erected in 1909 with enormous government and industrial funding. Dire warnings about overoptimism have an eerily reminiscent ring for us. Even Hansen cautioned against the dream of a new alchemy promising wealth and the indefinite prolongation of human life.

The meaning of zymotechnology would be incorporated within biotechnology, and nurturing its descendent, its institutions would provide a continuity as important as the close intellectual lineage. The Institut für Gärungsgewerbe was, for instance, a key player in bringing biotechnology to Germany in the 1960s. Zymotechnology therefore constitutes a vital stage in bridging the gap between biotechnology's ancient heritage and its modernist associations. The references to alchemical dreams and uncertain boundaries suggest, in addition, a rather different relationship. Many of the forces that shaped zymotechnology would also be important later. The subject came to sustain small consultancies which carried out special research projects meeting the needs of a variety of fermentation industries. As well, therefore, as an ancestor, zymotechnology was an important precedent for biotechnology.

The central role of alcohol production in their history has been recognized by biotechnologists pointing to the importance of Louis Pasteur who proved the microbial origins of fermentation. To his biographers and, perhaps, even to his French colleagues, Pasteur had, personally, established the science of microbes, microbiology, as the unique lens through which fermentation and its industrial applications were to be viewed.[4] Such claims are too much for the historian Bruno Latour, who has suggested that one man could no more be the unique source of a discipline than a general's genius the sole cause of victory in battle.[5] Many of the hygienic measures, for instance, that are associated with the great man, it is argued by Latour, have an older history. Biotechnology, too, is more than just the descendent of the applications of a single 'pure' science. A variety of skills and forms of knowledge were seen to contribute to the control of fermentation. Moreover, outside France, a Frenchman was less able to insist upon respect. In Germany, particularly, chemistry maintained a significant claim on the fermentation processes.

Even Pasteur was trained in chemistry, one of the largest of nineteenth-century sciences, and its lessons and skills could be combined in various ways with his claims for the special properties of whole, if microscopic, organisms.[6] So, against the model of an all-encompassing discipline of microbiology were other disciplinary formulations such as bacteriology, immunology, technical mycology, and biochemistry. Uncertainty in the relationship between chemistry and biological sciences was not unique to this front. It was manifest along a whole range of medical and physiological issues and, indeed, is not resolved to this day. Besides, whether the emphasis be on chemistry or microbiology, using achievements in basic science as benchmarks of progress in such industries as brewing overemphasizes the 'science-push' aspect of change. One ought to take into account too the demand-pull from manufacturing experienced in the new research institutes. Based on the practical application of all relevant sciences and skills, including chemistry, microbiology, and engineering, 'zymotechnology' described the cluster of skills and knowledge which responded to the pressures of supply and demand.

The chemical roots of zymotechnology

Zymotechnology as an ensemble of several disciplines and skills was a phenomenon characteristic of the last century.[7] It was also clearly descended from German chemistry of the Age of Enlightenment, the eighteenth century. Whether one looks to institutional roots, intellectual content, or the aspirations with which zymotechnology came to be attached, chemistry's promise to both explain and control provided a reliable precedent for the new dispensation. Strangely, though, while a generation of sophisticated historical study has elucidated the relationship of chemistry with biology and microbiology, scholars have looked more at the consequences for medicine and physiology than to the industrial skills of zymotechnology, and the small consultancies have, hitherto, been overlooked. So the question remains: How was an eighteenth-century vision exploited in a nineteenth-century compromise between chemistry and biology to sustain new forms of industrial research?

Despite its antique appearance, even the root '*Zymotechnia*' is not ancient.[8] It was coined by the German father of the discipline of chemistry, the Prussian court physician Georg Ernst Stahl (1659–1734) in his book *Zymotechnia Fundamentalis* published in 1697, at the very end of the century of the 'scientific revolution'. Condemned by the error of his burning principle, phlogiston, Stahl was once consigned to the prehistoric side of the later chemical revolution of the eighteenth century.[9] Since then, he has been, partially, rehabilitated; for Stahl identified the nature of chemistry in its union of empirically based scientific analysis and practical application. In Germany, in particular, he came to be seen as the father of the modern discipline. The medical chemists of his age, associated with bizarre and ineffective elixirs, Stahl denounced as charlatans and quacks. Unlike them, he rigourously separated the study of inert chemicals from the complexity of vital living matter.

Stahl argued that the study of practical fermentation, *Zymotechnia,* would be the basis of Germany's key industry of *Gärungskunst* – the art of brewing. Fermentation, to us the very expression of life, was amenable to scientific analysis, because it was associated with movement and because it occurred (like putrefaction) when the spirit was missing from living bodies. As an interpretation, this has its own roots and can be traced back to the Englishman Thomas Willis earlier in the seventeenth century. However, *Zymotechnia Fundamentalis* can be seen not just as an explanation, but also as the first bid for intellectual authority over related commercial processes and thus as the original work of biotechnology. Even if such claims are somewhat arbitrary, this one locates the subject's foundation in the period when its defining characteristics, the process of fermentation and the potential of science–technology relationships, were themselves being delineated. Stahl expressed, for the first time, the abiding hope that an understanding of the scientific basis of fermentation could be harnessed to the improvement of commerce.

Like his slightly older contemporary, Isaac Newton, Stahl was fortunate that others interpreted and popularized his frequently arcane ideas, expressed in a mixture of German, Latin, and Greek. While physician to the Prussian king, he found a protégé in the court apothecary, Caspar Neumann (1683–1737), a widely travelled and popular teacher.[10] Neumann had worked in England, the center of advanced brewing, and brought together Stahl's interest in the principle of zymotechnics and practical teaching of brewing. Moreover, the Latin original of *Zymotechnia Fundamentalis* was translated into German in 1734, making it accessible to practical men. So, zymotechnics survived the obscurantism of Stahl, even becoming an internationally recognized concept. The leading intellectuals of France, including Venel and Macquer, were much taken with Stahl's theories, and in 1762 the word '*zymotechnie*' entered the exclusive dictionary of the Académie Française.

From zymotechnology to organic chemistry

Incessantly, the Stahlians emphasized the distinction between their science and the charlatanism of the alchemists who had pursued the philosophers' stone that would bring the millennium – wealth through gold, and health through the instant cure of any disease. Nonetheless, chemistry did take the mantle of a science that would provide wealth and health. Its potential was a cliché of the nineteenth century. In her novel *Frankenstein*, published in 1817, Mary Shelley, a member of England's radical intellectual elite, managed to capture, but also slightly to extend, conventional claims of the day, expressed by such a well-known chemist as Humphry Davy at London's fashionable Royal Institution. In the mouth of Frankenstein's teacher, Professor Walden, she put the following words expressing a recurrent admiration for the results of chemistry:[11]

> 'The ancient teachers of this science,' said he, 'promised impossibilities, and performed nothing. The modern masters promise very little; they know metals cannot be transmuted, and that the elixir of life is a chimera. But these philosophers, whose hands seem only made to dabble in dirt, and their eyes to pore over the microscope or crucible, have indeed performed miracles. They penetrate into the recesses of nature, and show she works in her hiding places. They ascend into the heavens: they have discovered how the blood circulates, and the nature of the air we breathe. They have acquired new almost unlimited powers; they can command the thunders of heaven, mimic the earthquake, and even mock the invisible world with its own shadows.'

Despite the passage of time, this was still a legitimate fulfillment of the Stahlian dream. However, the pupil then defied the categorical chasm between the quick and the dead. Through chemistry, Professor Frankenstein creates a monster which pursues him, and finally, with his life, he pays the price for hubris. Beyond science fiction, medical men, particularly in France, were increasingly exploring the physiology and the chemistry of the sick. Historians have identified a profound shift during the early nineteenth century, as medicine, led by chemistry, moved from a concern with treating patients to treating diseases.[12] The idea that living matter was endowed with a divine spark was being gradually eroded, beginning with doubts as to whether a special aspect of life extended to products of the body. The discipline of chemistry therefore became a dominant means of linking the interpretation of living processes with their technological exploitation.

The 1828 synthesis of urea by Friedrich Wöhler is a useful benchmark in the erosion of the distinction between natural and chemical products, if not in itself the moment at which it disappeared. Wöhler, professor of chemistry at the liberal eighteenth-century University of Göttingen, showed how urea, a natural product extracted normally from the urine of snakes, could be made artificially.[13] The implications were explored by his friend, the brilliant chemist, teacher, and publicist Justus Liebig.[14] Sharing both Stahl's general faith in practical applications – he even called chemistry the 'philosopher's stone' – and his sense that it could not explain the potential of the living spirit, Liebig gradually went beyond his predecessor of the previous century in his conception of the science's scope.[15] At first even he argued for the existence of a vital force, but through the 1850s this concept came to be progressively secularized into just another force, rather like gravity, and, after that, it was silently dropped.

Having abandoned debates over pure chemistry in the late 1830s, Liebig came to be progressively more identified with the chemistry of agriculture and physiology. A radical reduction of physiological processes, be they fertilization or digestion, to the transformation of chemicals was his hallmark. Thus, though his main interest was in chemical inputs and outputs, with Stahl, he asserted that fermentation was the result of the movement of atoms and was transmitted from unstable bodies such as yeast, to susceptible victims such as sugars.[16] Liebig therefore affirmed the claim for control of fermentation by the chemist – deploying both explanatory theories and the key skills of temperature measurement, hydrometry, and analysis. Nonetheless, as a Europe-wide academic discipline, chemistry did not follow these later interests of Liebig; instead it was increasingly dominated by his earlier love, organic chemistry.

Organic chemists sought to replace the laborious and expensive extraction of natural products by laboratory synthesis. Extracted from the bark of the cinchona tree to fight malaria and other fevers, quinine was an outstanding challenge. Liebig's pupils August Hofmann and James Muspratt boasted in the introduction to an 1845 paper: 'If a chemist should succeed in transforming in an easy manner naphthaline into quinine we should justly revere him as one of the noblest benefactors of the human race'.[17] In fact, quinine itself withstood the chemists' charge until Woodward's final victory a century later, but dyestuffs submitted much faster. Attempting to synthesize quinine in 1856, Hofmann's pupil William Perkin obtained the first synthetic organic dye, Mauveine. This entirely new product was followed in May 1869 when Perkin himself, and, in Germany, Caro, Graebe, and Liebermann, simultaneously found commercial syntheses for a major natural dye, alizarin, the red colouring matter found in the madder root. Adolf Baeyer, who became Liebig's successor at Munich, created a research school based on the study of natural products. The chemistry which disowned Stahl but sanctified Liebig became one of the greatest success stories of the nineteenth century. The rapid growth of the great German chemical firms seemed to vindicate the vision of a science of the artificial.

The power obtained by the discipline of organic chemistry was celebrated as awesome. At the same time it sometimes seemed that, compared to the complexity of nature, the achievement was puny. Towards the end of the nineteenth century, as understanding of natural products grew, so did respect. With an intense interest and admiration for the subtlety of the chemical processes that could be found in living creatures, Emil Fischer, Baeyer's greatest pupil, explored the carbohydrates and proteins. His attitudes can be compared to those of the pupil who defers to the work of a master.[18] The 'laboratory of the living organism' (in the words of English biochemist William Foster) was increasingly looked up to, and the limits of human chemistry correspondingly appreciated, just at the moment when its potential was also most dramatically demonstrated. By 1925, the German chemical combine IG Farben, complaining about the training of traditional organic chemists, pointed out that synthetic organic chemistry was exhausted and that scholars were turning towards research on natural products.[19] Even a chemist who had been one of the first to develop a replacement for quinine as a reducer of fever, Wilhelm Koenig, is credited with a song whose third verse freely translated runs:

> Synthetic the Coffee, Synthetic the Wine
> Synthetic the Milk and the Butter Gloss Shine
> On top of it all, even beer is not pure
> Natural nutrition, you won't find to be sure.
> Then do let the devil take it for free
> That wretched synthetic-made Chemistry.[20]

The biological alternative

In its claims on the living world, turn-of-the-century chemistry confronted newly dynamic disciplines of biology and physiology with the result that zymotechnology transcended its origins as an aspect of applied chemistry. Instead, it became a loosely understood umbrella concept for theories and techniques relating to fermentation, drawn pragmatically from chemistry, microbiology, and engineering. Why did such a synthetic zymotechnology become a key interface between science and industry rather than the more coherent applied biochemistry? Perhaps surprisingly, the historian Robert Kohler has found that, despite the vitality of German academic life, biochemistry found it hard to find a purchase in German universities. He puts this down, in part, to the vigour of organic chemistry, and he suggests too that the programme of Liebig was incorporated, on the medical side, by the vigorously imperialist discipline of physiology. Just as chemistry was claiming more and more of living processes to its own, so physiology, driven by powerful medical interests, was becoming more reductionist. Typically, the interface was accommodated by institutes of physiology, such as that of Du Bois-Reymond founded in 1877 in Berlin, incorporating an integral chemical section as well as groups dealing with higher levels of biological organization.[21] The tendency in physiology to reduce life to chemistry and physics was complemented by other perspectives that emphasized the ecology of living organisms. This emphasis on the living being was expressed by the science of 'biology'. As was appropriate for an admirer of Stahl, Treviranus, one of the first to use the term, introduced his 1801 book *Biologie oder Philosophie der lebenden Natur für Naturforscher und Aertzte* with the words 'Only the exploitation and not the occupation makes the value of a treasure'.[22] He continued by emphasizing the value of biology when seen in conjunction with pharmacy and economics.

Nonetheless, it had only been with the gradual opening up of chairs and positions at the end of the century and the identification of its own generic problems that the science of biology acquired an existence separate from subsidiary disciplines such as zoology and botany. The issues that characterized the new discipline at the beginning of the twentieth century may not have appeared to relate to practice. Meetings were dominated by the new science of the cell, evolution, and the mechanism–vitalism debate.[23] Yet even these issues raised the problems of biological education and discipline building which, it will be argued, would underlie much of the development of biotechnology in the twentieth century. Repeatedly, the example of chemistry would highlight the attraction of claims for practical significance.

Individual biological sciences such as botany had already established practical uses. Increasingly, an ecological view of plants provided a characteristically botanical perspective complementing their reduction to chemical constituents. The distinguished German professor Julius Wiesner, author of *Die Rohstoffe des Pflanzenreiches* (*Raw Materials of the Plant World*) argued that an understanding of botany would provide an opportunity to find exotic materials in the tropics and improve agricultural production at home. In the tradition of Treviranus, Wiesner began his work with a quote from Helmholtz: 'To know alone is not the task of man on this earth. Knowledge must also be used in life.'[24] In a manner characteristic of his time, he argued that a technological approach to agriculturally produced raw materials could and should be a discipline represented in the technical highschools. Just as chemical technology was the go-between chemistry and its industries, so his *'Rohstofflehre'* should be the mediator between natural history and technology.[25] The boundary between the divine essentials of life and the secular province of technology was perforated by microbiology in a way that would be similar to Wiesner's *Rohstofflehre.* It too made strong claims for utility, making particular claims for fermentation. Since Stahl, chemistry had offered a vision of ever-expanding technical possibilities, an overarching theory, and a series of valuable techniques for monitoring the progress of particular fermentations. Yet compared to the detailed development of organic chemistry, fermentation chemistry was a marginal, empirical, and messy province. By contrast, Pasteur created a discipline that put fermentation at the centre of its concerns and, with its emphasis on microscopy, explored the process as well as inputs and outputs.[26] In 1857, Pasteur demonstrated that lactic acid fermentation was the result of the action of live microbes. Through the next decade, he debated increasingly hotly with Liebig who insisted on the purely chemical origins of fermentation phenomena. Pasteur constructed a new scientific discipline based on his understanding of microbes, 'microb-iology'. Where chemistry was characterized by the balance, the new science had its own central instrument, the microscope. His contribution thereby defined such subjects as brewing, wine making, and hygiene as central areas of application of this discipline. An institute named after the hero was established in Paris in 1887 and other Pasteur institutes followed later elsewhere. Pasteur's own contributions to the salvation of the French silk and wine industries have passed into legend. Such achievements prompted even an Englishman, Thomas Huxley, to assert that Pasteur's discoveries alone had reaped sufficient wealth for France to repay her enormous reparations to Germany after the 1870 war.[27]

In France, Pasteur's influence dominated his institutes, but elsewhere microbiology grew to live with other disciplines in institutions that emerged in response to local pressures. Thus, in Germany, the lessons of microbiology were assimilated in practical contexts, such as that of a medically oriented bacteriology whose father, Robert Koch, seemed to promise a solution to problems of such diseases as cholera and tuberculosis. Of course, the import of his breakthroughs, such as the growing of bacteria on solid media and the consequent ability to develop a clear morphological taxonomy, had far more than medical impact. They were applied in industries ranging from brewing to leather tanning, but important as they were, they were not sufficient in industry. Successful brewing, for instance, required a careful chemical control of the malt as well as botanical characterization of the yeast. The profession and discipline of chemistry were therefore still powerful and influential. Just as physiology acted as an umbrella within medicine, in industry, zymotechnology would act as a tent for chemistry to cohabit with the rather younger upstarts of various biological disciplines – microbiology, bacteriology, mycology, and botany.

Agriculture

Like human cohabitation, disciplinary mating often has an institutional base. In the case of zymotechnology, this grew out of the rich industrial and research system sustained by the search for rapid development in agriculture. Brewing was a major use of barley and hops, as well as a source of refreshment to the workers; and the integrated development of the vast network of associated industries was often seen as a key to agriculture's prosperity. Whereas the origins of modern industrial research have often been seen in the laboratories of the dyestuffs and electrical companies, agricultural research has a singularly strong claim to long-term significance; and it certainly provided the model for the advance of brewing.

The production and processing of agricultural produce became a fiercely competitive industry, worldwide, during the nineteenth century. In Europe, new sources of wheat supply – the United States and Russia, as well as Australia, South America, and other regions – brought to the consumer by ever more effective transport systems, threatened established domestic markets. The volume of U.S. wheat exports increased nine times between the early 1860s and the end of the century. The traditionally greater quantities exported by Russia were overtaken, but even they doubled in the last four decades of the century. At the same time, a rapidly growing population, particularly in urban areas, meant greater dependence than ever on a secure food supply. The great famines of the late 1840s, in which 700,000 Irish people died, had had searing effects across the whole of Europe, with revolutions spreading across the Continent in 1848. So, in the West, there were both the memories of scalding shortages and the need to compete with low-price imports.[28]

A hundred years later, there is still no consensus on how to resolve the tension between accommodating surpluses and ensuring against shortages. In the nineteenth century, intensive farming, economizing on land and labour inputs, was looked to, particularly in the Netherlands, Denmark, and the German states. Of these, Denmark took the most conscious measures to move 'up-market'.[29] Rather than producing wheat in competition with low-cost imports, the Danes moved to intensive livestock production and milk technology. They pioneered agricultural cooperatives as technology was coming to make centralization attractive. The development of the centrifugal milk separator in the 1880s made the extraction of cream from milk a factory operation, rather than a time- and space-consuming farmyard settling process. The factories yielded butter but also equivalent amounts of unpalatable skimmed milk. New uses included the casein glues, and the plastic known in Britain as 'Erinoid' was also produced from casein. While it could be stored as condensed milk, the bulk of this milk lake was dried. It was found to be a good pig feed, and all over Europe and the United States pig rearing flourished. The number of pigs in Denmark doubled between the mid-1880s and 1900 and doubled again to over 2 million by 1910. Denmark was not alone. In Germany, between 1873 and 1913, the number increased from 7 to 26 million, almost the same as the entire country's human population.[30] Scientific husbandry was further developed in the United States, where their feeding and fattening was brought to a high art. In 1903, 3 billion pounds of skimmed milk were produced there.[31]

Such productivity increases were often attributed to the growth of agricultural teaching and research institutions particularly in Germany. There, in the aftermath of the Napoleonic wars, the rulers of the still firmly agricultural states had been threatened by the emergence of major cities such as London and Paris, prone to civil disturbances, and by Britain's newly productive industry. Academic leaders argued that they should play their part in helping a development of the society that would avoid the division into a declining agricultural sector and an impoverished industrial proletariate.

The first German agricultural college was established in the year of Prussia's humiliation at the hands of the French, 1806, by a practical agriculturalist much impressed by British achievements, A.E. Thaer.[32] His academy at Möglin was combined with the newly established University of Berlin in 1810. Largely inspired by Thaer's example, twenty agricultural colleges were founded in German-speaking lands between 1818 and 1858. The development of trades traditionally closely associated with agricultural development would also enable organic change. In the Vienna Polytechnic, the founder and first director of the polytechnic school, J.J. Prechtl, argued for the importance of '*Gewerbeindustrie*'.[33] When in 1816 the polytechnic founded a chair of '*spezielle technische Chemie*', it offered the specialized courses of brewing, leather preparation, soap making, and dyeing and printing. So far the emphasis was on teaching. However, all over Europe research followed.

In France, Boussingault founded his private agricultural research

laboratory at Bechelbronn in 1835, and Lawes and Gilbert e
their laboratory at Rothamsted near London in 1842. These
inspired, in Germany, the foundation of a research laboratory, at Möckern, in 1851.[34] Two years later, another followed in Chemnitz. By 1863, there were seventeen and, by 1877, fifty-nine so-called research stations in Germany. In the United States, the Morrill Act of 1863 and the Hatch Act of 1887 sustained the development of land grant colleges and associated agricultural research stations. Great universities such as MIT, Cornell, and Wisconsin owe their origins to these initiatives.[35]

Nonetheless, the relationship between chemists who favoured laboratory experiments and farmers working in fields was not one of science discovers, tradesmen apply. Though it would be wrong to argue the uselessness of the chemists' services (indeed, the increasing productivity of German soil is a testament to them), farmers were repeatedly disappointed. Artificial fertilizers often proved inadequate replacements for farmyard manure despite the teaching of Liebig.[36] The agricultural research stations therefore had to evolve their approaches in the constant interplay of theory and practice.[37]

Brewing

If brewing was an agricultural industry, as northwest Europe became increasingly wealthy and urbanized, so beer's centralized production and worldwide distribution became big business. The British produced 4.4, and the Germans 3.9, billion litres of beer in 1883.[38] The value of the brewing industry's production in Germany, as Mikuláš Teich has pointed out, was the same as that of the steel industry even at the end of the nineteenth century.[39] Its economic importance was further increased by the tax revenue raised for needy governments. Nor was brewing a simple technology to master reliably; despite its ancient lineage and traditionalist associations, it was a complex process where consistency was hard to achieve, and erratic or even poisonous results were all too likely. With increasing scale, even in the eighteenth century, had come the search for consistency and the ability to brew whatever the weather. England was the largest centre, and firms such as Truman Hanbury and Buxton had assets of hundreds of thousands of pounds by 1800.[40] Brewing chemists offered their employers means of monitoring the process of their craft: specific gravity, weights of the various products, temperatures, and water purity were commonly measured even in the 1830s. Many textbooks instructed brewers on the keeping of so-called Gyle books in which the process of brewing particular beers was meticulously monitored and recorded for reproduction. Expertise was well developed particularly in England, the world's brewing centre.[41]

In London, several of the leading chemists of the mid-nineteenth century were employed by brewers whose names still survive. The foundation of the world's first major chemical association, the Chemical Society of London, in 1841 was driven by the energies of Robert Warington, a chemist who had recently resigned from the post of chemist to Truman's. Warington was a so-called practical chemist whose skill lay in his experimental technique rather than in his theoretical depth. The society he strove to create would bring together the different strands of chemistry so that the academics, consultants, and manufacturers could help each other.[42]

The trade schools of the German-speaking lands had already come to provide a continuity between the craft transmission of the eighteenth century and academic training. As early as 1816, Bohemian brewers petitioned the director of the Ständische Ingenieurschule in Prague to found a school which would train experts for their industry.[43] They were even ready to cover the expenses of setting up the course. There was already a one-year chemistry course at the school. It was extended to two years with supplementary practical courses. Thus, it happened that the former practical pharmacist J.J. Steinmann (1799–1833) offered a course on fermentation chemistry from 1818, possibly the first in the world. In 1824–25, Steinmann was succeeded by the young Carl Balling who was to be a full professor for thirty-three years. In the career of Balling, we see the transition from the craft to a new form which was academic but still practical. Here was a man who saw himself employing chemical principles certainly, but also serving an entire industry. Closely influenced by Liebig, Balling looked to the history of his subject for inspiration, and Stahl offered him an ancestry, identity, and historic location. He espoused the term *'Zymotechnik'*, entitling the fourth volume of his classic text on brewing, *Bericht über die Fortschritte der zymotechnische Wissenschaften und Gewerbe* (*Account of the Progress of the Zymotechnic Sciences and Arts*).

Just as agricultural centres moved from a purely educational role to a greater influence on research, so this process could be observed in the special case of brewing. The first great centre, established in 1872, was at the school at Weihenstephan near Munich where brewing had been taught for more than twenty years. Its formation was driven by the entrepreneurial pharmaceutical chemist Carl Lintner, who within three years of arriving at Weihenstephan in 1863 had founded his journal, *Bayerische Bierbrauer*.[44] In the first volume, Lintner ran a

Medallion commemorating Max Delbrück, scientific director of Berlin's Institut für Gärungsgewerbe. Courtesy Institut für Gärungsgewerbe und Biotechnologie.

series of historical articles about the life of Balling, as the first of the founders of zymotechnics 'for future cultural historians'.[45] He ran occasional columns on current developments in zymotechnics. The word expressed what he felt was the distinctive academic character of the subject.

Weihenstephan served the south of Germany; in the north, the Berlin research centre of the Verein der Spiritus Fabrikanten (Federation of Spirit Producers) was founded in 1874, shortly after the Weihenstephan station, by the eminent chemist Maercker and his assistant Max Delbrück (1850–1919) – uncle of the pioneer molecular biologist.[46] It began as a two-person laboratory in the Royal Academy of Arts. From there it developed in close connection with agricultural research – first in the Agricultural Institute, moving in 1882 to the newly erected building of the Royal Agricultural High School. In 1883, the Research and Teaching Institute for brewing was established in Berlin. Later, this and subsequent institutions dealing with other agricultural fermentation issues – yeast, potatoes, and spirits – were each linked with Berlin Technical University. The Institut für Gärungsgewerbe obtained its title in 1897. By 1909, buildings worth 4 million marks (£200,000) had been erected (2.5 million coming from the Prussian government and 1.5 million from industry), and were occupied by eighty scientists, including chemists, biologists, nutrition physiologists, engineers, and domestic economists, consuming a budget of 2.5 million marks. A research maltery, yeast breeding centre, and liquor- and vinegar-brewing pilot plants all contributed to the Institute for Fermentation Technology.[47]

Attempting to wrest leadership from the Germans, Pasteur himself had published a classic study of brewing, *Etudes sur la Bière,* in 1876, and under his successor Duclaux, the Pasteur Institute established a research brewery. In Britain, Birmingham University's British School of Malting and Brewing was opened in 1899. Ten years later, a Scottish school was opened at the Andersonian Institution, now Strathclyde University. Copenhagen's Carlsberg Institute was built up in the same period, unusually as a private research station attached to a single brewery. It had two departments dealing, respectively, with chemistry and physiology. Both attracted distinguished scientists – the first head of the chemical department was Johann Kjeldahl, most famous for his method of nitrogen determination. On the physiological side, the second head, Emil Christian Hansen, made a breakthrough in 1883 which, for the brewing industry, was as important as Pasteur's. Showing that beer could be ruined by wild yeasts, rather than only through bacterial infection, he provided a key to successful brewing.

Hansen's achievement was soon recognized to be of great practical importance. It was significant too in terms of the structure of knowledge. A new biologically inspired discipline had come to the centre, entailing not just taxonomic knowledge, but also an understanding of natural selection, as the change in yeast strains was explained. The requirement for such expertise could most easily be met by large centres such as the Carlsberg Institute or in Berlin, where the addition of a new discipline could be accommodated. Even there, the disciplinary departments of industrially oriented research institutes had to be justified as providing the scientific bases of a new composite technology, not just as expanding a single science, such as chemistry or microbiology.

Hansen's programme was developed in Delbrück's Berlin laboratory which explored the difficulties of scaling up laboratory single-cell cultures to the different conditions of the manufacturing plant. Technologists there developed 'natural pure culture' techniques in which the acidity and other environmental factors were controlled to encourage the growth of just the favoured organism. Thus, whereas Pasteur was famous for emphasizing the purely scientific nature of his work, supposedly quipping, 'There are no applied sciences... there are only... the applications of science', Delbrück took pains to emphasize the role of practice as well as science in finding the truth.[48] He highlighted the work of '*Technologen*' such as Balling in identifying the role of yeast before Pasteur and proudly announced to the German brewing congress in 1884: 'With the sword of science and the armour of practice German beer will encircle the world'.[49] During the early twentieth century, Delbrück and his colleague Carl Schrohe went further to establish the identity of their technology. They explored its history through a series of great books and founded a society for the history and bibliography of brewing in 1913.[50]

The debate over the role of practice seems to have been reflected in the use of the word 'zymotechnology' rather than the French scientific term '*microbiologie*'. Here was a technology closely allied with chemistry and its expansive ambitions, and with a pragmatically assembled range of techniques. It represented too the ambitions of institutions to grow beyond dependence on a single industry. The emerging ability to control the progress and products of fermentation stretched further than beer. The production of lactic acid was developed in the Berlin laboratory, and to this day the organism's name, *Lactobacillus Delbrücki,* commemorates Delbrück. An analogous conflict on medical territory was being fought between Pasteur and Delbrück's Berlin colleague, the medically oriented Robert Koch, who preferred the term 'bacteriology' to Pasteur's 'microbiology'.[51] Not that Delbrück ever competed so publicly with Pasteur as did Koch. Indeed, when, in 1896, Buchner identified the transforming role of a chemical, zymase, in fermentation, Delbrück was slow to be convinced.[52]

Zymotechnics as trademark

The problems of the need for both chemical and biological paradigms were greater for the smaller institute which could not afford a discipline structure. The word 'zymotechnology' to describe a practically convenient, if intellectually heterogenous, topic proved particularly useful here. It was given visibility by Hansen's Danish colleague, the well-known consultant brewing technologist Alfred Jørgensen (1848–1925), who in 1885 founded the Danish journal *Zymotechnisk Tidende.*[53]

Jørgensen was an internationally famous entrepreneur and discipline builder. He established a trade in yeast cultures, sending dried

Alfred Jørgensen, popularizer of the term 'zymotechnics'. Courtesy Alfred Jørgensen Laboratory.

pure yeasts worldwide, and devised equipment for the commercial cultivation of pure yeasts, used, by the 1890s, in 160 breweries.[54] Beyond that, his expertise was in microscopy, and his interests extended throughout the range of industries in which this tool, so marginal to the chemist but central to the zymotechnologist, could be useful: milk technology, vinegar, and wines among them. He established a private laboratory for microscopical and microbiological research work and then a teaching laboratory. Attempting to describe his area of skill, Jørgensen called his institute a 'Fermentology Laboratory'. Respected throughout Europe, and advertized by Hansen, by 1903 he claimed to have had eight hundred students. A list of the fourteen students currently attending, published in November 1907, included not just Scandinavians but also a Spaniard and a Czech.[55] Jørgensen's international fame spread after 1889 when he published the first edition of his *Micro-organisms and Fermentation* which was to pass through many editions until 1948.

Hansen published many articles in Jørgensen's journal and employed the umbrella concept of 'zymotechnics' there. In his book, whose 1896 English translation was entitled *Practical Studies in Fermentation,* the close association between the word and the programme is to be seen in Hansen's concluding triumphant sentence:

> Nowadays it must be clear to every zymotechnologist who has made himself familiar with the results of recent investigation, that wherever fermentation organisms are made use of, the aim must be the same, namely to give up the old traditional method which depended upon mere chance. In this entire field a new era has now commenced.[56]

Thus, zymotechnology acquired a specific connotation in Copenhagen, and though Denmark is a small country, the work of its brewing pioneers quickly won international renown. Even Chicago, Illinois, 4,000 miles away, was culturally close. A favoured pupil of Hansen, the Danish consultant Max Henius, established a consultancy there with fellow Marburg student, the American Robert Wahl, in 1885. The two emulated the institute of Jørgensen, even taking the equivalent to the Danish name, Institute of Fermentology, and pioneered pure yeast practices in the United States.[57] Not only the name of the Dane's institute, but also his journal's title was an inspiration, and a year after the launch of *Zymotechnisk Tidende,* another Chicago consultant (incidentally born the same year as Jørgensen), John Ewald Siebel, started a journal entitled the *Zymotechnic Magazine: Zeitschrift für Gährungsgewerbe and Food and Beverage Critic* (sic). Born in Germany, Siebel trained there as chemist and then moved to the United States, working first in the sugar industry and then branching out into brewing in Chicago. With the city's traditional German ties, the brewers there called each other 'Herr Kollege' and even spoke German to one another. So, despite the geographic distance between America's Midwest and central Europe, developments in the two were closely linked.

In 1872, Siebel founded an analytical laboratory from which developed an experiment station and, in 1884, a brewing school entitled the Zymotechnic College.[58] Like Jørgensen whose own institute had been founded a year earlier, Siebel used the word 'zymotechnic' as a hallmark to connote a field wider than brewing. The *Zymotechnic Magazine* was a temporary retitling of his journal the *American Chemical Review* founded in 1880 to address 'Manufacturers of Sugar, Starch, Vinegar, Pickles, Soap and other Animal and Vegetable Products'. The range of interests was well described by the detailed subheads for topics relating to animal products, oils, fats, and so on: Valuation of Milk, Half Measures, Extracting Glycerine from Fatty Substances, Grease-proof paper, Preparing food for animals; Soap; Apparatus for Distilling Oil; Tanning with Sulpholeic acid; Manufacturing Glue; To Preserve Eggs; A Cotton Seed Oil Scheme; Milk Analysis. Although his school would begin with instruction in brewing, its potential scope was far wider: 'education in the arts based on fermentation processes and changes such as malting, brewing, distilling, yeast, vinegar and wine making. . . . All those articles such as meat, glue, cheese, are subject to fermentative influence and their preservation may well be counted among the fermentative arts'.[59] Siebel was quite conscious of the parallel between his role as a scientific brewer confronting a practical industry and that of the agricultural research stations then springing up.[60] From 1890, his consultancy was called the Zymotechnic Station.

In 1901, with the support of his more business-minded son, Fred, John Siebel incorporated his renamed Zymotechnic Institute, to address generally all 'Zymotechnic Arts and Industries especially of those for the manufacture of food and drink'. The logic was explained in the business plan he seems to have submitted in support of his application for incorporation. His company was one of a handful around the country servicing the 850 U.S. breweries producing over 10,000 barrels a year. He pointed out that the city's best-known establishment, the Wahl Henius Institute with a staff of twelve, had, by 1898, 294 member companies each paying a $125 annual fee, submitting 8,641 samples for examination. There was also a flourishing school teaching sixty-eight scholars at $300 a time. As a result, the company made a very satisfactory profit of $22,000 a year on a capital of $100,000. The competing house of Schwartz was roughly the same size.[61]

Siebel did complain that 'brewers among all classes of business men were the hardest people as a class to get to comprehend the improvements of modern times', a complaint that will resonate throughout this account. Yet there were so-called modernized brewers abroad to whom he could supply yeast and scientific instruments, as well as special reports. He had already helped brewers in Canada, Mexico,

The Zymotechnic Institute in 1911, showing J.E. Siebel at the centre with E.A. Siebel, later founder of the Bureau of Bio-technology, to his right. Courtesy J.E. Siebel's Sons and Co., Inc.

and Australia. Moreover there were people in other neighbouring industries – cider and vinegar making and distillers – who had already sought his services. In 1901, he took four students, and by the end of the decade he was graduating almost thirty students a year. Under the name Siebel Institute of Technology, the company carried through Prohibition, concentrating on baking technology. Still under family management, it has continued operating to the present day.

Not only the Siebel family company operated under the zymotechnic label. In April 1906, the Chicago brewing industry under the leadership of the Siebel family created a Zymotechnica Association. J.E. Siebel himself gave the first address on 'alcoholic beverages and nutrition', an attack on the prohibitionists. He was shortly after elected the society's first honorary member. Two years later, the society having proved a success, Harvey Wiley, the powerful head of the government's Bureau of Chemistry and founder of the Food and Drug Administration, was also elected an honorary member. Jørgensen from Denmark was also elected to honorary membership, symbolizing the close affiliation between zymotechnology in Chicago and Copenhagen. At the same time, the Zymotechnica Association decided to make a claim for a national society in competition with New York's American Brewing Institute. Hence it anglicized its title to 'Society of Brewing Technology'.

Siebel was widely respected and sufficiently renowned to be the focus of a 1933 *History of Brewing in America.* His Germanic background and context in Chicago enabled him to introduce the European concept that he shared with Jørgensen in Copenhagen. Both sought a pragmatic interpretation of zymotechnics: It entailed an eclectic mix of skills taken from chemistry, microbiology, and engineering, yet it had sufficient scientific roots to expand the area of their competence under the umbrella of zymotechnics away from brewing. Zymotechnology therefore worked rather as biotechnology would at the end of the century, with a useful vagueness that both implied a particular focus to a company's work and allowed it to cross conventional market boundaries. Moving from denominating a mere subset of chemistry as it had a century earlier, it indicated a technological skill that was rooted in a variety of sciences but had a practical character transcending mere applied science.

2

From zymotechnology to biotechnology

> *The author includes within the category of biotechnology all such work by which products are produced from raw materials with the aid of living organisms.* (Original emphasis)
>
> (Karl Ereky, 1919)[1]

Therefore, while the new science of microbiology may, in retrospect, appear to have offered the crucial breakthroughs for fermentation technology, the process of applying such a science was complex and only a part of the more general development of zymotechnology. Still, as a series of brilliant microbiologists and bacteriologists showed the scope of their science, the introduction of biological thinking into technology became more common, if still haphazard. To the early issues of hygiene and alcohol had been added, by the First World War, the manufacture of a variety of chemicals including organic acids such as lactic, citric, and butyric acids, the cultivation of yeast, and sewage disposal. Gradually, the microbiological industries came to be seen as somewhat of a genre offering an alternative to applying conventional chemistry, rather than merely as peripheral variants of brewing. With this new biological awareness came a new term, 'biotechnology'.

In the thinking about the application of microbiology, there was as much variation as in the more familiar case of chemistry. There, one found chemists who felt that industrial processes, with their complex mix of scientific, engineering, artizanal, and commercial considerations, should be studied in themselves. This approach was described as chemical technology. Others believed that here were the applications of the pure science, and that one should recognize the key importance of applied chemistry. If there were engineering skills to be taken into account, they should be the separate and traditional engineering specialisms. A third school, placing a different emphasis, saw a new synthesis of a separate and increasingly sophisticated chemical engineering.[2]

Though a specific engineering form lay in the future, the study of fermentation technologies during the early twentieth century encouraged developments parallel to chemical technology and applied chemistry. Fermentation was increasingly seen as the subject of an applied science, economic microbiology. Even if the professorial adherents of zymotechnics took a more rigidly technological attitude to those organisms (Delbrück began an 1884 lecture with the phrase 'Yeast is a machine'[3]), the technologists had to take an increasing account of the importance of biological perspectives, and as we shall see, zymotechnology evolved into biotechnology.

This evolution of the identity of zymotechnology from an aspect of agricultural technology into an application of science was mirrored by the evolution of the concept of a microbiological centre. As early as 1884, Král at the Technical University of Prague had established a culture collection, and this was followed by a Dutch collection of fungi in 1906.[4] Paul Lindner (1861–1945), pupil of Koch and head of botany at the Institut fur Gärungsgewerbe, suggested to the 1909 International Congress of Applied Chemistry that a biological reference collection be established in Berlin.[5] This suggestion, explicitly aimed at the needs of the world's technical biology communities, was encouraged by the meeting which agreed that the suggestion for further support be put to the Council of the Institute of Brewing. Lindner revived his suggestion after World War I, and citing the ways in which his collection of pure yeasts had proved useful to the chemist Emil Fischer, he compared it to a set of standard chemical reagents.[6] Although it did not become the unique international reference centre, Lindner's collection did become famous and was only dispersed in the chaos of 1945 Berlin.

In the aftermath of World War I, a special British chemical symposium was held in 1919, at which the brewing chemist Chaston Chapman addressed the question of microorganisms in industry. He called for the formation of a national centre for industrial microbiology that would be equivalent to the German Institut für Gärungsgewerbe and a worthy outcome of Lindner's 1909 call for an international centre.[7] It would be a source of teaching in the scientific and technological bases of fermentation industries, would offer advice on microorganisms and have a culture collection, and would carry out research. More than a mere service bureau for the miscellany of industries that employed microorganisms, it would also be a national centre for microbiology. Whereas the Institut für Gärungsgewerbe had been a kind of industrial research laboratory, Chapman was proposing a more academic institution. With this orientation came a definition in terms of a scientific discipline, microbiology, rather than a technology such as brewing or agriculture. In the event, Chapman's plan for a central organization and culture collection was not realized. The old technological pattern was for the moment continued. A new Imperial Bureau of Mycology was established at Kew the following year to advise on fungi of agricultural interest, and the Medical Research Council also created a collection of medical and industrial cultures at the Lister Institute.[8] Nonetheless, Chapman's plea, repeated in a lecture at the Royal Society of Arts, did represent a growing discipline identity for industrial, or as it came to be called, economic microbiology.

The Germans themselves were not so satisfied with their own arrangements. A 1921 critique published in the general scientific journal *Naturwissenschaften* pointed out that, uniquely, bacteriology existed only in an applied science form and was not even considered as the application of a pure science.[9] Philosophically it might be seen as a subset of the science of botany. However, whereas a pure chemist could get a position as the head of a dye factory or a physicist as the head of an electrotechnical factory, one could not imagine a botanist becoming the head of a hygiene institute. The author, Otto Rahn, pointed out that, here, Germany lacked the leadership it had assumed in chemistry. The Institut für Gärungsgewerbe made a poor private comparison to the state-supported Carlsberg Institute. The American Bacteriological Society had devoted a third of its 1921 proceedings to theoretical bacteriology with a lecture on the importance of abstract bacteriology from the head of a dairy institute in Washington. The eminent Dutch microbiologist Beijerinck had also spoken on the need for theoretical bacteriology. Thus, already, the regrettable German situation was changing abroad. Increasingly, the technologies were being seen as applications of basic science.

This gradual shift from brewing to a greater emphasis on science, whether it be microbiology, bacteriology, or biochemistry and the variety of their possible applications, underlay the evolution of zymotechnology into biotechnology. The broadening meaning can be seen in two sites on the German periphery where zymotechnics had been most actively promoted: Chicago and Copenhagen. Though the transitions were probably separate, the pressures and opportunities in the two places were analogous. It is particularly easy to see in the Copenhagen context how the specific interest in zymotechnics was translated into a general interest in biotechnics.

Orla-Jensen

The close ties between fermentation studies and agricultural research seen in Germany were also explored in turn-of-the-century Denmark which was then the world's most advanced agricultural country. Typically, the rector of the Copenhagen teachers' school who converted his institution into a polytechnic was concerned to develop new agricultural industries.[10] In 1907, the polytechnic decided that the associate professor of agricultural chemistry should be made a full professor in a new subject: fermentation physiology and agricultural chemistry. Establishing the academic validity of the new applied science, the university *Yearbook* explained that fermentation physiology was now so weighty that it should be separated from chemistry, and yet taught to all aspiring chemical engineers stepping in the footsteps of Denmark's heroic Emil Christian Hansen.[11] So far it seemed as if the polytechnic was merely transferring from Jørgensen's laboratory the concept of a discrete zymotechnics. Orla Jensen, the professor, had been a pupil of Jørgensen and was an expert on the microorganisms responsible for cheese making, so he was a most appropriate choice for such a transfer. Still famous as a pioneer compiler of the lactic acid organisms, he is regarded as one of the great microbiologists in his own right. Jensen himself had worked at the Pasteur Institute and then for some years had lived in Switzerland where he became head of the central cheese institute. With this breadth of experience, he had a wider perspective and philosophy than merely fermentation physiology. This enabled him to make a major break in 1913 when the polytechnic made a further change: Jensen changed his title to professor of biotechnical chemistry.

The widening from pure zymotechnics was a deliberate act as can be seen in the introduction to his lecture notes of 1916. First, Orla-Jensen (he hyphenated his name in about 1914) defined 'biotechnical chemistry' as concerned with the 'foodstuffs and fermentation industries and as a necessary basis also the physiology of nutrition as well as fermentation physiology'.[12] Underlying these were living processes, which he attempted to define as the metabolism of proteins. However, Orla-Jensen was not the type to allow philosophical problems to impede him on a practical quest. In his 1934 *Lidt anvendt Filosofi*, he would reflect on the nature of applied science. He argued that it could not just be seen as the application of pure science; rather, the relationship was the other way around: 'Botany developed out of the search for plants with healing properties, chemistry out of the search for the philosopher's stone etc.'[13] Orla-Jensen's course of biotechnical chemistry, fulfilling this vision, linked treatments of proteins, enzymes, and cells with the analysis of particular foods such as milk, margerine, and chocolate.

Siebel

Orla-Jensen had developed the concept of biotechnical chemistry out of zymotechnics. In Chicago, with American linguistic inventiveness, the word 'biotechnology' was derived from the same root. The work of Siebel was carried on by four of his sons. Three of them continued their father's institute, and it operates to this day, but one of them, Emil, left, and set up on his own account in 1917.[14] This was the year that prohibition finally passed through Congress, and Emil, who had been working on a 'temperance beer' since 1908, concentrated on providing services to fermenters of the new nonalcoholic soft drinks. With the end of prohibition in 1932, his interests reverted to those of his father, teaching brewers and bakers and providing a consultancy for them.

At first Emil Siebel (1884–1939) established a consultancy under his own name. Then, apparently fairly quickly, his school acquired a title slightly different from that of his father's – rather than Zymotechnic Institute, he called his the Bureau of Bio-Technology.[15] The dilution of the brewing reference, implicit in the term 'zymotechnology', was doubtless wise in the era of Prohibition, and Siebel boasted of his good relations with Federal inspectors.[16] The title 'bureau' might well have been a pompous allusion to the federal government's powerful Bureau of Chemistry whose chief Harvey Wiley had been elected in 1908, with Jørgensen, as one of the first honorary members of Chicago's Zymotechnica Association.

Clearly, Emil Siebel's title was a trademark rather than a discipline. In academe, it had no perceptible impact. On the other hand, it did, possibly, have an influence in English commerce. When the chemical analytical house of Murphy decided to set up a microbiological consultancy in Leeds in 1920, it took the title Bureau of Bio-Technology, just as in Chicago. The English bureau had slightly greater academic ambitions than did its Chicago counterpart and published a bulletin recording the results of consultancies. This was sent to academic libraries, for instance, to the British Museum (Natural History). Nonetheless, the *Brewers Journal* rather sniffily referred to the 'transatlantic' tone of its title.[17] Murphy's consultant, the editor Frederick Mason, published articles which emphasized the significance of the microscope to his findings across industry.[18] Thus, while its origins in the zymotechnics of brewing remained clear, the bureau seems to have worked as much on the microbiological aspects of leather tanning, and it was even reviewed in Italian leather journals, thus introducing the word *'biotecnologia'* into Italian.[19]

Ereky

Nonetheless, the key source of the word 'biotechnology' was not in the United States or even Britain, but in Hungary. Its progenitor was a Hungarian agricultural engineer, Karl Ereky, seeking to transform his country into the equivalent of a Southern Denmark. Just as Denmark served as the agricultural supplier to its industrial neighbours Britain and Germany, so Hungary, with its vast feudal estates, had emerged as the specialized agricultural centre for the Austro-Hungarian empire. Whereas the country had been generally backward, Budapest was a very rapidly growing modern city. Until 1900, its corn mills were the largest in the world. (In that year Minneapolis took the lead.)[20] Its historian John Lukacs interestingly suggests that among the world's great metropolises, only the growth of Chicago could be compared with Budapest's trebling population in the last quarter of the nineteenth century. The countryside beyond the metropolis may have been scorned by the citizens, but amidst its backwardness it did boast several of the most sophisticated intensive farm projects in the world.[21] Hungarian cattle-breeding practices stimulated international interest – before the Great War, several hundred foreign experts visiting the cattle-breeders association included a party of eighteen veterinarians from the United States making a 'local' visit while attending a London conference. The intensive fattening of pigs was a Hungarian specialty. As early as 1894, the Köbánya pig-fattening plant on the rundown outskirts of Budapest handled 622,000 pigs within the year, though that particular centre failed with an epidemic among the animals.

Ereky coined the term *'Biotechnologie'* as part of a campaign intended to supersede the backward peasant. In the years 1917 to 1919, he wrote three testaments of faith. The last was given the German title *Biotechnologie der Fleisch-, Fett- und Milcherzeugung im landwirtschaftlichen Großbetriebe* (Biotechnology of meat, fat and milk production in large-scale agricultural industry).[22] Clearly, Ereky was no abstract intellectual: After the war, he would be the minister of nutrition in the counterrevolutionary Horthy government.[23] Later he would try to pioneer the conversion of leaves to protein and tried to raise British

Karl Ereky, father of the term 'biotechnology'. This photograph appeared in *Budapest közuti vasúti közlekeddésénrek fejlödése, 1865–1922 és a BSzKRT tiz évi müködése, 1923–1933* (Budapest: Budapest Székesfövárosi Közlekedési Részvénytárság Igazgatósága, 1934). Courtesy Library of the Hungarian Parliament.

interest; he wrote fluently in English.[24] It is, however, to his work before the death of the empire that we should now turn.

In 1914, Ereky persuaded two Hungarian banks to support an agricultural enterprise on an industrial scale.[25] There would be a slaughterhouse for a thousand pigs a day and also a fattening farm with space for 50,000 pigs, turning over 100,000 pigs a year. The enterprise was enormous, becoming one of the largest and most profitable meat and fat operations in Europe after the war.[26] Large pig farms were on the increase, as is testified by a burgeoning German literature, but even the largest of the German complexes had 27,000 pigs spread over three sites.[27] The scale of Ereky's plans can be compared with the entire set of 22,000 German pig farms with an area greater than 100 acres, which together held only 880,000 pigs.[28] Even today, the scale of Ereky's endeavour would count it among the very largest of pig operations.[29] His factory-farm covered fifty hectares, somewhat more than a hundred acres. The pig houses, forty-five in number, were 100 metres long, and 36 metres wide. Through the estate ran an 18-kilometre-long narrow-gauge railway. The newly built mainline spur carried annually 5,000 wagon loads of pigs, 8,000 of fodder, and 7,000 of dung.

In the first of his 1917 pieces, Ereky described his motivation.[30] Hungary was still an agricultural country with, in 1900, 31% of its population on the land. Now, he saw the traditional peasant economy as an anachronism in the age of a growing consumer culture of telephones, theatres, and power-based industry. Ereky envisaged its replacement by capitalistic agricultural industry based on science. He had visited and been very impressed by the Danish cooperatives and described their history and achievements in detail, but felt that the individualism of Hungarians would prevent such developments in his own homeland. Instead, the limited company was the key to the creation of industrial living standards on the land. As an economist, Ereky was not especially original. One can see the roots of his thinking in the ideas of German agricultural economists such as Theodor Brinkmann, whose well known text *Die Dänische Landwirtschaft* of 1908 had reported on his own pilgrimage to view the Danish agricultural miracle and the transformation of materials in particular.[31] More generally, the issues of capitalism and its tensions with technology and size, and indeed the catchword 'Americanization', were major themes in post-war Hungary.[32]

Ereky's expression of his vision through the giant pig-fattening station was described in considerable detail, together with expressions of a grand ambition for an industrial revolution in agriculture and the abolition of the peasant class in the second of his 1917 contributions, an article for the German Agricultural Society's transactions.[33] The difference between an industrial and a peasant approach to pig rearing, he reflected, lay not in his use of electrical pumps and automated feeding. It lay instead in the underlying scientific approach which Ereky labelled *'Biotechnologie'*. To him the pig was a machine, converting carefully calculated amounts of input into meat output. Indeed, he described the pig as a *'Biotechnologische Arbeitsmaschine'*.

In the third of his contributions, the book entitled *Biotechnologie*,

Ereky further developed a theme that was to be important and to be reiterated through the twentieth century. This book began with the same sentence as his first. Food shortages had been a crippling problem in the central states during the war. For Ereky, the great chemical industry could come to the aid of the peasant through its endorsement of '*Biotechnologie*'. As the quotation at the head of this chapter shows, by analogy with chemical technology, he chose his new term to indicate the process by which raw materials would be biologically upgraded. Ereky not only introduced this new classification, he also expressed great hopes for it: Just as previous eras had been defined by the use of stone and iron, a new biochemical era would follow if only the will were there.[34]

Certainly the parallels between the concepts of Orla-Jensen and those of Ereky were obvious to others. Unlike the Anglo-Saxon pioneers of biotechnology, Ereky was himself an influential leader in a centre of intellectual life. His ideas were noticed and the coinage of the new word was, for instance, highlighted in the review of Ereky's book that appeared in the general science journal *Naturwissenschaften.*[35] Paul Lindner, who had called for the new culture collection before the war and was now editor of the *Zeitschrift für Technische Biologie,* also reviewed Ereky's book favourably, following his notice with an editorial in which he praised Orla-Jensen's new book on lactic acid bacilli and described it as the quintessence of what '*Biotechnologie*' should be about.[36] Through Lindner's interest, the word had made a decisive shift into the heartlands of intellectual respectability. But its meaning had also been subtly changed; for by associating it with Orla-Jensen's work, Lindner had focussed on microorganisms, which had not been Ereky's intention. Thus, by the late 1920s, *Biotechnologie* had been incorporated into great German encyclopaediae – *Meyers Lexikon* and *Der Große Brockhaus* – each emphasizing (to a different degree) the particular significance of microorganisms in production.[37]

In German, there is a fundamental difference between the words '*Technik*' and '*Technologie*', which does not translate well into English. The latter refers to a branch of scholarship, whereas the former is an activity. The difference was explored by a 1920 article in Lindner's journal *Zeitschrift für Technische Biologie.* The author, Hase, worked in the still new Kaiser Wilhelm Institute for Biology in Berlin.[38] Reflecting on the immense range of applications of biology during the recent war, including the development of pesticides as well as fermentation technologies, he too felt there needed to be a new collective noun for the applications of biology. Pondering the use of the word '*Biotechnologie*', he rejected it on the grounds that so much of what was still done was a craft and therefore should be better described by the word *'Biotechnik'*.

Lindner's redefinition of *'Biotechnologie'* and the enthusiasm of Hase reflected an industrial phenomenon. By World War I, zymotechnology had yielded a panoply of opportunities. Even in the 1890s, Vienna's Professor of Fermentation Physiology and Bacteriology, Franz Lafar, had published a wide-ranging *Technical Mycology: The Utilization of Micro-organisms in the Arts and Manufactures* prefaced by Hansen's exclamation of enthusiasm, but also a warning not to look for immediate financial return.[39] An English chemistry professor, Raphael Meldola, overcame his anxieties about vitalism to publish a formulary of biologically derived compounds in his 1904 *The Chemical Synthesis of Vital Products and the Inter-relations between Organic Compounds.*[40] On an even larger scale, biological methods of sewage degradation were being developed and, during World War I, the activated sludge process was introduced in Britain.[41]

Three manufacturing processes developed during World War I achieved special prominence and helped to suggest a great future for manufacture by fermentation. In Germany, glycerol and food yeast and in Britain acetone became vital parts of the wartime infrastructure. All three were the result of prewar work. At Berlin's Institut für Gärungsgewerbe, the interest in yeast as a source of alcohol was transferred to yeast as a food in its own right. Max Delbrück published a pathbreaking article in 1910 on the uses of yeast as animal food.[42] Having come to see yeast as more than a machine, he proclaimed its versatility as an edible mushroom. Now, one year after Haber's famous demonstration of ammonia synthesis, Delbrück was looking forward to the day when humankind's protein needs would be met by nitrogen fixed directly from the air. It was a dream that would occur again and again in the history of biotechnology. With Germany starving and grain no longer available, Hayduck and Wohl developed means of growing yeast on molasses enriched with ammonia produced through the new synthetic processes. Ordinary brewers' yeast and the specially cultivated yeast together provided 60% of Germany's fodder protein needs during the First World War.[43] Hayduck saw this as the beginning of the ultimate solution to recycling, though he is supposed to have proclaimed, 'Not until man is in a position to convert his evening newspaper so rapidly into sugar that the protein produced therefrom can be consumed the next morning at breakfast will one of the greatest problems of this century have been solved'.[44] The use of yeast to prevent famine in Germany was a dramatic outcome of the new zymotechnology. Others were the replacement of materials hitherto produced chemically from raw materials that were no longer available. Lactic acid became a major project.[45] Another was the use of fermentation glycerol, which is particularly interesting because of its roots in basic biochemical research. In 1911, at Berlin's Kaiser Wilhelm Institute for Experimental Therapy, Carl Alexander Neuberg believed he had found a key stage in the process by which starch was fermented, when he demonstrated the presence of acetaldehyde. Further experiments showed that significant amounts of glycerol, which happened to be a crucial ingredient for explosives, could also be produced. These were taken up in 1915 by Connstein and Ludecke of Berlin's Vereinigte Chemische Werke. They found the optimum conditions for a sugar solution in the presence of a bisulphite solution and enabled Germany to produce more than a thousand tonnes a month of glycerol.[46]

The 'Weizmann' process

The British counterpart of this German enterprise was the acetone–butanol process. Though its development by Weizmann, the future first president of Israel, and its supposed role in engendering the British offer of Palestine to the Jews, has often been told, it is worth emphasizing two often overlooked factors.[47] Weizmann's own process, economically significant as it indeed proved to be, was the second generation of a prewar development in which several of Britain's leading chemists had played a part, together with France's Pasteur Institute. Second, in playing a key role himself, Weizmann was motivated by the wish to develop an industry that would be an outlet for Palestine's agricultural production.

A shortage of rubber in 1907–10 highlighted to the industrialized world the industrial importance of a monopoly of wild rubber held by the Brazilians and the increasingly important role of British plantations in Malaya, commemorated by Somerset Maugham.[48] Forced up by a Brazilian cartel, prices reached 12 shillings and 4 pence at their peak in April 1910 before collapsing to 6 shillings a pound by October. Rubber was a strategically crucial material because it was, literally, the basis of the motor car industry. Its hardened form, vulcanite, was also used throughout society from pipe stems to the lining of chemical vats. The obvious solution to those countries at the mercy of the Brazilians and the British was the development of synthetic rubber. After all, it had been known for over thirty years that the chemical isoprene would spontaneously react with itself to make a rubbery substance. What was needed was a source of isoprene and a way of speeding up the reaction.

And even the British were interested, for they were still smarting from the humiliation of losing the indigo market to the German synthetic product, and indeed the loss of the dye industry in general. The fiftieth anniversary of Perkin's discovery of mauve had been sadly celebrated in 1906 with a symposium on the variety of mistakes responsible for the subsequent German lead, ranging from inadequate provision for chemical research to tariff and patent legislation. On this occasion, a race ensued. By 1909, two groups in Britain and three in Germany were working on similar lines. Accounts in Britain that the feared German company Badische Anilin- und Soda-Fabrik (BASF), whose synthetic indigo had destroyed the Indian plantation industry, was now working on synthetic rubber raised national anxieties to fever pitch. Though it would take several decades to achieve the elasticity of natural rubber, in 1910 it still appeared that a breakthrough was imminent. In a process of consolidation, the British teams came together, with the consulting firm of Strange & Graham contracting the Manchester professor of organic chemistry William Perkin, Jr (son of the discoverer of the first synthetic dye), whose large department included the young Russian immigrant Chaim Weizmann.

The first problem of speeding up polymerization was solved simultaneously in Germany and Britain in 1910. Sodium was found to act as a catalyst. The problem then became one of finding a source of isoprene. A 1912 article, presented by Perkin, but actually written by Strange, carefully described the team's logic.[49] The price of natural rubber could comfortably fall to 2 shillings and 6 pence or 3 shillings a pound: 'It was useless, therefore, to consider any raw material which did not offer at least a possibility of the production of rubber at 1s. or even less. The only substances fulfilling these conditions seem to be wood, starch, or sugar, petroleum and coal.' Perkin then showed that wood was not a reliably cheap and plentiful source. Turpentine was subject to speculation and was not produced on a large enough scale. Coal and oil were abundant and could possibly prove satisfactory, but it was not yet known how to convert them. Starch seemed the obviously attractive feedstock at a penny a pound. This could, for instance, be fermented to lactic acid which could be converted to isoprene, but the conversion was complex. Higher alcohols were then investigated. Fusel oil, a long known fermentation derivative, contains iso-amyl alcohol. From that isoprene was derived, but the world's

The Department of Chemistry at Manchester, showing Perkin (first row, fifth from the left) and Weizmann (first row, second from the right). Courtesy University of Manchester, Department of Chemistry.

production of isoprene was only 3,500 tonnes a year, and the price was rising to £140 a tonne or over a shilling a pound itself.

It was in 1910, when both lactic acid and amyl alcohol seemed possible targets, that Weizmann suggested that the consortium consult Fernbach at the Pasteur Institute for advice on how to produce a reasonably priced starting point. The connection with Fernbach was to be Weizmann's key contribution to the first process. The idea of consulting the Pasteur Institute for help with lactic acid manufacture was easy to sell in 1910, only four years after the identification of *Lactobacillus Delbrücki* in Berlin. Still, the origins of the interest in fermentation of the chemist Weizmann, a pupil of Carl Graebe of alizarin fame, should not be taken for granted. Instead, they can illuminate the fundamentally agricultural nature of zymotechnology, for Weizmann's obsession at the time was the building of a new Palestine. He had visited the country in 1907 and came back feeling that far more investment was required. The Austrian millionaire Kremenetzky, proprietor of the world's largest lamp firm, promised him resources if he could show that agricultural-chemical production could be rendered economical. Weizmann had suggested that 'we start with an oil factory, then add a small lemon-[processing] factory and distillation plant'.[50] During 1909, relations between Kremenetzky and Weizmann worsened as the Zionist leader failed to provide information.

Having first visited the Pasteur Institute to learn bacteriology in 1909, Weizmann returned in 1910, bringing two problems set by Strange, first, of how to utilize the currently wasted hulls of rice produced as a by-product in the manufacture of starch and, second, of how to produce amyl alcohol by fermentation.[51] The rice problem was a pressing concern of Strange's business partner, Reckitt, and seems to have been given equal billing with amyl alcohol. Starch was produced by treating broken rice with soda. However, in the process 20% of the rice was lost in a stream of thousands of gallons of waste per hour.[52] The two problems proved to be interrelated, and by December 1910, progress was being made with converting glutine from the rice waste into lactic acid which in turn could be converted to isoprene.[53]

Subsequent research yielded a bacterium that produced a fusel oil rich in both amyl alcohol and a related chemical, butanol, from starch. Fernbach's process had been carried out under reduced pressure. When, in early 1912, the process was transferred to Strange's Rainham works and operated under industrial conditions, using oak casks and no vacuum distillation, the fermentation proceeded differently. Now, butanol and acetone were the principal products.[54] This result created tremendous excitement. The relevance of butanol to synthetic rubber manufacture was just becoming clear, as Harries in Germany published an article showing that a superior rubber could be produced from butadiene. The conversion of butanol to butadiene was analogous to that from isoamyl alcohol to isoprene on which the Manchester team had already put much work. So, between them, the Pasteur Institute and the Manchester team had generated a new route to synthetic rubber, with a raw material of butanol at £40 a tonne, or 4 pence a pound. Moreover, and even more exciting, the acetone was valuable in its own right as an important, and hitherto imported, ingredient in the manufacture of explosives, while isobutyl acetate was the best solvent known for the increasingly important plastic nitrocellulose.

This work therefore had enormous commercial potential significance, and the scientists, far from being unworldly, exploited the breakthrough to the hilt. Their enthusiasm for profiting from both technology and the public's willingness to believe established a fascinating precedent for the future of biotechnology. Within a couple of months of the discovery of acetone, it was decided, with the encouragement of the eminent chemists Ramsay and Roscoe, to float a company to develop the process for Britain (despite the fact that the key microbiological input was French). Though personally he had done little work on the project, Perkin, the distinguished Manchester professor, presented a dramatic account of its work to the Society of Chemical Industry in May 1912.

Perkin's lecture created intense interest, for here appeared to be a threat to Britain's burgeoning rubber interests. The London chemistry professor Henry Armstrong reflected on the potential of the discovery, and on the rumours of BASF's commitment of a million pounds to research on synthetic rubber, in a long article, published by *The Times Engineering Supplement,* entitled 'The Production of Rubber: With or Against Nature'. In case readers might miss this, the message was emphasized and endorsed by an editorial the same day. Would the BASF investment lead to the same disastrous results for colonial agriculture as synthetic indigo a decade earlier? In a sentence whose sentiment would echo through the twentieth century, Armstrong had pondered, 'We are competing with Nature in many directions at present and it is very desirable to discuss whether in the future it will be either desirable or possible to work so much against her'.[55] The theme was that chemists could make rubber only using other scarce raw materials – petroleum or starch, which would be more useful for other purposes. On the other hand, the rubber tree was itself a most powerful and effective machine for the manufacture of rubber. As he wrote, 'Ethically we shall probably be making a mistake in not availing ourselves to the full of the activity of the plant; but, apart from this, it may well be that, when everything is taken into account, the plant is able far more effectively than man to make rubber from starch'. The conclusion was the importance of careful scientific agriculture. Here, therefore, zymotechnology and agricultural technology were coming to be seen as alternatives, rather than one being an aspect of the other. The editorial had translated this message into a question of the British culture. 'The evidence in the case of the rubber industry cannot be accepted as overwelmingly in favour of stoically following Germany as well as nature: Nature is enough'.[56]

On 29 June, following the expenditure of £7,000 on newspaper advertisements, a prospectus for the new company, Synthetic Products Ltd, was launched. Though at their most ambitious the proponents had hoped that £500,000 would be raised, they were not too disappointed to raise, in the face of determined opposition from the plantation rubber interests, £75,000, a not inconsiderable sum when the net assets of Britain's largest chemical company, Brunner Mond & Co., amounted to £5m. In 1990 terms, it may be roughly equated to £3m or $6m. The prospectus implied that both potatoes and maize had been successfully fermented. This was false, since success had been achieved only with the more expensive potatoes. Nonetheless, the success enabled Perkin later that year to challenge the German Carl Duisberg, at the Eighth International Congress on Applied Chemistry in New York, each claiming to the Americans that his team had been the first to synthesize rubber.[57] Meanwhile, Weizmann who was technically an assistant to Perkin, tried to do a deal directly with Fernbach, was found out, and fired. He then continued experiments on his own account.

From the outset, Weizmann was fundamentally interested in creating an industry, rather than merely solving a scientific problem. On 1 January 1911, when fermentation seemed a plausible, but as yet unproved, solution, he wrote to Strange from Paris that he was hopeful about the fermentation route: 'I should like again and again to impress upon you the necessity of enlarging our field of activity in the sense of manufacturing chemicals and building up a new English industry. I am constantly thinking about it, have plans which we will discuss when we meet'.[58] He did not see the new process as a threat to agriculture; rather Weizmann was hoping for an industry that would transform agricultural produce into valued chemicals. Instead

This Prospectus has been filed with the Registrar of Joint Stock Companies.

The Subscription List will open on Monday, the 1st July, and close on or before Thursday, the 4th July, 1912.

The Synthetic Products Company, Limited.

(Incorporated under the Companies (Consolidation) Act 1908)

(For Manufacturing Acetone and Fusel Oil and for making further experiments in developing Synthetic Rubber).

CAPITAL - - - - - - - £500,000

Divided into 475,000 Cumulative Participating Preferred Shares of £1 each and 25,000 Deferred Ordinary Shares of £1 each.

The whole of the Deferred Ordinary Shares and 25,000 of the Cumulative Participating Preferred Shares are to be allotted to the Vendors as fully paid up.

The Cumulative Participating Preferred Shares (hereinafter called "A" Shares) confer the right to a fixed Cumulative Preferential Dividend at the rate of £6 per cent. per annum, together with two-thirds of the balance of profits referred to in the next paragraph.

The Deferred Ordinary Shares (hereinafter called "B" Shares) confer the right, subject to the payment of the £6 per cent. Dividend above mentioned on the "A" Shares, to a sum equal to the total of the £6 per cent. Dividend paid on such Shares, and any balance of profits available for distribution is to be divided between the holders of the "A" Shares and the "B" Shares respectively, in the proportion of two-thirds to the holders of the "A" Shares and one-third to the holders of the "B" Shares.

On a distribution of assets the "A" Shares will be entitled to receive the amounts paid up thereon. The "B" Shares will next be entitled to receive a sum equal to the total amount so paid to the "A" Shares, and subject thereto the surplus is distributable amongst the holders of the "A" and "B" Shares in the proportion of two-thirds to the "A" Shares and one-third to the "B" Shares.

The "B" Shares, the whole of which will be held by the Vendors, will not, for three years from the date of the Incorporation of the Company, be transferable without the consent of an Extraordinary Resolution passed by the Company in General Meeting.

Issue of 450,000 Cumulative Participating Preferred Shares at par, payable as follows:—

ON APPLICATION ... ... 2s. 6d. per Share. ON ALLOTMENT ... ... 7s. 6d. per Share.

and the balance of 10s. per Share in such instalments, not exceeding 5s. each, and at intervals of not less than three months, as may be determined by the Directors.

DIRECTORS:

SIR WILLIAM RAMSAY, K.C.B., LL.D., 19, Chester Terrace, Regent's Park, London.
PHILIP BEALBY RECKITT, J.P., East Mount, Sutton, Hull, Director of Reckitt & Sons, Limited, Hull and London, Manufacturers of Starch, Blue and Black Lead.
ERNEST JOHN HUMPHERY, M.A., 22, Bloomsbury St., London, Manager of the Platinotype Co.; Director of J. W. Brooke & Co., Limited, Lowestoft.
HENRY ARNOLD AVENEL VAN SOMEREN, 6, Throgmorton Avenue, London, Barrister-at-law.
FRANCIS EDWARD MATTHEWS, Ph.D., Ash Lawn, Blackheath, London, Director of Strange & Graham, Limited.
EDWARD HALFORD STRANGE, M.Sc. 7, Staple Inn, Holborn, London, Director of Strange & Graham, Limited.

BANKERS:

THE UNION OF LONDON & SMITHS BANK, LIMITED, 2, Princes Street, London, E.C., and Branches.
And their Agents: MANCHESTER & COUNTY BANK, LIMITED, Manchester;
BRITISH LINEN BANK, Edinburgh, and Branches;
BELFAST BANKING COMPANY, LIMITED, Belfast, and Branches.

CONSULTING CHEMIST:

SIR WILLIAM A. TILDEN, D.Sc., The Oaks, Northwood, Middlesex.

SOLICITORS:

To the Company, ASHURST, MORRIS, CRISP & Co., 17, Throgmorton Avenue, London, E.C.
To the Vendors (Research Syndicate, Limited), NICHOLSON, GRAHAM & JONES, 24, Coleman Street, London, E.C.
To the Vendors (Organic Products Syndicate, Limited), CLAPHAM, FRASER, COOK & Co., 15, Devonshire Square, London, E.C.

BROKERS:

MARKS, BULTEEL, MILLS & CO., 31, Threadneedle St., London, E.C.

AUDITORS:

W. B. PEAT & Co., Chartered Accountants, 11, Ironmonger Lane, London, E.C.

SECRETARY & REGISTERED OFFICES:

H. EDWIN COLEY, 50, City Road, London, E.C.

PROSPECTUS.

THIS Company has been formed to purchase from the Organic Products Syndicate, Ltd., and from the Research Syndicate, Ltd., certain exclusive Licences under British, Colonial and Foreign Patents and other Rights for new and economical processes for the manufacture of Acetone, Fusel Oil, and Synthetic Rubber, owned or controlled by these Syndicates. The two former substances, Acetone and Fusel Oil, are valuable, quite apart from the fact that they form raw materials for the manufacture of Synthetic Rubber. The Estimate of Profits hereafter given is based solely on the manufacture and sale of Acetone and Fusel Oil, whilst the Synthetic Rubber section may be looked upon as carrying great possibilities of future profits.

The Company is also to take over, under certain conditions, all improvements and further inventions connected with Acetone, Fusel Oil, and Synthetic Rubber included in the Agreement for Sale which may be made during approximately nineteen years from the present date by Prof W. H. Perkin, of Manchester University, and by Prof. A. Fernbach, of the Pasteur Institute, Paris, and by Strange and Graham, Limited, Technical Research Chemists, of 50, City Road, London, E.C. On page two of the inset there is given a list of patents and applications for patents connected with the processes to be taken over by the Company, in respect of which the Company will (without further consideration than that mentioned in the Agreement for Sale) be entitled to exclusive licences for the purposes of the processes sold. These stand in the names of one or more of the following: H. J. W. Bliss, H. Davies, A. Fernbach, W. R. Hodgkinson, F. E. Matthews, W. H. Perkin, C. A. Pim, E. H. Strange, and C. Weizmann.

It is proposed to commence the acetone and fusel oil processes in the first place by erecting a comparatively small plant of 10 units of 1,000 gallons capacity each at a cost not exceeding £5,000, manufacturing works already partially equipped being available for immediate occupation. This will occupy about three months. Then, after reporting to the shareholders, it is proposed to extend the Company's works on a suitable site by expending a further £145,000, making in all £150,000 expenditure on the Acetone and Fusel Oil sections.

It is also proposed to employ a sum not exceeding £25,000 for experiments in developing the manufacture of Synthetic Rubber on a commercial scale.

The 1912 Prospectus for Synthetic Products with annotations by Chaim Weizmann. Courtesy Weizmann Archives.

of emphasizing the English location of that future industry, one can speculate that this letter was a reflection of elation over the potential for his Jewish Palestine.[59] Certainly, Strange felt Weizmann was using the consortium's efforts for his own ends.[60]

Immediately, however, it appeared that the focus of Synthetic Products would be redirected from replacing British rubber to substituting for acetone, imported traditionally from Germany. A factory using Fernbach's organism was built in 1913 at Rainham, and shortly after, another was opened in King's Lynn, Norfolk. With the declaration of war, acetone as an essential solvent for explosives was in short supply. The government's attention was drawn to the new plant, and after a demonstration, a contract was placed with Synthetic Products' new plant at King's Lynn. However, the process for converting starch was cumbersome, requiring two steps, first the chemical hydrolysis of starch and only then a fermentation. Regrettably, the plant, performing poorly and exclusively reliant upon scarce potatoes, never realized the hopes expressed in the 1912 prospectus.

Meanwhile, Weizmann himself, now Manchester's Reader in 'Biochemistry', had left the consortium and, striking out on his own, identified a bacterium which could ferment another starch source directly to acetone and butanol. Weizmann wrote to the explosives manufacturers Nobels about his result, and such was the need that they picked up the process and commercialized it quickly. In April 1915, he was introduced to Winston Churchill, First Lord of the Admiralty. He asked for a factory. By July a pilot plant at Nicholson's London gin distillery was complete. Further research, carried out at the Lister Institute during 1916, led to the construction of a plant at the Royal Naval cordite factory at Holton Heath and the cooption of the poorly functioning King's Lynn plant. Production was increased from less than half a ton of acetone a week to more than 4 tons.[61] Maize rather than potatoes could be used. Even chestnuts were collected by English boys, on mysterious government instructions, though it eventually proved difficult to completely dehusk them, and their use was declared impractical.[62] The solution to the shortage of maize in England was, more easily, to transfer production to the prairies. Distilleries were purchased first in Toronto then in Terre Haute, Indiana, for the manufacture of acetone.

The so-called Weizmann process proved to be special not just because it used a new organism and had significant economic potential. It also required a new degree of microbiological sophistication in manufacture. Whereas most industrial fermentations till that time did require cleanliness, they were not aseptic. The Weizmann process required a laboratory standard of sterility at an industrial scale. The system required frequent purging with superheated steam and was carried out not in traditional oak casks but in the new aluminium vessels made by his friend Seligman's Aluminium Plant & Vessel Company (now known as A.P.V.).[63] In Britain, its development engendered a new professionalism and sophistication among men such as A.C. Thaysen, T.K. Walker, and L.D. Galloway who made it work. Walker went on to found a new department of biochemistry at Manchester College of Technology (now UMIST) while the bacteriological expertise of Thaysen and Galloway was employed by the government for decades, first at the Admiralty Laboratory in Holton Heath and, from 1931, at the Chemical Research Laboratory.[64]

After the war, the German innovations of microbiologically produced glycerol and food yeast proved uneconomic, and the plants were shut down. Meanwhile, the Weizmann plants in the United States and Canada were reborn, for the original product of butyl alcohol proved useful in a new consumer's world. It was the ideal solvent for the cellulose lacquers needed to paint the motor cars that rolled off production lines. In the United States, the Commercial Solvents Corporation applied Weizmann's patents, and by 1930 production was almost a hundred times the Synthetic Products' level of 1914.

The acetone–butanol process has since been associated with Weizmann. In part, this was because a major law case of 1926 establishing patent rights showed that, technically, the process developed by him was different from the earlier Synthetic Products process. Nonetheless, from a historical rather than a legal standpoint, there clearly was a continuity between the two. Moreover, as the full story shows, it was a much larger enterprise than that of a single man, involving, as early as 1912, fifteen chemists and bacteriologists. Its inception illustrated the faith placed in the economics of an agricultural product whose supply could be multiplied indefinitely, just by planting more cereals. Its success established the importance of the principle. It vindicated the elaborate infrastructure of research laboratories established at the beginning of the century and was interpreted as justifying a further development.

Renewable resources

Interest in the biological production of chemicals remained high after World War I. Francis Garvan, head of the Chemical Foundation, which took over the German chemical industry in the United States on behalf of the state, would become an influential enthusiast of zymotechnology. The British chemical community's leader and eminent wartime patron of Weizmann, William Pope, made a passionate plea for the proper use of biological resources in 1921. Speaking in Toronto to the Society of Chemical Industry, he returned to the theme explored by his own former professor, Henry Armstrong, and by Meldola before him.[65] In a brilliant prospect of the future, he asserted that Britain should not follow Germany's lead in attempting to replace natural products. The chemical industry, which had been dependent on brute-force, high-energy reactions, was now more elegant. It would in the future become even more intelligent by emulating the mild conditions of nature. The best way to do this would be to use nature's reactors. Rubber, he predicted, would be better grown than made, and he reflected on how that might be true for many other products. Thus, even Pope, an organic chemist, portrayed the use of natural organisms as an alternative route to traditional chemistry. It is intriguing to see the nationalistic colouring to this differentiation and how he, following Armstrong, highlighted this as the British way against the German chemical route. Germany, without imperial possessions, was condemned to using coal to supply energy and raw materials. The British Empire could use sunlight and the power of plants. After all, Pope was giving his lecture in Canada, where the Weizmann process had been brought to draw on the almost limitless supplies of maize.

Pope was irritated too by the wasteful use of petroleum which was merely burnt instead of its complex chemistry being valuably employed. This belief that oil and coal should be processed rather than burnt (in furnaces or cars) was a common theme of the time. Oil seemed to be running out and, in any case, had to be imported by European nations, while agricultural produce was in surplus. Several countries insisted on alcohol being added to fuel in the 1920s, and Pope's speech was cited in support. Thus, it was quoted approvingly by the great Dutch microbiologist Albert Jan Kluyver in his inaugural lecture at Delft in 1922.[66] In fact, the two feedstocks Pope highlighted, petroleum and agricultural products, were in competition, and petroleum proved, time and time again, to be the cheaper. Nonetheless, the belief that somehow renewable resources must enable an elegant approach to a new biochemical industry has proved irrepressible.

Chemurgy

When, in 1930, the American journal *Industrial and Engineering Chemistry* carried out a survey of industrial fermentation, the lead article, entitled 'The Chemical Approach to Fermentation', still generally discussed its subject as zymotechnology.[67] Nonetheless, industry, now recruiting professional biochemists, had a greater degree of disciplinary identity than in the previous generation.[68] Smyth and Obold's classic 1930 work entitled *Industrial Microbiology* is dedicated to 'a more complete use of the microbiological processes in industry'. The classic 1936 text *Gärungschemisches Praktikum* of the German-Czech biochemist Konrad Bernhauer began with an affirmation of fermentation chemistry's new affiliation with the chemical industry, rather than its traditional home in brewing. It then went on to lay out the scale of production of its various uses.[69] Though most of the products were ancient, new techniques were being used as in alcohol production now worth RM 12 b. Organic acids, including 6,000 tonnes a year of lactic acid and 8,000 tonnes a year of citric acid, were among the new products of modern technology. Lactic acid, an important chemical in the textile and leather industries, had been made industrially from 1894. By 1909, Germany exported 1,500 tonnes of lactic acid in addition to its domestic production.[70] During the war, its compounds had been used to replace the scarce glycerol in hydraulic brakes, and whereas previously its production had been a German monopoly, a variety of countries created their own industries in the face of wartime restrictions. In the 1920s, lactic acid became a favourite food flavouring in lemonade. In Britain, it is still familiar as the acidifying agent of Lucozade.

In the United States, where domestic refrigeration was available and soft drinks were used to make up for the lack of beer banned under Prohibition, there was a growing market for cooled carbonated drinks, and for them, citric acid was preferred. This had been observed as a microbiological product of sugars in 1890 by Wehmer, and four years later he took out broad patents. Manufacture from *Aspergillus Niger* was demonstrated in the United States during World War I, and Pfizer laboriously commercialized the sensitive process. Complex fermentations such as citric acid presented complex problems. As late as 1930, an article in the *Industrial and Engineering Chemistry* survey concluded: 'That the problem is no small one is attested by the fact that apparently very few industrial concerns have succeeded in producing citric acid by fermentation in quantities large enough to be noticeable in the markets'.[71]

Nonetheless, when declining prices of agricultural produce brought the United States to experience agricultural crisis in the 1930s, it too turned to the solutions that had been favoured in Europe. Recalled by John Steinbeck's *Grapes of Wrath,* the disaster, engendered by overproduction and low prices, was driving farmers to desperation and leading to the destruction of nutrients in a starving world. The Republican Party's research division, reporting the fall of the corn price from 121.2¢ a bushel in May 1937 to just 52.7¢ a bushel a year later, reflected on the necessity of a new approach. There was no possibility of a prosperous agriculture 'unless farm commodities are developed as raw materials for the manufacturing industry.'[72]

Characteristically, despite the similarity between American and European concerns, a new term was coined in the New World. The author was the research director of Dow Chemicals, William J. Hale, a felicitous inventor of new words and fisherman for new ideas. Witnessing the tragedy inflicted by the problems of farm surpluses and pondering the potential of using them to produce chemicals, he chose the term 'Chemurgy' meaning chemistry at work, from 'chem' and the Greek 'ergon'.[73] In fact, Hale did not distinguish too fastidiously the boundaries of chemistry and used his word rather as others had employed 'zymotechnology'. He explained his meaning in 'The Farm Chemurgic': The principal message was the potential of producing alcohol to replace petroleum, from fermented starch. This product Hale entitled 'agricrude', explaining in his 1939 work *Farmward March*, 'the development of an *agricrude* alcohol industry . . . will end all unemployment, provide unlimited extension of mechanical power and end all international trade in organic chemical material (farm produce).'[74] Oil-starved European countries had already made the use of 'power alcohol' established practice. (Many Britons will remember Cleveland Discol made by adding *Dis*tillers alcohol to petr*ol*). In the United States, with its enormous domestic gasoline production and powerful petroleum interests, the commercial challenge was greater, success less immediate, and the spark of enthusiasm not ignited until 'gasohol' in the 1970s. Hale's own interest was not, in any case, to use alcohol to burn, instead he saw it as the feedstock for the organic chemicals, such as acetic acid, essential for his beloved plastics.

Chemurgy would become a theme of repeated meetings during the late 1930s. These were held over many days and were widely attended. Financial support came from Henry Ford who had never forgotten his farm boy background. A famous picture shows Ford wielding an axe on his experimental car made from soya bean–sourced plastic to demonstrate its strength. Appropriately, the approach was intensely practical: particular uses for particular agricultural products. Although, as the word 'chemurgy' implied, the focus of the movement was on chemistry, it thus also led to a new institutional framework for biology.[75] The U.S. Department of Agriculture responded to the

Henry Ford demonstrating the strength of a car body made from soya bean–sourced plastic in 1941. From the collections of Henry Ford Museum and Greenfield Village.

new movement, and to the glut of agricultural materials, by creating four great regional research stations, each of which would concentrate on a local surplus. Thus, in Peoria, the Northern Regional Research Laboratory focussed on the surplus of corn products, and of how to turn to profit waste such as the starch-rich corn steep liquor. Echoing the solution of Fernbach to Reckitt's surplus rice hull problem a generation earlier, the Peoria scientists tried to use the liquor to grow useful microbes, in this case ones that would produce gluconic acid. They developed the art of fermenting oxygen-breathing moulds in submerged culture, which would become vital in penicillin production a few years later.

Nonetheless, fermentation chemicals such as the organic acids still lay on the margins of the developing chemical industry. The great corporations of the 1920s, IG Farben, Du Pont, and Britain's Imperial Chemical Industries (ICI), placed little emphasis on biological processing. Aromatic chemical compounds could be extracted from coal tar, and increasingly, aliphatic compounds such as the alcohols and glycols could be made, particularly in the United States, from petroleum derivatives.

Despite slow commercial development, the words and speeches of the period around World War I evoked many of the sentiments that would be familiar through the twentieth century: in particular, the possibility of a distinct biotechnology. This was being described as the technology of the future by Ereky, while others saw the production of chemicals through microorganisms as an alternative to chemical manufacture which was focussing on the high-energy transformation of coal and oil. The use of agricultural surpluses was repeatedly expressed. Yet again and again, there was the familiar problem of competing economically with chemistry. Beyond food products and sewage, biotechnology was, largely, small beer.

3
The engineering of nature

Biology is as important as the sciences of lifeless matter, and biotechnology will in the long run be more important than mechanical and chemical engineering.

(Julian Huxley, 1936)[1]

Introduction

During the years between the two world wars, biotechnology acquired another connotation apparently quite different from its chemurgic interpretation. On the basis of almost mystical concepts of the integration of biology with engineering as central to an entire new phase of human civilization, the focus of the composite subject came to lie in the application of biology to humanity. Health would be not just a matter of the occasional medical intervention, but also the result of an environment harmonized with the needs of society. Eugenic issues had their place here, but so did nutrition and nonpolluting manufacturing technology based on renewable natural resources. Hence, despite the apparent divergence from the zymotechnic tradition, the two came together during the 1930s in the vision of a new healthier technology. To the idealists, the image of a new technology for a new age was startlingly bright, even if this rendition of biotechnology perhaps resembled a perversely shunted goods train, with milk tankers hitched to coal trucks. In its apparent confusion of hitherto distinct categories, it resembled other aspects of European culture that were being radically reformulated in the wake of the disaster of World War I: art, music, physics, and even philosophy itself. In later chapters, we will deal extensively with institutions and technologies. Here the emphasis is on the formulation of increasingly complex and interrelated ideas of biotechnology in books and lectures. As was true of ideas of those times in so many fields, their relevance would endure long after the details of knowledge had changed unrecognizably.

At the heart of these concepts was the category confusion engendered by biotechnology and exemplified by the 'robot' – is it man or machine? The very word was introduced to the world by the Czech playwright Karel Čapek in a 1920 play *R.U.R.* Today the term has a mechanical connotation, but this was not the author's intent: His robots were made from 'catalytics, enzymes, hormones and so forth'. Čapek explained to London's *Saturday Review* of his inventor: 'Young Rossum is the modern scientist, untroubled by metaphysical ideas; scientific experiment is to him the road to industrial production. He is not concerned to prove, but to manufacture.'[2]

This expresses the view of machines as alien, linked only by commercial profit to humanity. The concept of biotechnology would, by contrast, integrate the contemporary ideal of manufacturing with visions of humanity and its environment, and a faith in the privileged view of life. Though it is perhaps surprising to find the metaphor of the machine relevant even to organic philosophers, for them the machine became a symbol of the system that was more than the sum of its parts and had an irreducible character of its own. Thus, both the severely mechanistic models of biological organisms comparing life to a chemical machine, and neovitalist models according to which biological systems were endowed with some special organizing or developmental character not available to man-made systems, yielded concepts of biotechnics. Though these were apparently far more remote from our first impressions of modern biotechnology than the zymotechnic tradition, and are apparently sometimes bizarre, they did become important constituents of later ideas.

To practical people, the attraction of the biological realm lay in the omnipresent raw material and its challenge to practical exploitation. To others, the idea of technology was still fraught. Frightful images of factories and the despoliation of the landscape were only the most visible aspects of the disruption of traditional ways of life put at technology's door. For a century, technical change had shaken the Western world, stimulating a variety of emotional, literary, political, and cultural responses. Emile Zola described the new debauched world of the worker. Socialist and Communist parties emerged throughout the industrialized world. New forms of art expressed an attempt to assimilate technology. An American visitor to Germany, Dean Randall of Columbia University, noted the ironic contrast between the feudal pomp of the 1902 opening of Munich's new museum, reassuringly entitled the 'Deutsches Museum', and devotion to modern technology.[3] In this process of domesticating technology as part of human history, and indeed human natural history, biotechnology itself would be constructed.

Biotechnics: biology and engineering

Hosting the 1901 World Zoological Congress, as the twentieth century dawned, Berlin provided an appropriate construction site for biotechnology. A German participant, Gustav Tornier, pointed out how many papers dealt with the analogies between biological and mechanical systems. In his own contribution, he went on to propose that the category of technology, which he formally defined, could be applied to living organisms in general, described as *'Bionten,'* and the process of modifying or using them technologically could then be known as *'Biontotechnik.'*[4] Tornier's paper was noticed and its new word was included in Roux's dictionary of 1910.[5]

Though Tornier's wordsmithing was innovative, his biology was conservative. The analogies between living beings and human technology had already been highlighted for decades by studies showing how human bones, joints, and organs could be considered as sophisticated machines. In 1877, Ernst Kapp had generalized this by describing technology as the result of an 'introjection' in which tools were the extension of the hand.[6] His work had transformed the rather specialized physiological literature into a philosophy of technology, with a momentum at the beginning of this century it would find hard to recapture.[7] It provided a reference for people trying to make sense of a wide variety of different concepts and issues such as the community attempting to combine evolutionary thought with mechanistic models of the body.

The mechanical analogies between skeletons and machines were deepened by the growing study of embryology. The rumbustious biologist Wilhelm Roux discovered that the development of organisms could be altered by removing one cell from a fertilized egg. Roux coined the term *'Entwicklungsmechanik'* to describe the machinelike quality of the growth of the embryo. He was a powerful teacher and inspired – and indeed coopted – a generation of biologists. In an autobiographical note, he went so far as to list the members of his school.[8] The young émigré to the United States Jacques Loeb was influenced in his expression of hope for a *'Technik der lebenden Wesen'*, a technology of living matter.[9] Loeb himself made an apparent breakthrough in 1899 when he managed chemically to induce an unfertilized sea-urchin egg to divide. The lay press were sure that this technique of parthenogenesis would soon lead to laboratory produced babies. Loeb himself was cautious, yet to him, as to Roux, the lines between the animate and the inanimate were arbitrary. His engineering approach to biology was to influence deeply a generation of disciples including Hugo Muller and Gregory Pincus who would themselves be influential into the 1960s.

French and German schools of physiology have been distinguished by Owsei Temkin, the historian of medicine, in a brilliant epigram: 'To the vitalistic materialists in France man was but an animal. To the mechanistic materialists in Germany he became but a passing constellation of lifeless particles of matter.'[10] A similar distinction informed the debates over the relation between culture and technology.[11] Man could be 'reduced' to a machine, but the creation of machines could also be seen as a characteristic aspect of the human animal. Among the French pioneers of such studies was Jean-Jacques Virey, a popularizer of science and contemporary of Lamarck, one of those to coin the very word *'Biologie.'* He developed a theme that was to become very important among evolutionary thinkers: that man has had to develop technology to make up for the loss of natural instincts. The term used by Virey to describe this innate human trait in 1828 was *'biotechnie'*, and although the links of this word with later coinages cannot be documented, it was this expression which was probably the first.[12] Virey's coinage introduced not only a word but also an ambition that was to recur. He was the editor of a journal urgently concerned about the status of biology, and his word identifying tool making as a biological phenomenon coopted the whole discipline of engineering as a subset of his own professional expertise.

If Virey himself was soon forgotten, his idea that intelligence is man's replacement for instinct was echoed by his far more distinguished successor Henri Bergson, a French intellectual superstar. So influential were works of Bergson such as his 1907 *Evolution creatrice* (*Creative Evolution* in its 1912 English translation) that their reputation hangs over much of early twentieth-century European philosophy, whether or not individual ideas can be attributed to others. By 1912, 417 books and articles on Bergson had been published in French alone, and *Evolution creatrice* itself had passed through sixteen editions by 1914.[13]

Three themes, underpinning ideas explored throughout this chapter, were developed by Bergson: life, the evolution of consciousness, and technology. He is perhaps best known for the special characteristic he saw in living beings, the *élan vital.* Less well remembered is his philosophy of work, shared with many of his contemporaries. Man could constantly recreate himself through work, and thus the use of tools was a part of the spiritual expression of humanity.[14] Bergson suggested that life was special, because its infinite potential was denied to the inanimate which is constrained by the circumstances of manufacture.

Seeing technology as a symbol of human ability to create new worlds, Bergson identified man as *Homo faber,* man the maker. Following the idea of early nineteenth-century archaeologists who had defined ancient man by his tools, he then classified more recent historical periods by their characteristic technology. With Bergson, therefore, the actual technology (and not just the relations of production as Marx had suggested) defines a historical era. Our current technological state defines our state of consciousness:

> In thousands of years, when seen from the distance, only the broad lines of the present age will still be visible, our wars and our revolutions will count for little, even supposing they are remembered at all; but the steam-engine, and the procession of inventions of every king that accompanied it, will perhaps be spoken of as we speak of the chipped stone of the pre-historic times: it will serve to define an age.[15]

Although the concept of the 'industrial revolution' dates back to Toynbee's use in 1881, it seems to be to Bergson's concept of a self-recreating *Homo faber* that we owe the concept of the third industrial revolution variously reported to be associated with the information age, the nuclear age and, of course, with biotechnology.

This conception of periods as defined by their technological style led, even in Bergson's time, to a vision of biotechnics as a definition of the current age. This judgement, at a time so removed from molecular biology, was based on interweaving with Bergson's historical model two other strands of biological debate: the problem of evolution and arguments over whether biological products were just like man-made ones.

Lamarckians

According to Darwin's formulation of evolution by natural selection, species developed randomly, and had then to await passively whatever brute fate decreed. His French predecessor, Lamarck, had suggested that animals evolve during their lifetimes in response to their environment. The neo-Lamarckian biologists, influential in Germany, suggested that there was the possibility of enhancement. Biological systems could thus be considered as machines with the special ability to improve themselves, making them particularly worthy of respect and emulation. Thus, the theory of 'orthogenesis', popular in the early twentieth century, suggested that evolution did not occur randomly but took a specific direction.[16] In human terms, the implication was that technological progress and human evolution could go hand in hand.

For the proponents of organicism, moreover, the body could not be reduced simply to the sum of its parts since an additional principle had to be taken into account. The most charismatic of the group was the passionate long-haired mystic August Pauly, author of *Lamarckismus und Darwinismus,* who was to influence a generation of embryologists and whose ranks of other admirers included Sigmund Freud.[17] For Pauly, evolution could only be explained by an extra, psychic principle characteristic of each cell. His friend, the Hungarian-born botanist Raoul Francé, editor of the journal on psychobiology, agreed that living organisms could be considered mechanistically but only when this extra principle was taken into account. As early as 1910, the circle was put into historical context by one of its members, Adolf Wagner, a lecturer at Innsbruck.[18] Wagner differentiated between the old vitalism, for which, he generalized, there was no bridge between the animate and the inanimate, and the new vitalism for which there was just the need for an extra step.

Wagner paid deep attention to a small group of modern thinkers, closely associated with the new *Naturphilosophie* inspired by Wilhelm Ostwald. Although most often remembered today as the founder of physical chemistry, Ostwald saw his contribution as far more than merely the creation of a scientific discipline. He articulated a new philosophy in which energy was the basic and unified component of the universe, explored in his journal *Annalen der Naturphilosophie.* A leading member of the Ostwald circle was the Austrian sociologist-novelist Rudolph Goldscheid. Separately, he and Francé coined the word *'Biotechnik'* during the second decade of the twentieth century. Eisler, author of an authoritative multivolume dictionary of philosophy, granted the first use of the word to his friend Goldscheid's 1911 terminology.[19] Francé's formulation followed in 1918, and though its meaning was slightly different, Wagner's book testifies to the common culture behind them. Interesting in themselves, these coinages showed how the word grew out of philosophies that were to be popular in many lands between the wars. We shall explore how both Francé and Goldscheid can be seen as precursors of separate, but interrelated developments in English thought during the 1920s and 1930s.

Goldscheid

Rudolf Goldscheid, the first of the biotechnic philosophers, was a respected novelist and social critic who entered sociology, as the discipline was being formed, with a fervent commitment to the betterment of humanity.[20] He even led the formation of the Viennese sociological society, though at the cost of alienating Max Weber for whom objectivity was all. As a socialist torn between the economism of some and the idealism of others, his solution was to believe that by making the best of people, both their happiness and economic good would be served. A member of that remarkable and close Viennese culture which included Mahler, Freud, and Wittgenstein among its luminaries, to him disciplinary boundaries counted for little and philosophical mastery for much. His studies of public finance published at the end of the Great War have been reprinted in modern times as a classic claim for the sociological grounding of economics, and they proved a spur to his fellow Viennese, the distinguished economist Joseph Schumpeter whose 'The Crisis of the Tax State' was a direct response to Goldscheid.[21] Among the many contributions for which Schumpeter is remembered is his theory of long-run technological trends and their relationship to society.[22] These ideas have been related by historians to the very visible industrializing developments in the hitherto agricultural Empire of his time, and such arguments could be applied equally to Goldscheid.[23]

In 1911, Goldscheid published his great work *Höherentwicklung und Menschenökonomie (Higher Development and Human Economy)* in which he argued from Lamarckian principles towards improving the lot of present generations to the benefit of the future. The work was widely noticed: London's *Eugenic Review* allotted it a four-page review, among the longest that journal ever published, though, as the generally favourable commentator noticed of the book, 'It is interminably long and diffuse'.[24] As a self-conscious exercise in metaphysics, it cites few references and less data. It was well known but rarely cited.

The Darwinian image of the survival of the fittest was portrayed here as wasteful and derogatory of the human being. Weismann's recent genetic interpretation of Darwinian evolution according to which people inherited immutable germ plasm from their parents was a particular focus for disdain. By contrast to this apparently 'liberal' and 'individualistic' philosophy, Goldscheid believed fervently that improved conditions of life would improve one's genetic legacy. Even the average person ought to have an economic potential. Like several other socialists of his time, he felt that the eugenics would serve to raise the level of the average. This was expressed by *'Höherentwicklung'*. Such new words were beloved by Goldscheid. Most associated with him personally would be *'Menschenökonomie'*, but he also pioneered the word *'Sozialbiologie'*, as the title of the section of the Austrian Sociological Society, and finally *'Biotechnik'*. He showed unbounded enthusiasm about the potential and importance of the latter, a concept whose vagueness regrettably matched its apparent significance. Reference to *Biotechnik* pervades *Höherentwicklung und Menschenökonomie*, and in its introduction Goldscheid proclaimed his belief that this would be the most important technology of the twentieth century.

Whereas *Biotechnik* was Goldscheid's original idea, its roots in his concept of *Sozialbiologie* were more conventional.[25] The idea of social medicine can be traced back to the eighteenth century, and quarantine provisions had been especially strict in Austro-Hungary with its borders on the Turkish empire, where the plague was endemic. An interest in medical care systems was spurred by the radicals of 1848. It is, nonetheless, to a constellation of doctors at the end of the century, most notably the Berlin physician Alfred Grotjahn, that the twentieth century owes the inspiration of a benevolent combination of genetic and social controls.[26] In work carried out after 1905, Grotjahn had identified a range of social factors such as alcoholism that contributed to illness and whose study would complement the bacteriological concern of the biological hygienists. He was careful to argue that he was concerned not just with the current preoccupation with underpopulation, but with every aspect of society and hygiene. His *Soziale Pathologie* whose first edition appeared in 1911, described meticulously the meaning of *'Soziale'*, coming from the Latin *Socius* and connoting the entire range of social phenomena, those related both to the state and to the community.[27] Grotjahn was himself impressed with the work of Goldscheid and the concept of *Menschenökonomie* (human economy) first explored in his 1908 work *Entwicklungstheorie, Entwicklungsökonomie, Menschenökonomie*.[28] Correspondingly, use of Grotjahn's prefix *'Sozial-'* would differentiate Goldscheid from the racist *'Gesellschafts-Biologie'* promoted by his opponent Schallmeyer.

Goldscheid saw *Biotechnik* as implementing the programme of *Sozialbiologie*. One of its foundations would be a solution to the problems of human reproduction dealing with the transmission, adaptation, and selection of traits. In addition, there was the need to understand the instability of reproducibility and factors behind variable fertility. Application of such knowledge would enable humanity to move from an erratic process of quantity production, with the attendant evils of alcoholism, prostitution, and low opportunities for the many, to quality production. Goldscheid often used the analogy between factory production of quality goods, which increasingly employed science-based technology, and society's production of people. The idea of *Technik* and references to the machine, was not to be taken as an insult – rather they represented his respect for the organic wholeness and functionality of machines which went far beyond the individual component's mechanism. Goldscheid suggested that *Biotechnik* was based on a respect for the living organism, the human being, and the wish to develop a complementary technology which was organic, spiritual, and ethical.[29]

The details of *Biotechnik* were never clearly spelled out by Goldscheid, though he was clearly impressed by the multifold developments in embryology, developmental biology, and bacteriology around him. He seems to have been particularly influenced by the work of an associate in the Ostwald circle, the charismatic young Lamarckian zoologist Paul Kammerer. Possibly best remembered through the biography of another Viennese, Arthur Koestler, Kammerer became infamous for his apparently fraudulent claims that he could alter the genetically transmitted markings of the midwife toad, but before his unmasking and subsequent suicide, he was a popular thinker.[30] Kammerer and Goldscheid were particularly close – Kammerer became secretary of the social biology section in the Viennese sociological society established by Goldscheid. He defended his friend's ideas against the right-wing eugenist Schallmeyer in the main eugenic journal and used them in his 1918 book *Einzeltod, Völkertod, Biologische Unsterblichkeit*.[31] During the early 1920s, he published two pamphlets that made claims for the radical historical importance of *'organische'* or *'biologische' Technik*, as the key twentieth-century science. He reminded readers that their sociological implications had already been explored by Goldscheid. As early as 1910, he had interpreted Lamarckian biology as implying the possibility of improving the human race.[32] Thus in the 600 pages of *Höherentwicklung und Menschenökonomie*, Goldscheid brought to Grotjahn's ideals of social hygiene a particular technological vision. This may have been based on the practical Lamarckian *Technik* of his friend Kammerer, but also on the sense of differentness about this technology, expressed by Bergson's vision of life as something beyond matter and evolutionary sense of consciousness, and the sense of the imminence of the transformation from mechanistic thinking found in Ostwald's work.

If Goldscheid himself was an obscure and unlikely father of biotechnology, his concerns were widespread. The interwar years saw increasing worries about the health of populations in terms of their numbers, their quality, and their welfare. Debates over eugenics, feminism, and birth control and the implications for nutrition and food policies of the new concepts of vitamins ensured that social hygiene was a key issue of the interwar years. Most unpleasantly, this is remembered through the Nazi perversions, but Goldscheid's *Menschenökonomie* also impressed Julius Moses, the campaigning health spokesman of the Socialist party in the Weimar Republic, and the thoughts he expressed, were, in different forms, popular throughout Europe.[33] In France, where 'quality with quantity' also became a call, there was talk of the importance of human zootechny.[34] The right-winger Sicard de Planzoles addressing a 1934 meeting at the Sorbonne, with the minister of health present, suggested that human zootechny would be the end stage of hygiene: Echoing Goldscheid, he suggested, 'Let us think of man as an industrial material, or more precisely, as an animal machine. The hygienist then is the engineer of the human machine'.[35] In the Soviet Union, the revolution was followed by a short-lived enthusiasm for an emphasis on public health. Kammerer himself was also briefly popular; a movie, *Salamandr,* was even based on his life, and Ivanov's *zooteckhnika* inspired Serebovskii's *antropotekhnika.*[36] It is therefore not necessary to argue that Goldscheid's particular formulation was solely responsible for later interpretations. Rather, it crystallized the variety of ideas of a new technology suggested by the conventional concepts of social hygiene.

Francé

Ultimately, Goldscheid was inspired by the primary importance of the biological over the more conventional inorganic obsessions of engineers. This provides a common basis for Goldscheid's vision of biotechnics as a means of changing nature and the model developed by his contemporary, Raoul Francé, in which biotechnics described man's attempts to emulate nature. Sharing the belief in the intrinsic worth of natural phenomena, Francé also ascribed transcendent importance to biotechnics. Trained as a botanist, he found himself frustrated by narrow science, materialism, and careerism.[37] Increasingly, he saw himself as a philosopher, inspired by Pythagoras, Kant, Nietzsche, and Schopenhauer. From his conception of the unity of life, he would argue that the activity of living things was controlled by more than mere chemical and physical laws.

Francé's ideas were framed within an intellectual circle vividly described in a series of essays published by his widow.[38] The couple was

Raoul Francé. This engraving, by Rudolf Engel Hardt, appears as the frontispiece to Francé's 1927 autobiography, *Der Weg zu Mir*. Courtesy Alfred Kröner Verlag.

friendly with the followers of Madame Blavatsky, and with Hans Meyrink, the Prague intellectual and author of the 1915 rewrite of the medieval Jewish tale of earth made into man, 'The Golem', and, later, with Oswald Spengler. Francé's psychobiology, developed before the First World War, informed but did not explicitly intrude on the theory he expounded in a series of works at the end of the war that used the word *'Biotechnik'* without ever defining it too tightly. His friend and devotee, Adolph Wagner, in a festschrift of 1925 pointed out that *Biotechnik* 'is only an idea, a word, but one of those which unifies an entire world'.[39] He pointed out that the word *'Technik'*, unlike a simple mechanics, incorporated the idea of design for a purpose. It therefore found for Francé's panpsychism a practical implication.

In the second half of 1918, as the old world was being overturned,

Francé published an article in the *Mitteillungen des K.K. Technischen Versuchsamtes.*[40] He began by suggesting that biology with its insight into the natural phenomena could be a teacher of technology with its own immense territories. His axiom was that life could be seen as a series of technical problems for which living organisms represented the optimal solutions. The relationship could be described technically as he did by pouring out examples. His first example in the 1918 paper, of a lock, should not be interpreted as mundane, for the real message was spiritual. It was first developed in the technical work *Die Technischen Leistungen der Pflanzen,* then in Francé's massive philosophical treatise, *Bios: Die Gesetze der Welt (Bios: The Laws of the World),* and finally in a range of popular works, one of which, *Plants as Inventors* (*Pflanze als Erfinder,* 1920), was translated into English:

> It was my thesis that we can conquer not only by the destruction of disturbing influences, but by compensation in harmony with the world. Only compensation and harmony can be the optimal solutions; for that end the wheels of the world turn.
>
> To attain its aim, life; to overcome obstacles, the organism – plant, animal, man or unicellular body – shifts and changes. It swims, flys [sic], defends itself and invents a thousand new forms and apparatuses.
>
> If you follow my thought, you will see where I am leading, what the deepest meaning of the biotechnical tokens. It portends a deliverance from many obstacles, a redemption, a straining for the solution of many problems in harmony with the forces of the world.[41]

So Francé had established his word as combining the two concepts of a new kind of technology, more harmonious than before with the natural model. He portrayed himself as a prophet of this new world, a leader. His magisterial two-volume *Bios* describes, in detail, the history and prospects of the concept of *Biotechnik,* and he professed that he had had no doubt that at the moment in 1917 when the idea first came to him, however incompletely, it opened a new era in cultural development.[42]

Perhaps Francé was pompous. Nevertheless, his ideas did prosper in Germany. His popular works won an immense circulation: His widow suggests a total printing of just his twelve *'Kosmos'* volumes of 3 million copies, each (in those pretelevision days) having five readers, giving a total of fifteen million readers.[43] Certainly, they are still to be easily found in second-hand bookshops, and the library of the distinguished architect Mies van der Rohe counted forty of them. Indeed, Francé's thoughts provided the solace that the Bauhaus spirit would live on in the wake of Hitler's victory.[44] Recently, he has been described as 'probably the most important inspiration for most European avant-garde artists and architects intrigued by the analogies of natural and technical form'.[45] *Bios* could also be celebrated by those closer to the Nazis. In 1939, Alfred Giessler, head of the *Biotechnik* group at Halle, brought out a work with the title *'Biotechnik'*, which reiterated, with some acknowledgement, the arguments of Francé, but as 'German' science.[46]

Stylistically, the philosophies of Francé and Goldscheid might have seemed remote from the pragmatic, even technocratic, concepts of Orla-Jensen and Ereky. Yet they did have much in common. Both were means of expressing a new constellation of ideas around biology that transcended particular techniques. The new science seemed to be the basis for technologies that were in their infancy and could be described better by their potential and collective significance than by their individual value. Moreover, they shared the metaphor of a new industrial revolution or new industrial age. At the very time that World War I and the Russian Revolution were bringing the era of the nineteenth century to a close, they provided a technological expression of the new century. The audiences were also analogous: Neither type was addressing peers engaged in detailed bench science, as in the normal scientific paper. Instead, their concepts were intended for a more or less lay public: Orla-Jensen was talking to students, Siebel and Mason to customers. The others, Ereky, Francé, and Goldscheid, were self-consciously communicating the meaning, as they saw it, of current change, to a nontechnical audience. They were going beyond the normal meaning of 'popularization', in which ideas standard within the scientific community are watered down for the public, for theirs were ideas that did not have currency within science. Analogous too were the reasons for addressing such audiences: professional and commercial self-respect. Though they found themselves doing important and significant work, to the rest of the world their work was peripheral – it was neither chemistry nor physiology. It did not have status within such traditionally highly respectable spheres. Biology was a new and poorly comprehended category. These men – editors such as Lindner, Goldscheid, Francé, and Mason, or the professor Orla-Jensen, and the consultant Emil Siebel – could use the new world and the idea of a new science-based revelation to give significance to themselves and to their work. These were themes that would endure in the history of biotechnology.

Britain

Francé's word, *'Biotechnik'*, successfully crystallized ideas in Germany. Through translation into English, its meaning came to be blended with other similar concepts and, at the same time, acquired currency worldwide. The declining and industrially laggardly empire that formed the backdrop to the Austro-Hungarian cultures of Francé and Goldscheid had its analogue on the other side of Europe in the anxious centre of the British Empire. There too, new sources of vitality were sought. Since before World War I, philosophies which elevated the natural and living, and condemned the dead and the grey, had pervaded British society. This was to be seen not only in London's romantic and evocatively named Hampstead Garden Suburb, founded in 1906, or in D.H. Lawrence's revulsion for his native Nottingham, expressed in the once-notorious novel *Lady Chatterley's Lover*. As Armstrong's 1912 article had shown, traces were also to be found even in *The Times Engineering Supplement.*

Of course, Britain, as the only long-established, industrial country which was also a global empire, had its own distinctive problems. It suffered palpably from smoke that destroyed the 'green and pleasant land', and city planners as well as novelists called for a new environmental awareness. Modern analysts have diagnosed an antiindustrial spirit in the response to industrial desolation. More immediately, the lungs, as well as the conscience and the eye, rebelled against the 'black country'. Meanwhile, the agricultural base of the empire as a whole still provided a green alternative. The response to the synthetic rubber issue made it plain that, in England as in Austria, nature and life were seen as answers to the chemical challenge posed by Germany.

In practice, then, the economic development of agriculture and the rebuilding of England's shattered environment were often linked. The new profession of biologist was well placed to address such problems. Biology had become a university subject in the later nineteenth century, and there were increasing possibilities for advanced training. It was dominated by medicine, though even in microbiology, nonmedical fora were emerging. The Society of Agricultural Bacteriologists was founded in 1929, later (1944) becoming the Society for Applied Bacteriology.[47] The academic Society for Experimental Biology, established in 1925, provided a context for modern laboratory sciences. A distinguished department of general biochemistry at Cambridge, the Dunn Institute, achieved world renown under Gowland Hopkins. It would lay the basis for generations of scientific distinction in microbiology and later molecular biology. By 1925, the institute held more than fifty workers brought together by a common interest in biological form and function.[48]

Still, academic posts were few, and in industry, there were almost no jobs either. The Society of Economic Biology, founded in 1907, numbering less than 300 members during the 1920s, was but a tenth of the size of the chemical community.[49] Thus, whereas chemists were established and held a certain social status, enhanced by wartime success and practical importance, biology was too new to be recognized as an occupation. Writing and appealing to popular interests provided an alternative remunerative use of knowledge, which also helped build public respect for the nascent profession. An imperial premier, General Smuts of South Africa, made a unique contribution for a modern statesman by launching a major new scientific concept, 'holism'. Moreover, the explanatory model of evolution stood high in public esteem. After all, the country could claim Darwin, his 'bulldog' Thomas Huxley, and the social Darwinist Herbert Spencer, who had established the close connection between sociology and biology. Human evolution had become a special concern since Francis Galton, and a remarkable succession of geneticists, Pearson, Weldon, and Fisher, built a centre at University College London. Such bastions of the scientific aristocracy as the families of the Huxleys and the Haldanes maintained their distinction. Their distinguished youth of the 1920s included Julian Huxley, grandson of Thomas, and J.B.S. Haldane, son of the distinguished biologist J.S. Haldane, and nephew of the equally famous James Burdon Sanderson and the distinguished politician Lord Haldane. Both disdained academic convention, as perhaps only they could afford, to put considerable energy into communicating directly with the public. Huxley, for instance, resigned his Kings College professorship to concentrate on a popular survey of biology he was co-authoring with H.G. Wells. By contrast, their brilliant, but parvenue contemporary, Lancelot Hogben, was very cautious about putting his name to his first (though subsequently immensely successful) popular work, *Mathematics for the Million*, which he was to follow with *Science for the Citizen*.[50]

If the British were not known for their regard for intellectuals, they nonetheless were close intellectually to Germany. Language was a complex barrier. Although British scientists generally read German papers and books – sometimes more than English material – German philosophy was not well regarded in a nationalistic age.[51] Its heroes, Nietzsche and Schopenhauer, were suspect and its current writers unknown. Even on the verge of World War II, the famous Austrian author Stephan Zweig, with friends throughout Europe, found himself isolated in England.[52]

There was, in fact, far more cultural trade than often acknowledged, but ideas experienced complex transformations in its course. When the German biological philosophy of vitalism was translated to England, the sympathetic English biologist J.S. Haldane refused to call himself a vitalist. Lancelot Hogben, his opponent, who in Germany might have been called a 'mechanist', retaliated at the British Association's meeting in Cape Town in 1929 when he opposed Smuts' new concept of 'holism' with what he called his publicist standpoint.[53] Indeed, a foreign taint was a positive insult in Britain, and even the linguistically inclined Hogben complained in an irritated note to Joseph Needham in about 1936 about the linguistic imports brought by immigrant Jews.[54] Thus, one must decipher relationships between English uses of the words 'biotechnics' and 'biotechnology' and German precedents which were not vaunted by their British authors. Still, neither did the English claim parentage, so if one neglects the German ancestry, these words seem to appear simply out of the blue across the North Sea.

Complex and hard to diagnose as Anglo-German relationships were, in such marginal subjects as biotechnics, they were the key to intellectual survival (or physical survival as in the case of many refugees). The speculative thought of Francé did cross the English Channel in the shape of English translations of his popular works. Psychobiology was picked up by the English neo-Lamarckian E.S. Russell and was widely promoted. Francé's ideas of the analogy between engineering and biological forms were paralleled in England by the Scottish biologist D'Arcy Thompson in his *On Growth and Form*. Within a few months of the publication of *Bios: Die Gesetze der Welt* in 1921, Patrick Geddes, also a Scottish biologist, used the term, 'Biotechnic', though without attribution. Even later, as Geddes became more and more enamoured of the word, he never acknowledged its source. He did not, however, claim it for his own or even define it formally. This casualness, uncharacteristic of his treatment of his own words, together with the timing, is strong circumstantial evidence for its German origin.

Geddes, an older contemporary of Francé, both directly and through his followers, particularly Lewis Mumford, would have a significant impact on future thought in the English-speaking world. He was also, himself, a key transitional figure in the transformation of biotechnics, from a biological evolutionary idea to a characteristic kind of engineering. It is therefore worth exploring the intellectual transition made as the idea of biotechnics crossed the North Sea. This is not a straightforward task, because Geddes expressed himself through his personality and voluminous correspondence, rather than through his published works.

Born in 1854, Geddes was a middle-aged man by the beginning of the twentieth century. Still, he was an advanced thinker writing on sex and evolution linked more and more closely with the evolution of societies and evolution of the city. A botanist by training, Geddes had been a pupil of Thomas Huxley but had, against his master's wishes, attended the lectures given by Herbert Spencer, the father of social Darwinism. In the heart of old Edinburgh, he established his Outlook Tower in which the whole of man's knowledge could be brought to bear on the city's problems. This required an ever more elaborate grand synthesis of human knowledge, based on the tryptych of the French geographer Le Play – folk, work, and place.

The two interests of biology and sociology were too different for the increasingly professionalized twentieth-century world, and Geddes acquired two sets of collaborators: on the biological side, J. Arthur Thomson, professor of zoology at Aberdeen; and on the sociological, Victor Branford. With Branford he published a journal, *The Sociological Review,* and continually pondered a great work which never appeared. With Thomson, he did manage a magnum opus, *Life: Outlines of General Biology,* published in 1931, just before his death. Even though it is comprised of two massive volumes, it was no more than a pale and somewhat disorganized reflection of the master's thoughts.

As early as 1895, Geddes, reflecting on the emerging distinctions between the old and the new stone age identified by the archaeologist Evans as 'paleolithic' and 'neolithic', described the industrial age as equally divided. He recognized 'paleotechnic' and 'neotechnic' civilizations, partitioned by what, in 1915, he was calling the 'second industrial revolution'.[55] These two civilizations were characterized by different kinds of technology; the former was realized by steam engines concentrated around the filthy coal mines, the latter, by the clean, distributed technology of electricity. Geddes did not consider that such technologies by themselves drive civilization; rather they were to him expressions of it. Reasonably, one can compare his work with that of Bergson: Thomson gave the English translation of *Creative Evolution* a most favourable review in *Nature.*[56]

For Geddes, as for so many Europeans, World War I marked a fundamental divide. His beloved wife died in 1917, and three months later his son and potential heir was killed in the trenches. Before the war he had been much concerned with practical schemes of town planning and organized an exhibition that toured the world; it was lost on a ship sunk by enemy action in 1914. Thereafter, he became more interested in idealistic schemes. He was contracted to design the new University of Jerusalem; from 1920 to 1923 he was professor of town planning in Bombay. And from there he moved to a college

Patrick Geddes. Courtesy University of Strathclyde.

that he established himself in Montpellier, southern France. Geddes came to describe a future stage in man's technological development, the 'geotechnic', in which technology would harmonize with the earth's needs. The excitement of the impending millennium infected a young American regional planner, Benton Mackaye, who much later would report a 1923 conversation: ' "Geography", said he, Geddes, "is descriptive science (*geo* earth, *graphy* describe); it tells what *is*. Geotechnics is applied science (*geo* earth, *technics* use); it shows what *ought to be*." '[57] This would lead up to the 'eutechnic', which would characterise 'eutopia', that is, a practical utopia.

Though early public characterizations at first excluded biotechnic, this did not mean disinterest on Geddes's part. This can be seen in the evolution of *The Coming Polity* he published with Branford in 1917 and modified in 1919. The later edition had an additional chapter entitled 'The Post-Germanic University' which indicated lines of thought rather similar to those of Francé:

> Throughout the biological sciences and their arts as applied in the faculties of Agriculture, Hygiene and Medicine, we see then a definite transformation in progress. It consists of a turning round upon the mechanical, physical and chemical science and a deliberate harnessing of these to the services of life. From its former servitude to these preliminary sciences, life is not only escaping, but learning to apply whip and rein to its previous master.[58]

The word 'biotechnic' was available by 1921, after the publication of Francé's *Bios,* when he sketched a historical sequence giving biotechnics a key role in a letter to Mumford.[59] Geddes's transfer of Francé's word, which dealt with plant processes, to particular human processes, as seen in *The Coming Polity,* was confirmed in a 1923 letter in which Geddes defined 'biotechnics' as comprising agriculture, medicine, hygiene, and eugenics.[60] By then, the concept had been incorporated in a grand historical scheme, that Geddes would call 'Transition IX to 9'. This foresaw the translation of a society described by Comte and interpreted by Geddes as a nine-box grid (IX) describing 'wardom' with the key words 'militancy', 'state', 'individual', and 'industry' ('mechanotechnic') and its transformation into a future 'peacedom', also defined by nine boxes (9), described by the key words 'biotechnic', 'synergy in geotechnics', and 'folk', 'work', 'place in etho-polity'.[61] By 1931, Geddes was triumphantly seeing the dawn of the new age expressed in Krupp's investment in tractors rather than guns, and its use of the profits in garden cities.[62]

Despite its importance to his IX to 9 conception, Geddes first published the word 'biotechnic' in brackets and without explanation in a chapter heading to his and Thomson's short preliminary text, *Biology,* of 1925.[63] Again, although drafts make it clear that the concept was intended to be the climax of his tome *Life,* in the event it appeared rather haphazardly towards the end, without introduction or elucidation. So *Life* turned out to be a faint spark in the history of biotechnics, rather than an explosive affirmation. An article on biology written by Thomson for the *Encyclopaedia Britannica Supplement* of 1926 was probably a more important means of popularizing the concept.[64] Thomson attributed the coinage of the word 'biotechnic' to Geddes and explained that it meant the use of biological organisms for the benefit of man. If this appears a very simplistic rendering of Geddes's ideas, and certainly it was a long way from those of Francé, it does show how their very complex ideologically loaded terms were transformed into a more easily digestible form for the Anglo-Saxon market.[65]

So, while Geddes seemed something of an anachronism to the English professionals of the 1930s and *Life* was not academically respectable by the time it appeared, that is not to say that Geddes's work did not affect the intellectual environment.[66] A memorial supplement to the *Sociological Review* contained tributes by leading economists and sociologists and was paid for by a small group including R.A. Gregory, the editor of the all-powerful scientific journal *Nature*.[67] To understand the acceptance of his biotechnic concept, one needs to turn to the thoughts of his younger English contemporaries among the biologists.

Just as Geddes had promiscuously taken ideas from his predecessors, so younger biologists appropriated the concepts of biotechnics from Geddes, Francé, and Goldscheid. As early as 1914, the Cambridge student J.B.S. Haldane gave a lecture, which he was to develop in published form a decade later as a slim volume entitled *Daedalus or Science and the Future*. Although this paean to biological invention was first conceived before the war, its detailed expression allows us no room to doubt the significance of that catastrophe in the history of Western civilization. The book begins with an evocation of a battle on the western front:

> Through a blur of dust and fumes there appear, quite suddenly, great black and yellow masses of smoke which seem to be tearing up the surface of the earth and disintegrating the works of man with an almost visible hatred. These form the chief parts of the picture, but somewhere in the middle distance we can see a few irrelevant looking human figures, and soon there are fewer. It is hard to believe that these are the protagonists in the battle. One would rather choose those huge substantive oily black masses which are so much more conspicuous, and suppose that the men are in reality their servants.[68]

This experience, common to thinkers on both sides of the front, was perhaps even more important in shaping hopes for a more life-centred technology than any crystallizing ideas that they then exchanged.

Haldane goes on to suggest that, whereas in 1902, when H.G. Wells published *Anticipations*, flying and radiotelegraphy represented the limit of scientific ideas, they had, by the 1920s, been exploited commercially. The scientific problems of the present lay in biology. In the future, physiology would invade and take over mathematical physics, and Bergsonian activism would be the working metaphysical hypothesis of science and practical life. Haldane defined the 'biological in-

J.B.S. Haldane. Courtesy National Portrait Gallery.

vention' as, 'the establishment of a new relationship between man and other animals or plants, or between different human beings, provided that such relationship is one which comes primarily under the domain of biology rather than physics, psychology, or ethics.'[69] He suggested that every biological invention would begin as blasphemy, and indeed, the book is famous for its prediction of one process that is today being so described. Haldane described ectogenesis, which combined in vitro fertilization with the further complete development of the foetus outside the womb. In case the reader misses the moral, Haldane makes it explicit: 'I have tried to show why I believe that the biologist is the most romantic figure on earth at the present day.'[70]

Haldane's little book of less than a hundred pages is one of the most influential stimuli of science fiction ever written. The novelist Aldous Huxley made a conscious attack on its vision in *Brave New World,* but his brother Julian, himself an eminent biologist, was an enthusiast for human engineering. Possibly Julian Huxley was influenced directly by Goldscheid: His thought certainly took a remarkably similar path. In his 1926 Norman Lockyer lecture, Huxley referred to the categories of the quantity and quality of people which would be controlled respectively by birth control (a 'revolutionary biological invention' he called it, echoing Haldane) and eugenics.[71] Large families engendered the waste of child deaths, the 'massacre of the innocents'. In this lecture, Huxley willingly accepted the need to improve the environment to provide diet, bodily exercise, air and light, and mental interest, and also called for the restriction of reproduction of society's bottom 20%.

The relationship to the philosophy of Goldscheid was even clearer eight years later in an article Huxley unambiguously entitled 'The Applied Science of the Next Hundred Years: Biological and Social Engineering'.[72] Here he called for the planning of leisure (social engineering) and of the quantity and quality of population (biological engineering). In the face of anxieties about the role of science in the depression, he robustly answered: 'The cure for the ills of the Scientific Age is not less science, but more science. And the science that is most needed at the moment is that of biological engineering'.[73] In a sentence that seems to demonstrate his indebtedness to Goldscheid's *Menschenökonomie,* Huxley goes on to point out that human development would be a nation's most important process of manufacture. The use of the idea of biological engineering seems to have become quite common at the time. At a Workers Educational Association meeting in Cambridge in 1934, the socialist biologist Joseph Needham, discussing *Brave New World,* proposed to his work-a-day audience that the themes were basically biological engineering and world dictatorship, later reprinting his remarks without explanation in the popular magazine *Time and Tide.*[74]

Needham's conception of biological engineering owed more to the school of Loeb, than to the now disgraced Kammerer. Ectogenesis was an increasingly common idea. Polyembryony, by which a fertilized egg would be subdivided to produce several genetically identical offspring, again was already being explored with rabbits. The social implication that Needham suggested to his audience was a 'genetic strain of submission or capacity for routine work and staff a whole factory, not merely with brothers, but with identical "twins" all from the same egg.'[75] Practical, in principle, as this already seemed to be, its application to people was still in the realms of science fiction. At the same time, biological understanding was providing insights of a more immediately applicable social significance.

Decline in population and increasing evidence of malnutrition in Britain, even before the depression of the early 1930s, had provided a context for 'social biology' in Britain. Though the term can be found in nineteenth-century English, it was considered 'odd', perhaps because it had been popularized in Austria by Goldscheid.[76] Nonetheless, the radical London School of Economics, as a centre of social sciences, was well connected with German social hygiene and was given Goldscheid's *Höherentwicklung* in the early 1920s.[77] William Beveridge, the school's director from 1919 to 1936, worked assiduously to create a chair of social biology from the early years of his tenure.[78] Even before the First World War, he had been concerned with German solutions to the problems of social welfare. Now he worried about the problems of the population and in 1920 addressed the British Association on the dangers of underpopulation. The talk impressed Beardsley Ruml, the head of the immensely wealthy Laura Spelman Rockefeller Foundation. The two men talked, Beveridge putting forward a vision of social science based on the natural sciences that encompassed a range of anthropological, eugenic, nutritional, and psychological issues.[79] This vision, so similar to the *Biotechnik* visions of Goldscheid, impressed the Rockefeller Foundation, particularly in its reference to social biology, and over the next few years Beveridge persuaded his colleagues and the foundation of the importance of the new subject. Beveridge sought a biologist interested in economics and politics and finally recruited Julian Huxley's close friend Lancelot Hogben, currently at the University of Cape Town.[80]

Hogben was a bitter opponent of eugenics, and indeed of any unclear philosophizing about biology. Desperately, he sought acceptance as an eminent scientist, scheming with Huxley for election to the Royal Society. At other times, he would consider that promoting the public understanding of science would be even more important. The two interests sometimes almost tore him apart. He wanted people to recognize the importance of zoology and his importance within it. Hogben's militant friend, J.D. Bernal, had portrayed the biological level as a mere temporary holding operation until the world could be reduced to physical and chemical units in his 1929 satire of the Book of Common Prayer, *The World, the Flesh and the Devil.* Even to the

Lancelot Hogben on his election to the Royal Society in 1936. Courtesy Godfrey Argent.

cynical Hogben, this was too much. At a personal level, having moved from South Africa to the LSE, he found the polluted London environment intolerable for his family, which then settled in the rural charm of Devon.

The chair of social biology seemed to give Hogben an opportunity to promote the wider significance of his science. From being a rather sceptical hard-nosed embryologist, Hogben developed his interests in the social implications of biology. His inaugural lecture pointed to the by then well-established fear of depopulation and linked it to Haldane's idea of the biological invention: 'The declining birth rate has brought us face to face with the fact that we are entering upon the era of biological invention.'[81] Hogben devised a pregnancy test and recruited a demographer with concerns about depopulation, René Kuczynski, a refugee from Hitler's Germany.[82] In his *Political Arithmetic* of 1937, which summarized the work of the department, Hogben called for applications of biology as practical as the applications of chemistry.[83] So by the early 1930s, when both Geddes and Goldscheid died, social biology had achieved the status of university subject. Geddes may have been considered old fashioned in his personal beliefs, but the concept of biotechnics with which he was associated had become part of the cultural substructure of biology. As for Goldscheid, there were many routes through which his ideas might have entered English discourse – his 1929 visit to London to attend a sexual reform congress and his impact upon Beveridge, Hogben, and Kuczynski, not to mention his book readily available in the London School of Economics' sparse collection of biological texts.

It was not necessarily coincidental that on the first anniversary of Geddes's death, *Nature* published an editorial entitled 'Biotechnology'.[84] With the exception of Siebel's and Mason's commercially motivated use of that word, this seems to have been its first use in English. The author was the chemist and frequent contributor, Rainald Brightman, but the topic and title smack of the mark of the editor, R.A. Gregory, an old friend of Thomson and admirer of Geddes, and an enthusiastic technocrat and eugenist. The article argued that science-based 'technology' should advance humanity and that biology, through contraception and eugenics, could improve the quality, and make up for the shortfall in quantity, of the British population. Echoing Goldscheid's concept of biotechnics twenty years before, it was expressing sentiments similar to Huxley's when he spoke of 'biological engineering'. Confusingly, the roots of the title 'Biotechnology' in the thought of Huxley would not have been unambiguous, and would also have carried continued connotations of Geddes's 'biotechnics'.

An eclectic interpretation was again called for three years later. At a 1936 lecture given by Hogben at London's Conway Hall, entitled *Retreat From Reason,* the chairman, Julian Huxley, introduced his friend's theme by predicting that 'biotechnology' would in the future be as important as technologies based on chemistry and physics had been in the past, using the words quoted at the beginning of this chapter. From our vantage point, the statement seems remarkably forward looking. However, here he was using Hogben's word to repeat the claims he had made about 'biological engineering' two years earlier, which in turn had echoed the twenty-year-old arguments of Goldscheid and the decade-old promises of Geddes for biotechnics.

In the lecture itself, Hogben did more than merely reiterate those old ideas; he combined, amalgamated, and transformed them. For him, the ideal of modern society was to be found not in the factory, but in the country where he had settled his family. In a letter about his talk, evoking a 'Green and Pleasant Land', to his friend Joseph Needham, Hogben explained that he saw biotechnology as a socialist and aesthetic answer to the polluting mechanical technologies inherited by his generation. If only energy could be obtained from the fermented product of the Jerusalem artichoke![85] Humanity (of the English brand) therefore needed to be radically transformed. In his lecture, he announced, 'Social science can no longer accept the work of evolution as finished', echoing Goldscheid.[86] Not that he would have tolerated Lamarckism, but he did see the old opposition of the immortal germ plasm of Weismann as obsolete. Influenced by his wife, Enid Charles, Hogben reflected on the need to reverse population decline. At the same time, production on farms could not be allowed to continue its traditional slovenly ways: 'As a biologist, I am in favour of planned production based on bio-technology, because I cannot conceive how planned consumption is possible without planned production.' A final component of biotechnology was microbe-based chemical transformation: 'Biochemistry shews that we do not have to wait till Nature has converted green forests into Stygian gloom. I should be more impressed by the arguments of the professional economist, if he could convince me that he knows how easily mesitylene can be made out of acetone.'

Hogben's use of the word 'biotechnology' brought together the ideals expressed by Ereky, Pope, Goldscheid, and most obviously, Geddes. Biotechnology did not stand in isolation; it was an aspect of his vision of a 'bioaesthetic' utopia in which small-scale communities would be based on hydroelectric power, light metals, fertilizers, and the applications of biochemistry and genetics to the control of population.[87] Huxley's firmly eugenic interpretation as expressed in his introduction emphasized one aspect. There was also, in the belief in production through fermentation, a link to the zymotechnic tradition.

Hogben had written a caustic review of Geddes' *Life*, but he did have a tendency to express himself with considerable acid, and he had in fact moved away from particular biological inventions, in the Haldane sense, to the model of an entire new biologically sensitive agriculture, much more in the spirit of Geddes. Here he was clearly strongly influenced also by the Russian biologist and theoretician of natural sites for plants Vavilov, who had visited England in 1931. Two years after *Retreat from Reason,* Hogben reiterated his vision of biotechnology as an aspect of productive agriculture in his hugely successful *Science for the Citizen.*

In the meantime, he had moved to the Regius chair of Natural History at Aberdeen, brought there by the nutritionist John Boyd Orr, director of the nearby Rowett agricultural research station. Hogben, normally cautious in praise, said later that Orr, popularizer of the phrase 'the marriage of health with agriculture', had a major influence on his values.[88] Orr, a principal figure in the development of British nutrition policy and later founder of the United Nations Food and Agriculture Organization, aimed to solve the problems of both agricultural low prices and urban malnutrition through the promotion of free school milk which began in the 1930s.[89] Better agricultural technology could be seen, by Hogben too, as closely bound up with health and, therefore, higher birth rates.

Orr's work was but one element of a worldwide trend in the 1930s emphasizing the importance of nutrition. The importance of vitamins and balanced diets was being widely promoted within individual nations and through the League of Nations.[90] These emphasized the uses of biology in improving the health of populations and agriculture. Equally, new concepts in genetics were casting doubt on the simple certainties of early race hygienists. Population genetics, to which Haldane and Hogben were key contributors, was showing how complex were the links between the characteristics of succeeding generations. The new understanding of environmental issues and the growing complexity of genetic claims, as well as the perversion of Nazi interpretations of genetics, led to the 1939 declaration by leading Anglo-American geneticists, including the 'three Hs' (Haldane, Huxley, and Hogben), that the enhancement of the environment would be a key factor in the improvement of populations.[91]

These promoters of biotechnology were coming towards the social centre of British science as they were formulating their ideas during the 1930s. Geddes accepted a knighthood shortly before his death, and the fortunes of the younger generation were also transformed. In the late 1920s, they had still been marginal men, even if marked with talent and self-confidence. Haldane lost his readership in biochemistry at Cambridge on account of a sex scandal, and even when he was reinstated, he held parallel part-time positions at Rothamsted and at the Royal Institution. We have already seen Huxley leaving Kings College to work at the popular biology text he was writing with

H.G. Wells. Meanwhile, Lancelot Hogben was feeling marooned in Cape Town. During the early 1930s, the careers of the 'three Hs' blossomed.[92] Haldane won a chair at University College London in 1933. Huxley was elected to the Royal Society in 1935 and became secretary to the Zoological Society the same year. Even Hogben was elected to the Royal Society and obtained his Regius chair, with the active support of the Scottish secretary overwehelming the memory of his pacifism and imprisonment during World War I.[93]

At a popular level, Huxley and Haldane became stars of the new medium of radio broadcasting, while possession of Hogben's book *Science for the Citizen* came to define the 'intelligent layman' beloved of publishers. Their inspiration propelled increasing numbers of students and acolytes into a relationship with the great events of the day, with industrial change, and with malnutrition. Haldane's friend Pirie, at the beginning of the war, shifted his interest from Tobacco Mosaic Virus to a lifelong engagement with the production of protein from leaves, following pioneering work by Karl Ereky.[94] Another protégé of Haldane was the young refugee Ernst Chain who would be a key player in the development of penicillin.

The biologists Huxley and Hogben had coopted the chemists' enthusiasm for natural production and, by combining it with the ideas of environmental engineering as an influence on human nature, had amalgamated 'biological engineering' and 'biotechnology'. At least one influential chemist was willing to return the favour and to pay his respects to this expression of the potential of biology. The British chemist and industrialist Harold Hartley, himself one of the moving spirits, would often recite a story which he was told by Carl Bosch, head of Germany's IG Farben. Charles Steinmetz, the hunchback electronics genius of General Electric, echoing the vision of Pope a decade earlier, had suggested to Bosch, 'You can make indigo cheaper than God, you may make rubber cheaper than God, but you will never make cellulose cheaper than God.'[95]

Hartley's own son had taken part in a major expedition to the lush rain forests of British Malaya, and Sir Harold seems to have been impressed with the truth of Steinmetz's remark. In a much cited 1938 lecture on forest products to the Textile Institute, he reflected upon Hogben's praise of biotechnology. He was not impressed by the particular example of acetone, but the general point was crucially important. Hartley concluded rhapsodically: 'With the modern techniques of genetics and the closer association of the farmer and the manufacturer there is a fascinating prospect of new strains that will yield the ideal products for industry almost to a standard specification. Then indeed we should have Bacon's ideal of commanding nature in action.'[96] The concepts of biotechnology descended from zymotechnology and of *Biotechnik* were coming together. The next chapter will explore their translation and elaboration in the United States.

4
Institutional reality

Biotechnology is that phase of modern technology which integrates the theory and practice of engineering, medicine, and biology. It applies the principles of technology to the biological sciences and applies the principles of living biological material to technology. The contribution that each can make to the other cannot be measured. With each profession at its advanced state, it is becoming extremely important to coordinate and integrate the professions.

(H.J. Sauer Jr. and R.G. Nevins, 1965)[1]

Introduction

Hartley's euphoria had been stimulated by developments in the United States. There, engineers were already beginning to make the crucial moves towards embedding an alliance with biology in institutions and bureaucracy. The war's end in November 1918 had left a legacy for Europe of internal disorder, lost world political hegemony, millions of casualties, and bankruptcy. Haldane's evocation of the image of the tank, the biotechnics of Francé and Geddes, and the biotechnology of Ereky were thus aspects of a European cultural creativity born of disaster. For the United States, by contrast, the end of World War I, far from inspiring a sense of cultural crisis, had been a supreme triumph and confirmation that this was indeed the American century. Although anxious Europeans and their debates did come to the United States and isolated individuals did fret about the direction of their civilization in the 1920s, their thoughts were not widely shared among the rich or powerful. By contrast to Haldane's reflections on charges of blasphemy, microbe hunters were cast as popular heros by Paul de Kruif and Sinclair Lewis.[2] In 1918, 'America was thus clearly Top Nation and History came to a .' solemnly concluded the British historical satire *1066 and All That.*

By 1930, though, when that classic was published, crisis – economic and then cultural – was coming even to the United States, with the collapse of the stock market, the failure of banks, and the Great Depression. In threatening traditional optimism and values (the novel *You Can't Go Home Again* was Thomas Wolfe's warning), the depression shook American certainties about engineering. Despite their impeccable Yankee lineage, mass production and modern soulless technology were seen to be driving people out of work. Against this background biotechnics came to be reinterpreted in the United States as the promise of a new engineering intimately bound to the needs of the human and biological world.

From the dream of a new engineering was born a conception of biotechnology which was to flourish for two decades after World War II. Then, for the first time, biotechnology would acquire a precise sense, caught up in U.S. concerns about enhancing the status and scope of engineering as an academic discipline with published curricula and grant proposals in prestigious institutions. Distinguished universities including MIT, in 1939, and UCLA, in 1947, created units with the titles of 'Biological Engineering' and 'Biotechnology'. By 1960, the American Society of Mechanical Engineers had established a group with the title 'Biotechnology'. The singular significance of biotechnology in the United States, as a new stage in engineering, is brought out by comparison with conceptions held in Britain. There the phrase 'biological engineering' also acquired a currency, and it applied to much the same practices: Yet in Britain, its significance was the emphasis on applying engineering to the particular problems of the biological and medical sciences, rather than to a new stage in engineering itself. These two visions were perhaps only subtly different, yet they were negotiated openly in Sweden.

Protected by its neutrality from the most overwhelming challenges of the world wars, during the intervening years Sweden went through its own industrial and cultural revolutions. In these, engineering played a key role, coloured by British, U.S., Danish, and German influences, as the Swedes created their own, rather rigid, new institutional forms and gave bureaucratic reality to biotechnics and biotechnology.

Despair and chemistry

By August 1930, reflecting on the sudden onset of mass unemployment, an article in the *New York Times* lamented, 'It is often said that the machines have become the masters of men, that mass-production is the ruination of society'.[3] A few months later, the distinguished commentator Stuart Chase spelled out the new threat to American values: 'Technological unemployment is entering a new and terrible phase, a condition unknown during the entire course of the industrial revolution'.[4] Responding to such attacks on their assumed status, the engineers began to fight back. During the early 1930s, the 'technocrats' suggested that the whole price system needed to be reworked to give a proper representation of the energy costs of different products. Engineers and scientists turned to a vision of the control, harnessing, and encouragement of their *métier* as an alternative to its restraint. Whereas innovation was popularly seen as a threat, they portrayed it as a promise. Rather than being seen as the result of pull from the uncontrolled and fickle market, change could be controlled by the planned application of science.

Chemistry, once again associated with the horror of Frankenstein's monster brought back to life by the movies, was vigorously defended by the optimists. This science, so often associated with cartels and with war, had been under attack for a long time, and a rhetoric of 'boosterism' had evolved in defense.[5] In the aftermath of World War I, Edward Slosson published his immensely successful *Creative Chemistry.* There he showed how his science underlay the more superficial perturbations of modern times. Following Slosson's lead and drawing on the ideas of a new age propelled by new materials, sketched by Bernal in *The World, The Flesh and the Devil,* Dow's William J. Hale offered his *Chemistry Triumphant,* published for the 1933 Chicago Century of Progress Exposition. To Bernal's list of novel metals, Hale added silicon and plastics and prophesied the start of the 'Silico-plastic Age' – thereby inaugurating half a century of reflection of what came to be known more commonly as the 'Plastic Age'.[6] More generally, technological change based on the applications of chemistry would reverse the gloomy economic situation being compounded by the economists. 'Chem-economics' would show 'how these test-tube termites are nibbling the insides out of all the nicely arranged and seemingly solid timbers upon which *all* the economic theories of the Machine Age rest'.[7]

Biotechnics in the United States: Mumford and Wickenden

The sense of shock also found expression in the adoption of ideas of scientific change and cultural direction developed earlier in Britain, where planning and a purposive control of the direction of technology were already being sought as alternatives to *laissez-faire* capitalism.

British science commentators were brought to the United States: Geddes was about to visit when he died. Richard Gregory, editor of *Nature* and a leading British promoter of planning, made a well-publicized tour in 1938. Again, the belief in the power of agriculture acquiring currency there through the chemurgic movement drew on European authority: Christy Borth's popular 1939 *Pioneers of Plenty: The Story of Chemurgy* lauds both Hogben's 'Retreat from Reason' and Hartley's Mather Lecture.[8] At a philosophical level, Hogben's bioaesthetic vision was alluring in a United States which, as Arthur Thomson reflected in 1930, was particularly drawn to the 'natural' or, to use the then new word, the 'holistic'.[9] Geddes himself was more popular in the United States than in his home country, and his most devoted followers, such as Amelia Defries and Philip Boardman, were Americans. As historian Arthur Molella has written, reflecting on the founding fathers of the history of technology – the Swiss Siegried Giedeion and the Americans Abbott Usher and Lewis Mumford: 'All three believed they were about to witness a major cultural transition from a mechanical to an organic world picture.'[10]

Two immensely influential publications serving to translate European concepts into the U.S. context during the early 1930s were Lewis Mumford's *Technics and Civilization* and William Wickenden's final report on engineering education. Mumford (1895–1990), an anti-academic intellectual, had been a pupil at Columbia University of Edward Slosson from whom he learnt, in a course heavily influenced by Ostwald, of the importance of energy to science and culture. Impressed as Mumford clearly was, it was not chemistry alone that he would see at the root of things. Technology became important as he came under the spell of Geddes. This was a relationship to which he would turn, time and time again – writing in *Encounter* in the 1960s and, as recently as 1976, broadcasting on the BBC about 'the master'.[11] So although he did reject the role of Boswell, which Geddes had hoped he would play, Mumford's thinking was intensely affected by the prophetic Englishman.

In *Technics and Civilization,* published in 1934, Mumford adopted the Geddesian historical categories – the paleotechnic, the neotechnic, and the biotechnic.[12] For him too, the biotechnic age, in which things were made in a way respecting the workers' biological needs of clean air and light and designed in a way that would respect the customers' biological needs, was the next stage in design. Following Geddes, Mumford's greatest interest lay in town planning. There, above all, the distinction between dismal paleotechnic factory town and the modern green residential garden city of the biotechnic age highlighted the nature of progress. Refining and reorienting his ideas over the following half century, Mumford was horrified by the growth of the phenomenon he named the 'megalopolis', and later of the web of militarized power which so contrasted with his 1930s dreams of the biotechnic way out of the depression.

Mumford approached engineering as a talented journalist, and for all his interest in town planning, throughout his life Geddes had been a biologist looking at manufacturing in abstraction. So, despite their eulogies of technology, both looked on it from the outside. Biotechnics, however, also impressed professional U.S. engineers. As engineering moved, in the 1920s, from the vocation of the skilled craftsman to the profession of a graduate increasingly armed with a doctorate, so the image and conception of an appropriate curriculum had to change. It had grown up as a set of distinct and jealous specialities and professions. But the distinguished engineer and future president Herbert Hoover argued, after World War I, that this was now an obsolete arrangement and a coherent unified curriculum would be required. William Wickenden, an AT&T researcher who would go on to become president of Case Institute of Technology, was recruited by the Society for the Promotion of Engineering Education to undertake a fundamental reevaluation of the U.S. engineering curriculum. His voluminous reports published between 1925 and 1934 were key elements in the movement to have engineering recognized not just as a university discipline, but as a graduate career with an intellectual base equal to the traditional professions of science and medicine.

Wickenden was also concerned with the place of technology in human history. Indeed, he presented several lectures with the title 'Technology and Culture'.[13] By the early 1930s, the meaning of the depression, in which food was burnt despite a starving world, was deeply worrying him.[14] Surely, the ambit of the problem-solving engineer should encompass the production of food. Biology then should be progressively drawn on to sustain his vision of an expansionist science-based engineering.[15] The psychology of work would also become part of the management-oriented engineer's responsibilities, and with it 'psychotechnics' (a word coined in Germany also at the beginning of the century). So when, in his June 1933 final report, Wickenden called for consideration of the implications of 'biotechnics' and 'psychotechnics', he was furnishing engineering with a new science base, a more human orientation, and a way of exploiting the general 1930s interest in new biological findings.[16]

The work of implementing the Wickenden Report was given to the

Karl Compton, sponsor of biological engineering, pictured in the late 1930s. Courtesy The MIT Museum.

Engineering Council for Professional Development (ECPD), established in 1932. Historian David Noble has argued that this bureaucratic innovation was the report's most enduring outcome.[17] One man who made it so was the chairman of the ECPD Committee on Engineering Schools, Karl Compton, the president of MIT. Compton, perhaps the most academically distinguished patron of the chemurgic movement, was himself particularly interested in the biological aspects of engineering. It was a cause he championed at his own university, as he sought to revitalize old established programmes in food technology and bacteriology whose faculty was falling behind modern advances in biochemistry. Now he saw the potential for a new approach 'recruiting and training physical scientists for careers in the growing health industry'.[18]

Moreover, by emphasizing the human side of engineering, MIT could respond to sustained political pressure on the profession. In 1936, President Roosevelt had committed the unusual step of writing an open letter to the heads of schools teaching engineering (published in the *New York Times*) condemning the impact of science and engineering on unemployment. Enclosing a recent conservation pamphlet entitled 'Little Waters', he argued that engineers should recognize the problems resulting from the application of science, and engineering institutions should provide more social science education to their students. Compton replied, also publishing his letter in the *New York Times,* that new technology was the solution to any problems raised by the old. Whereas machines made through mechanical engineering may have replaced human labour, the application of the modern sciences of chemistry, metallurgy, and biology would now be the way to enhance human welfare.[19]

Biological engineering

At MIT, a plan for 'biological engineering' took definite form in 1936, and a combined research and teaching programme was implemented a year later. The name chosen after much thought, over alternatives such as biurgy and biotechnology, had had the backing of Compton's Vice-President Vannevar Bush, because of its resonance with engineering defined as 'the art of organizing and directing men and of controlling forces and materials of nature for the benefit of the human race'.[20] Not very far in the background, and a useful rhetorical weapon, was the precedent of 'chemical engineering', in which MIT had taken a world lead a decade earlier – once again chemistry had provided the model.[21] 'I know you don't like the name now', Bush admitted, 'but it will grow on you as you think it over'.[22]

The subject was defined as 'the art of applying to problems of human welfare knowledge obtained from the investigation of biological problems with the aid of physics, chemistry and other allied sciences.' Five sections were to be incorporated within the laboratories: bio-electrical engineering (concerned with x rays and cathode rays), electro-physiology, biophysics, microbiology (including mycological biochemistry), and nutritional biochemistry.[23]

The chemical precedent was highlighted in a 1941 symposium on the future of engineering held to celebrate the 150th anniversary of Rutgers University in New Jersey. The key speech was made by Vannevar Bush, who used it to promote the vision of bioengineering. What was required was not merely the provision of a smattering of biology for the businessman or economics for the biologist, but a new discipline. He foresaw applications ranging from agriculture to fermentation:

> The primary needs of man are for food, clothing and shelter. Into the first two of these the application of biological science enters very definitely, and it enters somewhat into the third. The ultimate field for the biological engineer is correspondingly wide, and his ultimate position in society will be correspondingly important.[24]

The commentary by Bush's MIT colleague, the biologist Detlev Bronk, concluded with a ringing call to bring to life 'biotechnics', so well described by Mumford.[25] However, MIT was an old school in which disciplinary demarcations were well entrenched. Despite the enthusiasm of Bush and Compton and their original insistence on an engineering approach, they employed a professor with ideas of his own, and biological engineering evolved quickly into plain biology.

Biotechnology

In California, where institutions were being established for the first time, new intellectual conformations were easier to entrench. When the University of California system decided it required a new engineering school in 1944, UCLA, the campus of the city of the future, Los Angeles, was chosen as its home, in recognition of the demands imposed by Southern California's formidable growth and growing role in high technology. L.M.K. Boelter, appointed to head the new school, was a thermodynamicist interested in heat transfer but also with the interface between people and machines. At Berkeley, he had investigated methods of devising improved car headlights, and from there his interests moved to vision as a whole, in order to create information useful to a doctor or optometrist. During the war, Boelter worked on problems of human performance in aircraft, analyzing problems of heat transfer to pilots to understand what happened at high altitudes.

Boelter, himself a friend of Wickenden, was determined to head a school that dealt with engineering as an integrated unified whole.[26] Within that, of course, specialities did appear, but these were considered as *technologies* which, taken together, added up to engineering.

His own wartime experience had highlighted the interaction of man and machine. This, Boelter and his associate Craig Taylor called 'biotechnology'. In a programmatic article in *Science,* they explained their thoughts, drawing upon Wickenden's use of the word 'biotechnic' and even quoting the paragraph from his 1933 report.[27] There was, they said, a need for a deeper preparation for the engineer with a human-centred approach to the discipline. As their colleague Myron Tribus has put it, 'By "Bio" we meant the things that biologists study but from a perspective of a person trying to do design'.[28] Thus, Tribus suggests, it would have been fair game in Boelter's course to have had a final examination in which you proposed how you would do experimentation and analysis to do an energy balance on pigs.

The UCLA programme was widely respected. Boelter's colleague, Craig Taylor, was featured in *Life* magazine; a photograph used in that article is shown opposite.[29] Although broadly philosophically grounded, Boelter's special interests were in the man–machine interface, and biotechnology became associated with the study of what was elsewhere known as 'human factors research'. This was a growth area, especially as military and then space engineers needed to optimize conditions for fast-reacting pilots. By 1962, about 500 engineers and psychologists in the United States alone were identified with the new discipline.[30] At the same time, the wish to broaden the intellectual base of engineering to include the life sciences was common to many educational institutions. Some fifty U.S. universities and colleges included life–behaviour science interfaces with traditional physical-science-oriented engineering education by the 1960s.[31] Although they did not all use the title 'biotechnology' and the MIT equivalents to UCLA's 'Machine and Systems Biotechnology' course were entitled 'Sensory Communication' and 'Man–Machine systems', 'biotechnology' was a popular word. Carnegie-Mellon University followed UCLA in offering seminars under that title. A 1964 textbook of this rendition, by Fogel, who had been a graduate student in the UCLA department, defines its subject in terms of a wheel. Geddes had used a very similar device, but the content of Fogel's wheel is, however, strikingly different from that of his English predecessor.[32] Motivated by the wish to develop a profession, this time engineering, rather than biology or sociology, lay at the centre. The new biotechnology had replaced biological with engineering categories. The term 'engineering' was being promoted more widely at the time: In 1959 the engineering professions testifying to Congress on their attitudes to a federal Department of Science and Technology, objected to the inclusion of the 'T' word in the name and sought instead 'engineering'.[33]

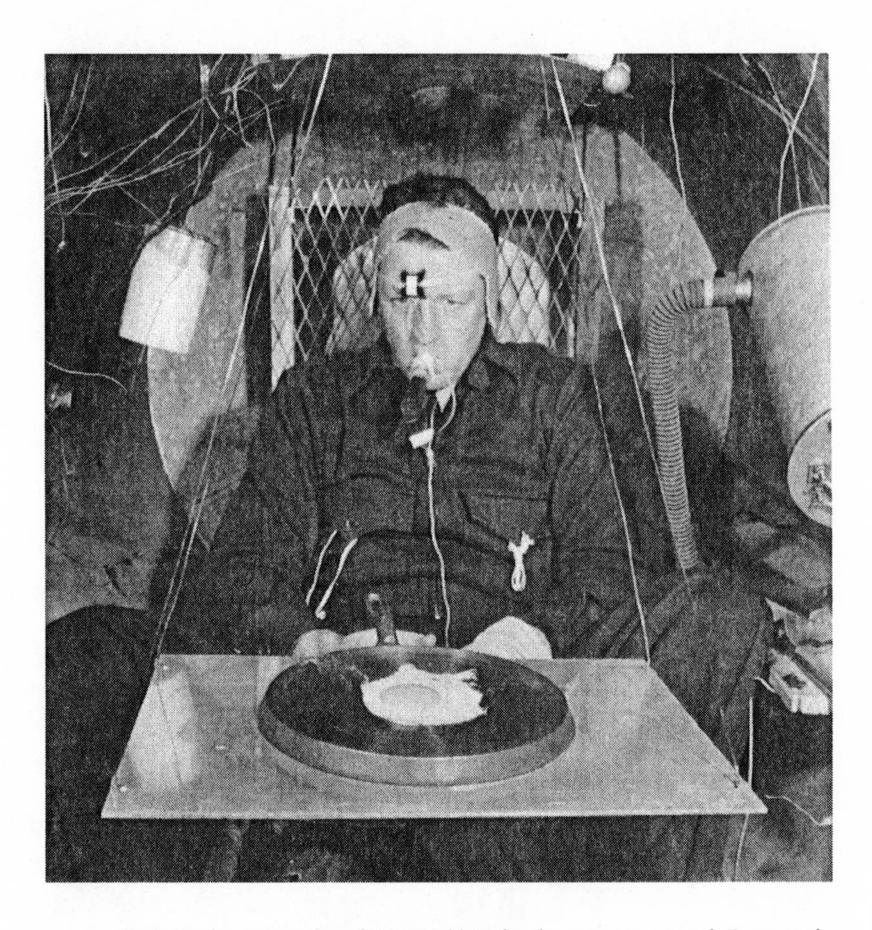

Craig Taylor, a founder of UCLA's biotechnology course, at work. Portrayed by *Life*, 9 February 1948. Photograph by Johnny Florea, *Life* magazine © 1948, Time Warner Inc. and Katz Pictures Ltd.

Boelter's vision of 'technology' even as a subordinate class was possibly too complex, and in 1966, the term 'bioengineering' was officially chosen by the Engineers Joint Council Committee on Engineering Interaction with Biology and Medicine, Subcommittee B, to mean 'the application of the knowledge gained by a cross-fertilization of engineering and the biological sciences so that both will be more fully utilized for the benefit of man.'[34] This sweeping category included a set of subjects little different from those construed by Vannevar Bush and familiar to earlier pioneers of biotechnology and biotechnics: medical engineering, environmental health engineering, agricultural

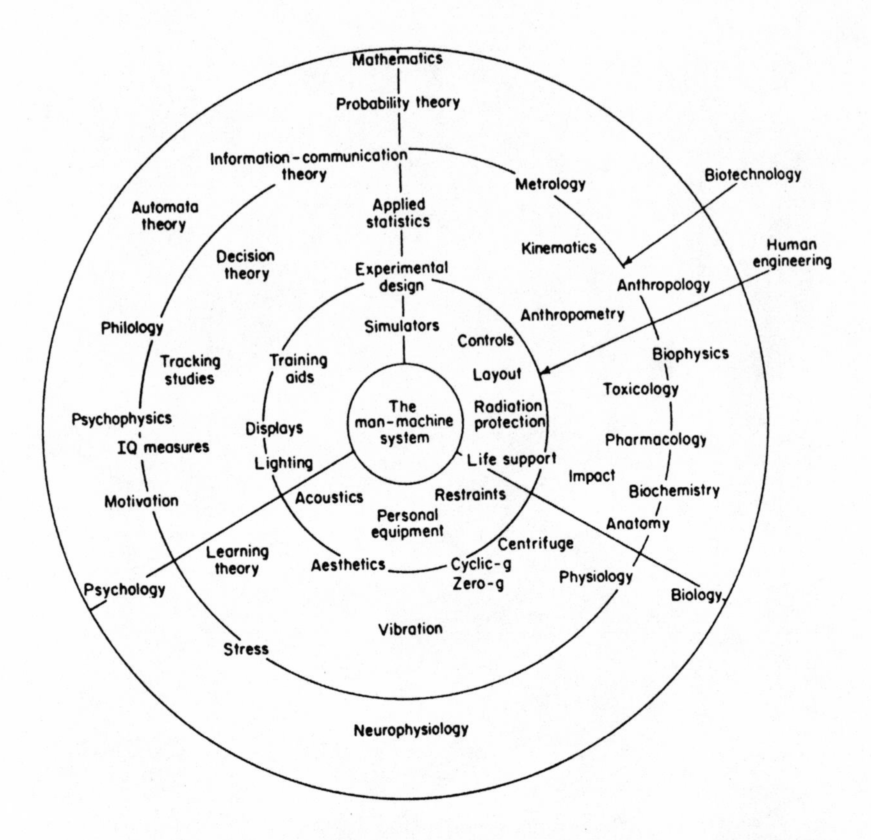

Biotechnology as seen by U.S. engineers in 1963. Lawrence J. Fogel, *Biotechnology: Concepts and Applications*, © 1963, p. 798. Reprinted by permission of Prentice Hall, Englewood Cliffs, N.J.

engineering, bionics, fermentation engineering, and human factors engineering. Bionics, for instance, is the English rendition of Francé's concern with 'the study of the function and principles of operation of living systems with application of the knowledge gained to the design of physical systems'.[35] Most recently, it had been rendered as 'cybernetics' under the visionary guidance of men such as Norbert Wiener, whose own parallels between machines and organisms themselves harked back to the era of Francé.[36] Agricultural engineering, 'the application of engineering principles to problems of biological production and to the external operations and environment that in-

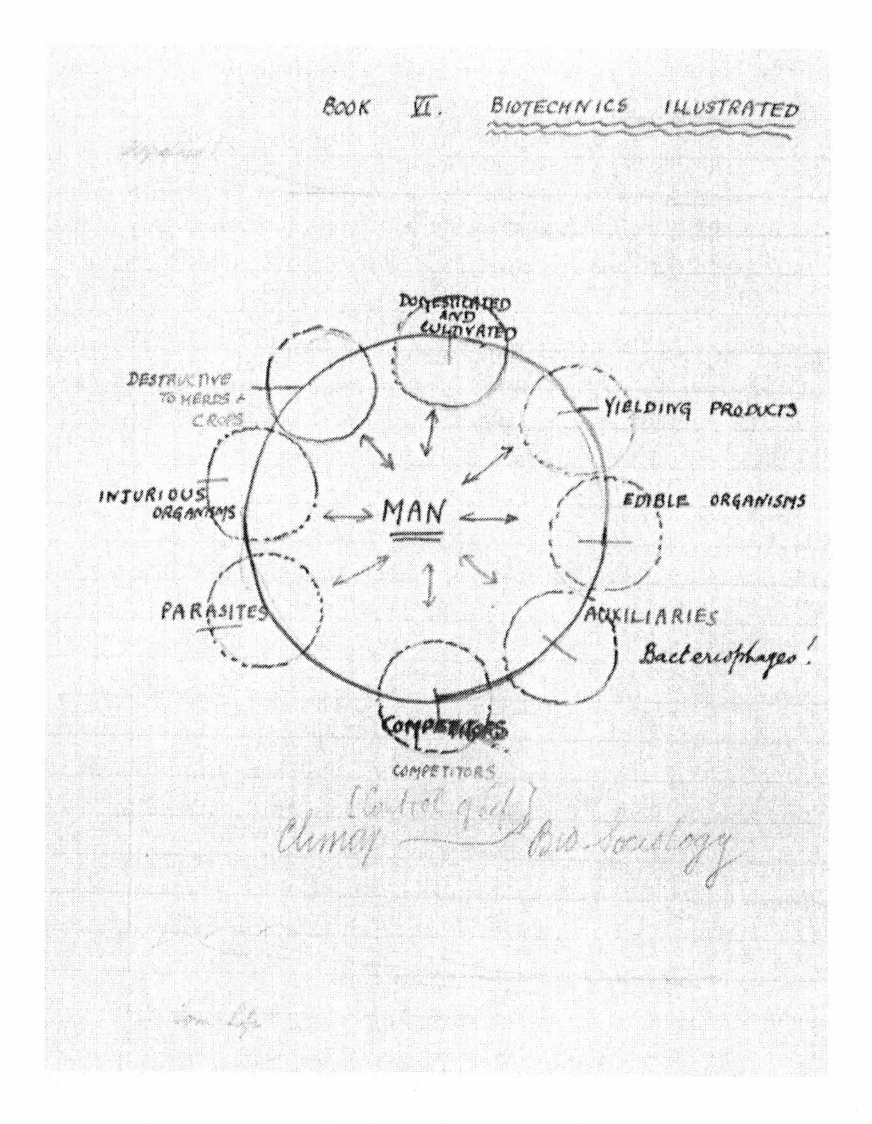

Biotechnics as construed by Geddes from a draft of his book *Life*, c. 1930. Courtesy University of Strathclyde.

fluence it', was of course the concern of Ereky. Fermentation engineering was the problem of Orla-Jensen, while human factors engineering was an alternative name for Boelter's original research interests.

From a research perspective, these subjects might have seemed diverse; equally, their commercial and technological significance and dynamics were quite different. However, they did share an educational basis and were held together by the belief that life sciences should play a crucial part in the education of engineers. The belief in the centrality of biotechnology to the future of the whole of engineering had been maintained. The quotation at the beginning of this chapter is taken from a classic 1965 paper, whose concept of biotechnology as a 'phase' in the history of modern technology would have endeared the European promoters of biotechnics a generation earlier. Moreover, whatever the variety of interpretations, as a 1968 UCLA report argued, 'There *is* a general consensus, however, that whatever content is taught should emphasize the unity of function in living systems and that wherever possible the material should be treated quantitatively.'[37]

In principle then, 'biological engineering', as it was understood in America around 1960, encompassed an indefinitely broad range of potential specialities. The debates over the proper name and character of biological engineering highlighted two characteristics. Bioengineering emerged as a single engineering category because, primarily, it was a vehicle for the introduction of the life sciences into undergraduate teaching. It was a teaching rather than a research category. UCLA, which had pioneered biotechnology, was in the 1950s the recipient of a ten-year $2m Ford Foundation Grant into the development of the engineering curriculum. Second, as MIT's argument made clear, this was a border subject getting its integrity and identity from its functional role as an interface between the historically quite separate disciplines of biology and engineering.

Biological engineering in Britain

By the 1960s, as interfaces between biology and engineering multiplied, available words were mapped on to these disciplines in a variety of ways. In Britain, the approach taken was different than in the United States. Human factors engineering, and the related human engineering, which Boelter had called 'biotechnology', were together termed 'ergonomics', and an Ergonomics society was launched in 1949. Whereas 'biological engineering' was a term with a very broad connotation in the United States, in Britain it was at first used for the single specialty that Americans called 'medical engineering'.

Physiology, of course, had a well-established engineering orientation of its own, and this was particularly well represented in the Biophysics Research Unit of University College London founded by the distinguished physiologist A.V. Hill (whose World War I work on antiaircraft gun aiming is a classic ancestor of cybernetics and who, Rutherford said, 'made a noise like a physicist').[38] Wartime experience, from prostheses to radar technology, had created a community of engineers working at the interface with the human body, and increasingly, individuals from these different backgrounds were working together. Beginning in the early 1950s, Heinz Wolff at the National Institute of Medical Research (NIMR) was describing his work as 'Bio-Engineering'.[39] Nonetheless, action at a national level was to be at the instigation of an American.

Since 1955, Vladimir Zworykin, who had earlier pioneered U.S. electron microscopy as well as his better-known innovations in television technology as head of research at RCA, had directed a medical electronics laboratory at the Rockefeller Institute (then led by Detlev Bronk). Finding the community working in isolation and often doing parallel work in ignorance of others, he decided to pull them together. In June 1958, Zworykin convened the first conference on medical electronics in Paris, leading to the launch of the International Federation of Medical Electronics in 1959.[40] The British were stimulated to create organizations that could be affiliated. The first was the Medical Electronics Section of the British Institution of Radio Engineers. In a speech launching the group, A.V. Hill complained that 'medical' was far too narrow a term. Instead, the broader 'biological' should be used to describe the applications of sensitive detectors and amplifiers that engineers were now designing.[41]

Of course, the issue was wider than electronics too. This was recognized in June 1960 when a meeting attended by doctors, physiologists, electronic engineers, mechanical engineers, and physicists led to the establishment of a new specialist grouping under the title of the Biological Engineering Society. Its purpose was to bring together members of the different disciplines from hospitals, research institutes, and industry and 'to further the application of engineering to biological and medical problems'.[42] The first scientific meeting held in October of that year at the NIMR conjured up the genie of 'a fusion of the two disciplines of biology and engineering'.[43] Thus, quite independently from the contemporary U.S. usages, 'biological engineering' was once again being chosen as the most appropriate umbrella concept. A 1964 inaugural address by the new professor of bioengineering at the University of Strathclyde, Robert Kenedi, conjured up prewar concepts in striking resonance with Harold Hartley's lecture on the potential of biological design, and indeed with Huxley's earlier concepts: 'Ultimately, in the more distant future, once genetic control is understood it would seem fairly certain that if the specialist fashion still obtains, genetic engineers or "genengs" will arise who will vie with each other in producing functional human body designs most suited for the career chosen.'[44] At a more mundane level, in practice the scope of concepts were similar on both sides of the Atlantic. After all, the philosophical bases of a subject are rarely a critical concern. British usage was recognized by the International Federation, and when in 1963 it launched a journal, the title chosen was *Medical Electronics and Biological Engineering*. Nonetheless, the work of biological engineers was not portrayed as a revolution at the heart of engineering education as it had been in the United States. The promoters were self-consciously specialists, bringing together diverse skills to animate a new area of research. Moreover, there was one significant respect in which the meaning of biological engineering differed: the place of fermentation technology. This could be incorporated, in principle at least, within the U.S. model of an integrated biology and engineering. The British emphasis on human applications distanced the topic. Perhaps the distinction might never have been noticed, but for the punctilious Swedes. The outcome of their careful debates was that the category of 'biotechnology' would be redefined and came to be focused upon microbes, not man.

The Swedish contribution

Sweden is an ordered, much more centralized country than the United States, Britain, or indeed Germany, so the processes of negotiation can be seen in their bureaucratic concreteness. Nor was this just petty politics made visible. The institution within which debate occurred was important and very sophisticated. Sweden, open to other cultures, provided an intellectual crucible in which different biotechnic traditions were blended. Curiously, the U.S. movements of biological engineering and biotechnology arrived separately and were independently interpreted, and their meanings were formally specified. 'Biological engineering' (*bioteknik*) emerged as a very wide term, and 'biotechnology' as a specific term relating to human factors engineering.

As early as 1942, the word *'bioteknik'* was first institutionalized within a national institution. The Royal Swedish Academy of Engineering Sciences (IVA) concerned with the provision of energy, food, and medicines amid wartime trade difficulties and shortages had turned to biological methods, and a new section of the academy entitled *'bioteknik'* was launched in December 1942. Two interests were clearly at work. On the one hand were practical men associated with the variety of biological food industries, such as brewing and baking, that were not properly represented in the organization. The initial proponent, Almgren, the former secretary of the Swedish Brewing Research Institute, came from this community. On the other hand were cosmopolitan, philosophical engineers who led IVA. Almgren died just after his initial suggestion had been given institutional blessing with the formation of a section on biotechnics. It was then taken up by the recently retired founder-director of the academy, Axel Enström.[45]

Enström was the epitome of the cosmopolitan engineer. In the aftermath of the First World War, he had been responsible for establishing IVA and later was responsible for its museum. It is interesting how many associated with this story, Geddes, Mumford, and Boelter, as well as Enström, were fascinated by science museums which illustrated the transformation of man's technological environment. Enström himself, like Schumpeter in Austria and Kondratiev in Russia, was absorbed by the problems of technological evolution, unemployment, and the business cycle. Here he led debates that were exercising the minds of many intellectuals in the tight-knit capital of this newly industrializing country.[46] Sweden suffered badly during the depression, and Enström, though an electrical engineer, had represented his country in meetings to discuss many technologies including agriculture. At a world energy conference in Berlin he reflected, rather like his U.S. colleagues, on the centrality of engineering in the solution to modern problems.[47] In 1940, Enström was succeeded as director by Edy Velander, an equally cosmopolitan engineer who had graduated at Harvard and MIT and kept closely in touch with his American contacts even during the war. During 1943, Velander would pay a formal visit to the United States. Again, like his American colleagues, he was interested in finding new scope for engineering and, developing a wartime interest, would launch a journal of food technology in the 1950s.[48]

Developments in Stockholm reflected the creative tension between these cosmopolitan engineers and the industrialists. At the initial meeting of the section on 19 June 1943, Velander tabled a preliminary paper he had written ten months earlier (when the section was being contemplated) about biotechnical research.[49] This began with general reflections about the contact between biology and engineering, reflected on the importance of food and hygiene, and suggested very general zones of interest. Intriguingly, it contains exactly the same evocation of clothing, food, and housing that Velander's old MIT teacher, Vannevar Bush, had summoned up a few months earlier at Rutgers, in calling for biological engineering – providing clear evidence of the use of an American model. The discussion, by contrast, centred on the varieties of food technology, and indeed on aspects such as hormones, which would be encompassed by the new section.

Literally, '*bioteknik*' translated Bush's 'Biological engineering'. Moreover, the word already existed in the departmental title (Bioteknisk kemi) of fellow-Scandinavian Orla-Jensen. Whichever was the more important influence, both sustained the belief in the centrality of *bioteknik* to engineering as a whole and in its practical implications. In the February 1943 issue of the official journal of IVA, Velander argued:

> During its development up to now, engineering technology has predominantly had to be based on natural phenomena, treated as far as science is concerned within the branches of physics and chemistry, including physical chemistry, and consequently falls within the inorganic world. However, for a few years now scientific research in the area of living organisms has produced a range of results of far-reaching significance to practical life, in medical and physiological respects on the one hand, and as regards the conditions for the production and use of agricultural and forestry products on the other hand. This has given and is still giving rise to problems of a technical industrial nature which increasingly require the cooperation of engineering science.[50]

It is worth following this with the formal definition on which the new section was based:

> Under this term ['*bioteknik*'] I [Velander] would tend to bring together applications which arise while one is learning to influence biological processes scientifically and exploit them technologically in an industrially organized activity, for example in industrial yeast cultivation, in the food industries for processing and improving the raw products as well as for the preparation and conservation of foodstuffs.

Just as in the United States, this development was seen as providing new opportunities for the engineer: 'A broad field is opening up here for industrial and agricultural technological work, in which the engineer has a major role to play by as usual being the interpreter who explains the scientists' theories and findings to the practitioner and the economist, thereby facilitating their practical assimilation.'[51] At first, the section was dominated by brewing interests not least because of the interest of Velander's successor Professor Bruholt, but increasingly, the activities of the section in Stockholm were influenced by medically oriented engineers, whose understanding of the word '*bio-*

Carl-Göran Hedén pictured in the 1991 IVA membership list. Courtesy IVA.

teknik' seems to have reflected the technology of physiology as connoted by 'biological engineering' in England. By contrast, in the late 1950s, the section also included the gentle, but zealous bacteriologist Carl-Göran Hedén, interested in fermentation technology and then assistant professor of bacteriology at Stockholm's Karolinska Institute (the Stockholm medical school). Hedén was, and is, not merely interested in bacteriology. For over forty years, he, more than any other biologist, has communicated visions reminiscent of Hogben and Geddes to succeeding generations.

Hedén

The Swedish school of biochemistry is well known for its instrumental tradition. In Uppsala during the 1920s, Svedberg developed the ultracentrifuge, and in the 1930s his successor Tiselius carried out the Nobel Prize–winning development of protein separation through electrophoresis. The cytologist Caspersson at Stockholm's Karolinska Institute also emphasized the use of automated instrumentation; later he was to pioneer the banding techniques with which human chromosomes would be mapped.[52] His student, Hedén, had become interested in the engineering aspects of microbiology as he struggled to build his own spectrophotometer for estimating nucleic acids in bacteria. From there, Hedén built a fermentation pilot plant that was one of the world's largest academic fermentation facilities. So although Hedén was not himself an engineer, his skills were thoroughly practical.

Certainly, Hedén saw *'bioteknik'* as dealing with problems of controlling microorganisms. In 1956, he persuaded the head of the technical nomenclature commission of his vision. Two years later, he encouraged the technical nomenclature commission of IVA to hold colloquia dealing with the meaning of *'bioteknik'*. The invitees to the initial colloquium would be strictly from the original category of industrially interested parties, and they were to be divided between those concerned with protein (microbiologists, plant physiologists, and brewers, as well as cellulose and forestry industry interests) and those more interested in fats (representatives of such industries as milk and margarine).[53] Though Hedén called his own special area of interest 'bacteriological biotechnics' to distinguish it from medical biotechnics, he much preferred the word *'bioteknologi'*, a similar, but distinct Swedish word.[54]

In Sweden, Hedén was frustrated because Boelter's word 'biotechnology' had been imported as an alternative to the new English term 'ergonomics'. This use was backed by the influential Professor Svent Forssman, of the institute for the study of work (ASTI). Moreover it was a subject of Swedish preeminence – indeed the first World Congress of Ergonomics was held in Stockholm in 1962. Contradicting this claim of the human engineers, Hedén wrote to the secretary of IVA, claiming that biotechnology 'has long' been associated with his interest in the 'Industrial Application of Biological Principles' (under which he put mostly microbiological examples) and biological aspects of technology (which included biomedical engineering).[55] He resented the way it had been claimed by the practitioners of human engineering and pressed the Technical Nomenclature Commission to accept the English word 'ergonomics' or some more Swedish variant – such as 'adaptation technology'. The use of the prefix 'bio-' to refer to just the human aspects of biology was repugnant agreed the purist Technical Nomenclature Commission.[56] Moreover, the suffix '-teknologi' was considered too similar to *'teknik'* to make the distinction between the broad-based combination of biology and engineering on the one hand, and the human focus on the other. However, Forssman's influence was too strong, and *bioteknologi* maintained its Boelter-derived meaning.[57]

Hedén can be seen to have been the first to have divided the issues of applied microbiology from the physiological aspects of bioengineering with which they had been combined for twenty years. Further, he reversed the U.S. interpretation of his subject as a purely engineering speciality common in the United States, towards a claim that 'the emphasis must now be regarded as lying on the biological plane'. Besides, with an international perspective, his influence was not limited by his own small country. Thus, it was an American rather than a Swede he managed to convert to his view of biotechnology. Through a conversation with his American friend Elmer Gaden, editor of *Journal of Microbiological and Biochemical Engineering and Technology,* founded in 1958, that journal changed its name to *Biotechnology and Bioengineering,* whose story will be recounted in the next chapter.

Conclusion

The vision of a newly vibrant engineering sustained the concept of an umbrella 'biological engineering' in both the United States and Sweden. Within that, enthusiasts sought to coopt the word 'biotechnology' for their favourite speciality: Boelter choosing what others would call 'ergonomics', and Hedén, biochemical engineering. The alternative visions of biotechnology as a boundary object whose control was sought by the engineers, physiologists, and microbiologists was clear enough, therefore, to be debated, at least in Sweden. Linguistic confusions, or at best niceties, engendered by these debates are still with us, so that the words 'bioengineering' and 'biotechnology' have quite different connotations in English. The fossilized remains of this debate are to be found in Sweden where the balance of power was different: The distinction was ironically inverted so that *'bioteknologi'* still refers to what in England would be called 'ergonomics'.

Confusing as these groupings were, they clearly indicate an ebullient upswelling of thought about the relations between biology and engineering. Such men as Mumford, Wickenden, Compton, Bush, and Boelter and their institutions in the United States and Velander and Enström working through IVA in Sweden provided the linkages with earlier concepts of the transcendent significance of *Biotechnik,* biological engineering, and biotechnology. However removed in time and in intent the ferment of practical thinking seems from prewar biological philosophy, that somewhat abstract earlier thought had provided a language and a framework for the postwar engineers.

5
The chemical engineering front

> Biotechnology embraces all aspects of the exploitation and control of biological systems and their activities. Some of these form the bases for old and well established industries. Industrial 'fermentations' and the isolation and purification of chemical products from natural products are examples.
>
> (Elmer Gaden, 1962)[1]

Although idealistic educators wished to promote an integrated vision, in practice engineering has been, consistently, specialized. In the years after World War II, individual disciplines stood proudly apart, and important as ergonomics and biomedical engineering were, it was the interface between chemical engineering and microbiology that seemed to have the most outstanding commercial implications on the front between biology and engineering. Chemical engineering, the youngest and fastest-growing engineering discipline, was dominated by the potential and problems of the petrochemical industry whose distinctive tall towers immediately conveyed its style: manipulating distillates, cracking with catalysts, and synthesizing polymers in enormous quantities using capital- and energy-intensive continuous processes. Polyethylene demand multiplied sixfold between 1954 and 1960, reaching 1.2 billion pounds.[2] Yet by the end of the 1960s, there was a sense that the boom times were ending. The great plastics were a decade old, and it did not look as if latter-day polypropylenes, polyethylenes, or nylon were emerging. An elaborate analysis of 1970 puts chemical breakthroughs on a timeline showing the relative sparsity of more recent achievements.[3] This perception, rather than any reality, is of course the key. Some chemical engineers began to look elsewhere. Food and drugs seemed to be areas in which demand would be on the same vast scale as plastics.

Even before World War II, Compton had envisaged that graduates in biological engineering would contribute to medical technology. The society that emerged after the war was yet more health conscious. In many countries, the Western world's revised priorities were indicated by government-backed guarantees of free or subsidized medical attention. It was even said that more Americans knew of the 1954 test of the Salk vaccine against polio than knew the name of the president.[4] This chapter shows first how a research programme and journal, *Biotechnology and Bioengineering,* were built on this prospect. Second, it explores the industrial and technological underpinning of the new discipline.

The close links between medical care and chemical engineering were consolidated by wartime discoveries of the microbial antibiotics, above all penicillin. These ensured that the problems of life would be grappled technologically through microorganisms, and their influence will pervade this chapter. Shortly after the war, in 1947, the journal *Chemical Engineering,* reflecting on Vannevar Bush's speech about the 'biological' engineer, suggested the analogous title of 'biochemical' engineer.[5] Any implication of priority was immediately rebuffed by the University of Wisconsin whose professor of biochemistry immediately wrote to say it had already created a course with that title. Keeping up the momentum, a few months later the 1947 Award for Chemical Engineering was bestowed on Merck, in recognition of its development of an industrial fermentation process to produce the antibiotic streptomycin.[6]

By the mid-1960s, 'biochemical engineering', taught in five graduate programmes (MIT, Columbia, Pennsylvania State, Cornell, and Minnesota State), was growing, its leaders enthusiastic, and industrial research in the large pharmaceutical companies and food companies ebullient.[7] Accordingly, the organizers of the fall 1965 meeting of the American Institute of Chemical Engineers decided that its theme should be the application of chemical engineering to biological processing. In advance, they published an article entitled 'Bioengineering – Is a New Era Beginning?'[8] Chemical engineering could provide a means of manipulating large quantities of biological materials. It could also be a means of 'inquiring into the nature of biological systems themselves'. Thus, contributions could range from studying gas absorption in biological systems, through enhancing the action of enzyme catalysts, to changing nucleic acids themselves. Already there seemed to the organizers the prospect of sequencing genetic material: 'Once this knowledge is available, the possibility that the sequence and, hence, the genetic properties of biological systems can be altered offers itself. The commercial and social implications of the availability of methods for such alterations are tremendous.'[9] Exciting as contemplating such prospects was for the chemical engineers just as for Kenedi a few years earlier, this prospect was still speculative.

The Minneapolis congress included sessions concerned with the more immediate problems of protein processing, natural products, and cryogenics in food processing, though time was also devoted to the growth of cells, biological catalysts, and engineering models in biological and medical research.[10] The treatment of a broad range of areas including the use of liquid nitrogen for shellfish protection and the fluidized bed characteristics of wheat flour, not to mention an 'analog model of a patient undergoing artificial kidney treatment', reflected the broad view of biological engineering enunciated by Compton in the previous generation. Nonetheless, the emphasis was on the manipulation of microorganisms and their products. Out of the twenty papers published in the proceedings, eleven dealt with microorganisms.

Strikingly, the professional relationship between microorganisms and chemical engineering was not reflected by the availability of journals. Although in the 1930s and 1940s biochemical engineering had access to the major U.S. journal *Industrial and Engineering Chemistry,* in 1953 the American Chemical Society formed a new specialist journal, *Agricultural and Food Chemistry,* and tried to divert the new contributions there – against the wishes of practitioners. There was therefore a sensed vacuum in publication opportunities that contrasted with the dynamism of the profession.[11]

The niche for a new publication was realized in 1957 by Elmer Gaden, a chemical engineer who had worked at Pfizer, the pharmaceutical manufacturers, before joining Columbia. As a teacher he has been credited as the father of U.S. biochemical engineering, shaping, together with his pupil Arthur Humphrey at the University of Pennsylvania, a generation of professors and pharmaceutical company executives.[12] As he pondered the potential for a new journal in the United States, a similar thought was impressing two academics in London, Donald and Crook, the former a chemical engineer, and the latter a biologist. Having been introduced to Interscience, they discovered that Gaden was already in negotiation, and although the two British scientists were recruited as advisors, Gaden became the editor of the new journal, launched in 1958 with the cumbersome title *Journal of Microbiological and Biochemical Engineering and Technology.* Three years later, Gaden and Hedén met and agreed to change the title and shape of the journal. The editorial board was to be reduced and tightened, and at their insistence, the title was condensed to *Biotechnology and Bioengineering.* As we have seen, the choice of the former word was the suggestion of Hedén, frustrated in Sweden. The latter reflected contemporary U.S. practice, for the word was still widely used to encompass biochemical processes as well as other interfaces.[13]

As a publication, the journal proved to be a great success, and the retitling launched the modern use of the term 'biotechnology'. Reflecting triumphantly after almost a quarter century, Gaden – editor and founder – wrote in 1981 to his publisher about his conception of the term. The letter shows how the microbiological vision he shared with Hedén had grown out of the earlier and more synoptic bioengineering that had entailed the whole of biology. Two major areas of activity are distinguished:

(1) The first comprises extraction, separation, purification and processing of biological materials. Extraction of oil seeds, soy protein preparation and processing, and paper making are examples. Most of these are long-established, traditional technologies in which the rate of progress is rather slow. Furthermore, many have their own professional organizations and publication vehicles.

(2) The second embraces utilization of complete biological systems (e.g. cells and tissues) *or* their components (e.g.) enzymes to effect *directed* and *controlled* chemical or physical changes.

This second aspect of biotechnology is often referred to as the technology of *bioprocesses* and is the main focus of current interest.[14]

Gaden's emphasis on the technology of 'bioprocesses' and the dominance of microbiology in the partnership with chemical engineering was sustained by the great continuing importance of two endeavours that had taken off in wartime: antibiotics and the response to the threat of biological warfare. The importance of the former in stimulating a new industry is well known, though the succession of other products does need to be laid out, but the significance of the latter has been perhaps underestimated. The perhaps surprising implications, for the industry as a whole, of military-supported research at Britain's Microbiological Research Establishment at Porton Down will become clear, through an account of the ambiguous development of continuous fermentation.

Penicillin

The story of penicillin has been told many times, and its main outlines need only be recalled here, though that brevity should not be misunderstood: Penicillin was to transform the industrial place of fermentation. The experience of its success provided the biochemical engineers with the memory of an industrial revolution which they would try again and again to recapture. Showing how a biological product could become an industrial superstar, penicillin acquired a market that grew even faster than that of other war babies such as computers, jet engines, and nuclear power.

During the late 1930s, a group of British scientists at Oxford picked up earlier reports of the action of the penicillium mould.[15] Alexander Fleming had described antibiotic properties in 1928, but during the early 1930s, the eminent chemist Harold Raistrick had failed to extract the mould's active principle. Within two years the Oxford group was successful in isolating a drug 'penicillin', in establishing protocols for its evaluation, in producing small quantities, and even in curing a few patients from fatal infections. Because wartime Britain lacked suitable resources, Howard Florey with Norman Heatley went to the United States in 1942 and there found the expertise, organization, and resources to guarantee success in large-scale production.

The market was dynamic, and the profits astounding. In 1947, production at 42 billion units was ten times that of late 1943, and though the price had fallen by a factor of ten since 1944, the industry's production was already approaching $100m – this for a product which was only seven years old. Still the world was crying out for more – as Graham Greene's classic *The Third Man* about trade in black-market penicillin in postwar Vienna reminds us.[16] By the time of the Korean War in 1950, production had increased another fourfold though prices had also fallen almost as much. Worldwide, 21 million patients a month were being treated with this material unknown a decade earlier. By 1955, still only a decade old, with 2 million pounds produced, the antibiotics industry was worth $268m.[17] Pharmaceutical companies built on prewar skills and reputation were to become major corporations. Thus, Pfizer, which had pioneered citric acid manufacture, became one of the leading penicillin companies. Squibb, Eli Lilly, Merck, and other familiar names provided the commercial basis for a new industry and a research-intensive context for technological development. The research expenditure of Merck leapt from $3.4m in 1945 to $21m just fifteen years later.[18]

Although it seemed that even such a complex chemical should be amenable to chemical synthesis, and though this was indeed achieved in 1947, to this day total synthesis is extremely expensive.[19] Instead, the fermentation experience of Peoria's Northern Regional Research Laboratory enabled it to scale up the microbiological process developed in Britain. Rather than growing the air-breathing moulds on the surface of shallow bowls of nutrient as the British did, the U.S. engineers managed to grow great tanks of mould fed with oxygen bubbled through the continuously stirred soup of corn steep liquor. They proved immediately that this was the perfect nutrient, and on a mouldy Peoria cantaloup they found a penicillium mould that would grow in submerged culture giving a high penicillin yield. This interaction between solid spores, liquid nutrient, and gaseous air was quite unknown. Some designers suggested that the large scale should emulate the gentle action of shaking laboratory flasks, which would not damage the mould growth as in current yeast manufacture. By contrast, Merck engineers experimented with violent circulation and high shear. Contrary to expectation, they doubled the yield.[20] Thereby, vast quantities of penicillin were being produced by the end of World War II. At the same time, a general system, the so-called Stirred Tank Reactor, had been developed for growing any air-breathing microbe. It required a new engineering style, very different from traditional chemical technologies. The first attempts by plant engineers tended to be disastrous: 'The converted chemical kettles they offered us invariably led to serious and crippling infections. The engineers then tried to solve this problem with steam seals at every possible point... until we ended up with a hissing and puffing contraption that resembled a noisy old steam engine more than a fermenter'.[21]

If penicillin did prove a great boon, it was not all-encompassing. Some bacteria, such as the tuberculosis bacilli, were immune, but the answer was found in the streptomycin extracted by Rutgers microbiologist Selman Waksman from soil bacilli.[22] In Paris, Dubos identified tyrothricin, and a host of different weapons were forged from microorganisms. By 1957, about 350 antibiotics had been introduced by an industry that, unusually, could produce products which actually cured disease, bringing bacterial infections to heel.[23]

The close association between penicillin and fermentation in the early 1950s is nicely illustrated by developments at the old established British company of Beecham. The company, a maker of over-the-counter medicines, was seeking to enhance its production of tartaric acid, an ingredient of its traditional remedy, 'Eno's Fruit Salts'.[24] Through its distinguished chemical director, Sir Ian Heilbron, the company was put in contact with Sir Ernst Chain, one of the original Oxford group that had pioneered penicillin. Sir Ernst was at the time investigating the development of semisynthetic penicillins in which penicillin was first produced by fermentation and then chemically modified. He suggested a plant which could produce both tartaric acid and penicillin, and from there he led the company into supporting his pioneering efforts in semisynthetic antibiotics, still, in 1992, a mainstay of the successor organization SmithKline Beecham.[25]

The model of penicillin and the other antibiotics was applied to other drugs, compensating for the maturation of the penicillin market itself by 1960. The 1955 value of $268.5m had increased to $385m by 1963 in ten plants.[26] Meanwhile, it was discovered that vitamin B_{12} could also be extracted from streptomyces. By 1955, sales of the vitamins B_2, B_{12}, and C were worth $42m. At the Upjohn Company in 1950, D.H. Peterson had shown how the new wonder drug, cortisone, used to tackle problems ranging from arthritis to insect bites and even, possibly, cancer, could be produced by hydroxylating the recently synthesized hormone progesterone with microorganism-produced enzymes.[27] Within five years of his discovery, sales of cortical steroid hormones had already reached $27.5m.[28] The technology of growing mould cells was applied to mammalian cells by the Wellcome Company from 1960. While the Salk and Sabin vaccines for polio were grown in live chimpanzees, foot and mouth disease vaccine was produced from baby hamster kidney cells grown in standard fermenters. The microbiology-based pharmaceutical industry, which had begun with penicillin, quickly moved beyond being a single-product industry.

Antibiotics grew out of work in Britain and the United States, but neither country was destined to dominate the fermentation industry. Instead, Japan became the most dynamic centre as early as the 1960s. An unusual number of traditional Japanese foods are based on fermentative processes. It is less often noticed that the key early years of Japanese industrialization and exposure to the West coincided with the beginning of systematic microbiological examination of fermentation. From 1875, Westerners became aware that the Japanese had evolved a parallel route to the Western use of malt to break down starch to sugars in brewing, using the kōji mould.[29] The traditional Japanese foods sake, soy sauce, and tofu were all the products of the action of kōji. Since the use of this fermentation was closely related to the utilization of Japan's agricultural produce, as in the West, the development of microbiology was closely related to agricultural development and found an institutional home in the Agricultural Chemistry Society established in 1924. So, for all the special features of Japanese culture, the concept of a microbiology harnessed to agriculture closely paralleled chemurgy in the United States. There is a strange irony in this, since the promoters of chemurgy were strongly nationalistic and particularly anti-Japanese.

In 1936, the key appointment of Kin-ichiro Sakaguchi as professor of agricultural chemistry at the University of Tokyo established the reputation of the nation's premier department of industrial microbiology. As he would suggest in 1970, there was a close relationship between fermentation and natural products chemistry. Biologically active chemicals – such as gibberilin, which stimulates plant growth – were discovered through examination of rice plant fungi.[30] Just as in the United States, the possibilities of power alcohol replacing scarce petroleum were attractive in Japan, and in 1940, an institute of applied microbiology was established to explore the technology.

In the postwar world, penicillin was a powerful attraction to Japanese industry, accustomed as it was to growing mould for profit. With the assistance of the Americans, particularly the Texan Professor Jackson Foster, a vigorous industry emerged – seventy companies were competing in the early postwar years. Though drawing upon foreign scientific discoveries, the strong industry that emerged provided the infrastructure for exploiting a major Japanese discovery, fermentation amino acids. As long ago as 1908, Kikunai Ikeda showed that the active principle of the favourite Japanese flavouring, konbu (a seaweed), was sodium L-glutamate. This was then isolated from acid hydrolysate of soya bean protein by the Ajinomoto company. In 1956, Shukuro Kinoshita and his team showed how to produce glutamic acid from bacteria. In the subsequent decade, a new industry producing amino acids emerged with Japan holding a dominant position worldwide.[31] The classic text on biochemical engineering – so important that its very possession has been said to define a biotechnologist – was the product of a 1963 Tokyo University course by Aiba, Humphrey, and Millis.[32]

Enzymes

Beyond pharmaceuticals and the Japanese manufacture of amino acids, there was but one genre of product which showed actual commercial growth, and was already beginning to offer the kind of industrial expansion that antibiotics had contributed: Enzymes, the catalysts used by living cells. These had first been separated right at the end of the nineteenth century, and a variety of uses for them had been developed in removing skins from hides, in producing artificial rennet for cheese making, and in clearing beer. Though chemicals, enzymes are derived from living organisms, and their manufacture involves the processes of fermentation and separation. The Danish firm of Novo Industri, which had emerged as a manufacturer of insulin from pig pancreases, became the world's largest enzyme manufacturer. The emerging market enabled expertise gained in antibiotic manufacture to be transferred back to the realm of agriculture-related industry.

However, enzymes are expensive and the market was restricted.[33] The 1960 U.S. industry was worth $25m. Even by 1975, sales of rennin (used in cheese making) had risen to only $15m, and papain (used in clarifying beer) to $10m. A more important innovation was glucose isomerase which enabled the product of an enzymatic transformation, high fructose corn syrup, to make major inroads into the market for a traditional product – sugar, particularly in soft drinks. The sweetener glucose, though easy to produce from materials such as corn syrup, is less soluble than its relative sucrose 'sugar' and less sweet, but chemically it is very similar to the highly soluble sweetener fructose. This is about twice as sweet as traditional sugar. As early as 1895, it was shown that glucose could be converted (isomerized) to fructose by heating in a solution of the powerful alkali sodium hydroxide. In the 1950s and early 1960s, a number of scientists particularly in Japan experimented with enzymes as means of effecting the conversion. With high sugar prices in the United States during the early 1960s, their work came to the notice of the Clinton Corn Processing Company, which in 1967 launched a complex enzyme system. It was commercially viable, but it was only in 1974 that Novo, and Gist Brocades in the Netherlands, developed cheap and effective methods. In the United States, this led to the widespread replacement of sucrose in the soft drinks industry. The 1975 glucose isomerase sales of $15m had increased to $40m by 1977 and $50m in 1980.[34]

The second major use of enzymes emerged in the detergent industry. Here enzyme preparations competed with a chemical product, detergents, winning because of their superior environmental quality.[35] Enzymes had first been used in cleaning before the First World War. Otto Röhm patented an enzyme preparation for washing in 1913, and his company Röhm & Haas marketed their presoak product 'Burnus' for about half a century. Shortage of soap during World War II stimulated further work in Switzerland which led to the launch of the product Bio 40 as early as 1959. Novo entered the field when it was confronted with the problem of developing a detergent to clean overalls stained with the protein products of fish and meat, which proved even harder to remove than the mud stains shown on television. By 1966, enzyme washing powders had a 2–3% market share, which rose to about 50% within three years. The subsequent history was slightly erratic as it turned out that the skin of some users was irritated by

Table 5.1. *Fermentation products, 1930s and 1960s: world production in tonnes*

Product	1930s	1960s
Beer	20 m	43.5 m
Spirits	2 m	1.8 m
Acetone and butanol	50,000	170,000
Acetic acid	110,000	950,000
Citric acid	8,000	60,000
Lactic acid	6,000	11,000
Pressed yeast	55,000*	79,000
Fodder yeast	100,000*	180,000

Note: For comparison purposes, all figures are given in tonnes. The production for beer and spirits and the 1960s acetic acid production are converted from hectolitres to tonnes using, for these purposes, adequate approximation of 10 hectolitres to the tonne. The 1960s figures for lactic acid are taken from H. Benninga, *A History of Lactic Acid Making: A Chapter in the History of Biotechnology* (Dordrecht: Kluwer, 1990), p. 449.
*Germany only.

the enzymes. However, the pressure for lower temperature washes and the unacceptability of phosphates kept the enzyme preparations on the market and enabled recovery.

Antibiotics, new drugs, high fructose corn syrup, amino acids, and enzyme-based detergents displayed the kind of growth that indicated not just the arrival of new products, but perhaps also an industrial revolution, to the believer. The application of the products was diverse, covering chemicals, foods, raw materials, and consumer products. All showed promise, even if none, beyond pharmaceuticals, were yet big business; and to the optimists, growth was just a matter of time. Against the hopes for these new products, however, had to be set the relatively slow growth of older technologies. A survey of the world fermentation industry in 1963 provides a quantitative comparison with Bernhauer's analysis of a generation earlier.[36] The production of many products had hardly grown (see Table 5.1). The manufacture of simple chemicals was now being disdained by industrialists. In part, this was because of competition from cheap petrochemicals, but it was also due to the price of nutrients such as molasses, ironically forced up by the demand from the antibiotic industry. So, growth in fermentation had to rely almost exclusively on new products.

By the late 1950s, even among the pioneers of antibiotics, there were pessimists feeling the need for a further infusion of talent and ideas. Jackson Foster, the Texan microbiologist who had been responsible for the building up of Japan's postwar antibiotics industry, reflected in 1962, 'Truly, industrial laboratories, with their superb teams of integrated technologists, are fuddled with respect to fertile new areas to exploit.' He suggested that the only really significant breakthroughs of the previous decade had been semisynthetic penicillin and the Japanese 'condiments'. Now he proclaimed a divide between 'the chemists, engineers, sales people, and above all, management, all evincing a sort of resigned skepticism' and the microbiologists whose 'euphoria not only sustains them, but overwelms all within range. One wonders whether, in the present context of things, this buoyancy of spirit smacks of whistling in the dark.'[37]

Foster suggested two solutions to this crisis: One was to attract more talent, otherwise bound for academe, into industry. The other was to develop that vigourous hybrid, bioengineering. In other words, if biological products were to be successful, their production needed to use the same principles as had been so cleverly developed by chemical engineers. Instead of the costly batchwise production taken over from the brewers, it came to appear that continuous production would hold the key to commercial prosperity. More efficient ways of using enzymes, whose application had long been constrained by their great expense, was an obvious avenue for chemical engineers. From the early 1960s, it appeared that enzymes could be immobilized. By sealing enzymes on to the surface of an inert bed, such as glass balls, titania, or cellulose, a chemical could be transformed just by passing it through the reactor. The two major applications were in the manufacture of semisynthetic penicillins and high fructose corn syrup.[38] Thus Malcolm Lilly at University College London and, independently, groups in Germany and the United States developed an immobilized penicillin amidase for stripping the side groups of penicillin to produce the core used for making the many semisynthetic penicillins. Today, the approach continues to stimulate new processes and has also been applied to the use of immobilized cells.

As a product, penicillin raised hopes for many more industrywide biological products. Equally, its production technique also seemed to presage a revolution. Chemical engineers have often been struck by the traditionalism and empiricism of fermentation industries, contrasting vividly with the scientific nature of the dynamic petrochemical industry. Typically, fermentation industries have operated batchwise, requiring constant interruptions to production, while petrochemicals are produced continuously. So, even before the development of immobilized enzymes and cells, chemical engineers, showing how new processes could supersede the batch-operated stirred-tank fermenter developed for penicillin, sought to develop continuous fermentation. Today, it is used principally as a research tool, but for twenty years it also symbolized hopes for a new process technology, appropriate to the microbiological industry. It raised hopes that were not dispelled even by the short-term disappointment raised by the commercial problems of continuous fermentation.

It is true that continuous fermentation had indeed been tried since the turn of the century, but with little success.[39] Nevertheless, the situation appeared to have changed; for, in 1950, Monod, in Paris, and Novick and Szilard, in the United States, published papers that analysed the bacterial growth under the conditions in which the rate of fermentation was controlled by allowing a plentiful supply of all the nutrients, but rationing the input of one.[40] Their work was fundamental and the apparatus, which Novick and Szilard called 'a chemostat', was still very small scale.

Porton Down

The opportunities highlighted by these fundamental studies were picked up at Britain's new Microbiological Research Department at Porton Down, described by an advisor of the early 1950s as 'the finest microbiological research institute in the world'.[41] Papers relating to its foundation are still closed and, perhaps because of such secrecy, the significance of preparations for biological warfare has been much less often recognized than the importance of penicillin. Nonetheless, such preparations did constitute an important wartime development, bringing together microbiologists and engineers in well-funded institutions with significant consequences for the future of biotechnology, particularly in Britain. The possibility of biological warfare had to be taken seriously in 1939. After all, the Germans had launched a surprise and, but for their own lack of faith, devastating, chemical attack in the previous conflict, and there were intelligence reports of work on biological weapons. In 1925, Churchill had warned of 'Anthrax to slay horses and cattle, Plague to poison not armies only but whole districts'.[42] Aldous Huxley's *Brave New World* 'featured' anthrax bombs. That was fiction of course, but the boundaries between nightmares and contingency planning were becoming blurred, and anthrax spores were indeed among the damaging agents that would withstand an explosion. In May 1934, the British Chiefs of Staff authorized the Cabinet Secretary, Maurice Hankey, to approach the Medical Research Council to test the possibilities of collaboration. The approach was rebuffed, but two years later Hankey persuaded the Committee of Imperial Defence to bring into being a committee on biological warfare. And in 1940, Hankey established a team to study defence against bacterial attack based at the Chemical Defence Experimental Establishment, Porton Down.[43] The head would be the distinguished microbial physiologist Paul Fildes, who had spent the previous decade studying the nutritional needs of bacteria.[44]

Fildes built up a team capable of both defensive and offensive roles. As a possible retaliatory weapon, 5 million anthrax-spore-filled cattle cakes were made. However, the technology was far less sophisticated than that being developed for penicillin at the same time, and a much more simple means of deep fermentation was developed using aerated milk churns.[45] An island, Gruinard, off the Ross-shire coast of Scotland was experimentally sprayed and remained contaminated for many years after the war. Of course, the Americans too developed biological warfare at Camp Detrick headed by George Merck, whose company was also a leader in penicillin manufacture.[46]

After the war, Fildes, and most of his senior staff, returned to academe; however, wartime experience had taught both the British and the Americans that biological weapons posed a considerable hazard. With renewed cold war anxieties over Russian potential, both countries built up their facilities again. Porton would always be much smaller than Fort Detrick, but because of its outstanding scientific standing and lack of such a great industrial infrastructure, its national significance would be far greater.

Fildes's assistant, David Henderson, was asked to run the postwar organization. Until Britain, still accustomed to being a top nation, acquired an H-bomb in 1957, biological warfare was the only weapon of mass destruction it could readily develop. At a time of austerity, Henderson was given a free hand, and he determined to build a proper laboratory. The building rising up next to the Chemical Defence Experimental Establishment (CDEE) covered 140,000 square feet and was the largest brick building in Europe.[47] Its benches consumed 90% of the teak imported into foreign-currency-starved Britain in 1947. There were strong central services, such as engineering, and there was access to the superb engineering facilities at CDEE next door. At the same time, the scientists benefited from a minimum of control. They all reported directly to David Henderson; as Keith Norris, a colleague, recalls, 'You didn't work for the Ministry of Supply, you worked for David Henderson.'[48]

Although Porton Down was technically supposed to be dedicated to biological warfare, the resources invested in it, the scientific advisory board, and the scientists employed permitted major contributions to civil science. More than 80% of its work was published. The distinguished alumnus John Postgate later recalled the atmosphere of this institution suspended between the collegiality of the army and the intellectual ambition of academe:

> Anyway, coffee and tea breaks were, for me, time for brisk interchange of ideas – political, social, sexual but mostly scientific – with my colleagues without, happily, the competitiveness which I have since met elsewhere. No-one seemed to fear that his ideas might be 'pinched' and used by another. Ideas were common currency: if someone used one of yours, you'd use someone else's.[49]

The scientific staff had close relations with academic science. There were links with Chain, then working in Italy. From him came an understanding of how to build beautifully designed fermenters, rather than the puffing kettles of earlier days, and key personnel, such as Chain's assistant John Pirt who moved to Porton in the early 1950s. Many of the staff had been trained at Oxford, where the Nobel Prize–winning chemist Sir Cyril Hinshelwood had come to be interested in cells as complex biochemical machines. His vision was biologically untenable. However, the emphasis on first-order reactions, in which rates of reaction were related to the amount of reactants, stimulated a sensitivity to kinetics and bacterial nutrition. The mathematics were similar to that shown to be characteristic of continuous fermentation by Monod. Thus, Hinshelwood's Oxford colleague A.C.R. Dean became a leading British promoter of continuous fermentation. NIMR, in the London suburb of Mill Hill to which several of the Porton team had returned after the war, provided another important link.

In 1950, Monod presented a lecture on continuous fermentation at the NIMR. Among his audience was an ex-Porton scientist, Dennis Herbert. Though stimulated by the approach of the young Frenchman, whose wartime doctoral thesis had only just been seen in London – the audience thought 'he would go far' – Herbert was not impressed by Monod's improvised and cumbersome rotating apparatus.[50] He himself was fascinated by gadgets, an aspect of Mill Hill's culture. Herbert was working in a team led by A.J.P. Martin who with A.T. James was then perfecting the gas chromatography.[51] So, when, shortly afterwards, his old boss David Henderson brought him back, Herbert took with him an interest in Monod's approach but with the wish to implement it better.

At Porton, Herbert created the world's leading team concerned with continuous fermentation. By the mid–1950s, there were already about a dozen scientists in the group, which was coming to combine a theoretical interest in controlling the rates of reaction with the development of practical pilot plants. Twenty years later, its members had published almost a hundred papers on various aspects of continuous culture.[52] The availability of engineering facilities, the development of automated oxygen monitoring, silicone rubber tubing, and a variety of disciplines, provided an unusual combination of resources. There were over a hundred craftsmen on the other side of the road at CDEE and several dozen at the Microbiological Research Department itself. As a scientist there, Charles Evans, remembers of the workshop, 'It wasn't just accurate, it was good'.[53] In the background were the needs of understanding biological warfare. The study of the dynamics of aerosols required large quantities of identical pathogenic organisms, and for their production continuous fermentation would be ideal.

Ironically, the Porton scientists had counterparts on the other side of the Iron Curtain, in Czechoslovakia, where Ivan Málek had a team.[54] An international congress was held in Prague in 1958. In the cold war atmosphere, awareness of potential advances being made in the Warsaw Pact was a potent factor supporting research at Porton Down. As tension thawed, the two teams established stronger contacts, with regular conferences on continuous fermentation. The understanding of the process appeared to be improving sufficiently well for industrial application. The equipment was actually very similar to standard batch fermenters. What was new was the monitoring and control that were required, as well as the standards of sterility necessary for fermentations that might be allowed to run for weeks.[55] There was another attraction for the biochemical engineers: The design required just their blend of biochemical and engineering knowledge. It was a matter of personal pride to be able to build a rig which would run for hundreds of hours without contamination, and, ideally, without deleterious mutation. National pride too was at stake, for it had been leadership in fermentation technology which had made possible U.S. control of the penicillin patents, to the enduring annoyance of the British.

The earliest industrial application to be explored at Porton was a contribution to the research of K.R. Butlin's microbiology group, at the Chemical Research Laboratory outside London, which was seeking a microbiological solution to a severe national shortage of sulphuric acid, a key industrial material. In the early 1950s, inadequate supplies of native sulphur from which it was produced were threatening Brit-

A typical continuous fermentation experimental setup, Microbiological Research Establishment, 1965. Courtesy Centre for Applied Microbiology & Research.

ain's postwar industrial recovery. Butlin's group were investigating the exploitation of an anaerobic bacterium, common in soil and water, called *Desulfovibrio desulfuricans*, which reduces sulphates to hydrogen sulphide, a product which could readily be used to make sulphuric acid. A team led by by P.S.S. Dawson, seconded to Porton, successfully overcame the problems of growing this awkward anaerobe in continuous culture, but other ways of making sulphuric acid came into industrial use, and the process was never used in Britain.[56]

The sulphur shortage might have been side stepped, but by 1955 continuous fermentation was being considered for the manufacture of acetic acid, penicillin, citric acid, and other products.[57] Again, competition from petrochemicals meant that acetic acid and other organic acids could be chemically synthesized more cheaply than they could be fermented, though continuous fermentation was applied in the production of vinegar for which petrochemical routes are not acceptable.[58] In penicillin, there had been a massive investment in batch production, and worries about mutation inhibited continuous manufacture, though there had been repeated small-scale experiments. The first area of manufacture was yeast production, but it was the continuous brewing of beer that was counted upon to provide sufficient impetus for widespread industrialization. Though this development too eventually proved a tragic disappointment, it illustrated both the drive to revolutionary change based on biochemical engineering and the problems of meshing with only slowly changing markets that would dog the development of biotechnology.

The Brewing Industry Research Foundation

At one time, continuous brewing appeared tremendously successful, and by the early 1970s 4% of British beer was produced in this way, with, apparently, prospects of further growth.[59] Instead, by 1980 none was produced continuously, and today the process is used only in New Zealand. The rise and decline can be illustrated by following just one project, that associated with the Brewing Industry Research Foundation, Arthur Guinness Son & Co., the brewer, and A.P.V. plant manufacturers.

Despite the zymotechnic enthusiasm of the late nineteenth century, the British brewing industry had no systematic organized research centre equivalent to Berlin's Institut für Gärungsgewerbe. Companies conducted research, and a trade organization, the Institute of Brewing, provided research grants to academics in departments such as Birmingham's British School of Malting and Brewing. The contrast with the chemical industry's great research laboratories was telling even during the interwar period and was drawn on in a 1930 proposal for a research institute put to the Institute of Brewing.[60] The promoter was Richard Seligman, the founder of A.P.V. who had devised and manufactured aluminium vessels for breweries and Weizmann's acetone–butanol work, and who had just introduced his pathbreaking plate heat exchanger. Though Seligman had little immediate impact, the upheaval induced by World War II gave him a new opportunity to promote advances in brewing technology.

In 1944, Seligman, as chairman of the institute's Research Committee, presented a major report on the future of the research programme.[61] Once again the principal recommendation was to set up a laboratory, but this time the idea was acted upon, and the Brewing Industry Research Foundation (BIRF) was established. The first director to take up post was Ian Heilbron, an organic chemist who had been engaged on wartime penicillin work and who was an advisor to Beechams. He had been recommended by Seligman despite his lack of any experience of brewing. By 1950, the new laboratory boasted an array of established talent. Heilbron had clear and specific hopes, often expressed, for his new laboratory. He saw it as the successor in the development of brewing science to the Carlsberg laboratory in Copenhagen. At last, claimed Heilbron, Britain had the equivalent of Berlin's Institut für Gärungsgewerbe that Chapman had envisaged half a century earlier. Heilbron would claim that the new laboratory housed the first systematic investigation of fermentation in the world.[62] Such rhetoric has to be understood in terms of his unstable position. The role of the research institute vis-à-vis the industry that supported it was continually difficult. It was assumed, from the beginning, the BIRF would not work on day-to-day problems.[63] These would be the preserve of individual companies. However, there had to be some relevance to the brewing industry.

Continuous brewing

In the light of the ambivalence about participation in the industry's normal concerns, the incorporation, within a wider research portfolio, of an interest in a completely new approach was a shrewd strategic move. Continuous fermentation had been investigated in the last century, and in the early 1950s Dominion Breweries in New Zealand and Labatt Breweries in Ontario were, for their own reasons, exploring the process.[64] Still, with British expertise, it seemed that the process could be put on a new scientific footing at BIRF. It would be both practical and scientific, and at the same time it would bring brewing within the broader development of fermentation technology.

Of course, fermentation is only the final stage of brewing. Breweries buy malt, produced from germinated barley. This is 'mashed' in a process in which the malt is heated with warm water to convert the starch to soluble carbohydrate. The resultant liquor, called the 'wort', is filtered through the bed of spent grains to produce a clear solution which is boiled with hops to obtain a bitter flavour. The hopped wort is then fermented with yeast. To obtain a truly continuous brewing process, each of these stages must be continuous. The first to be tackled at the BIRF was the mashing process, and by 1956 a continuous tube masher had been investigated at a laboratory scale. In order to develop this to practical brewery practice, BIRF recruited A.P.V. and Arthur Guinness Son & Co. to form a consortium. Another brewer, Courage Barclay, also joined.[65] The attitudes of these parties were interestingly different. For BIRF, as speech after speech made clear, continuous brewing exemplified the new integration of science and practice on which the institution's existence was predicated.[66] The scientists involved were chemists and biologists. To A.P.V., by contrast, the problem was one of straightforward chemical engineering. The company was at that time seeking to diversify from its narrow dependence on the aluminium vat and the plate heat exchanger. Its research director, Tony Dummett, was tireless in his reiteration that the application of chemical engineering to food industries such as brewing was the way of the future.[67] Brewers were more sceptical. When Dummett published an article in the United States in 1962, the following issue of the same journal, the *Wallerstein Laboratory Communications,* contained an attack on the idea that chemistry could define taste as recognized by an expert public.[68] A much better understanding of beer flavours was required before continuous brewing could be possible.

Difficulty in matching existing products proved to be the experience of Guinness, the third party to the original agreement. Their scientists had shown the most caution at first. Asked, in a 1960 patent suit, about their attitudes five years earlier, most said that they had only been marginally interested in continuous fermentation. The problems of sterility had seemed overwhelming: Excise regulations based on calculations that assumed batch fermentation had also seemed to militate against it. Nevertheless, there clearly had been some curiosity: One interviewee testified to having become interested from seeing an exhibition of the Porton Down work.[69]

Despite the initial scepticism, Guinness put in considerable effort during the early 1960s. Continuous brewing offered the prospect of cheaper, more compact, and less heavily manned breweries. Typically, a batch took ten days to brew, of which almost all the time was needed to allow the yeast to collect. Continuous fermentation would allow the time to be cut to a few hours. In other words, through biochemical engineering a plant would become an order of magnitude more efficient. This vision drove the company to overcome many problems. For example, the BIRF masher turned out to be unworkable. Guinness developed their own and went on to develop other innovations.[70] A.P.V. developed a process of continuous wort boiling which seemed to give good beer, but it seemed that after a few sips a strange flavour could be picked up, and thereafter it was very irritating. The fermentation process itself also proved very complex to render continuous. The rate of beer production is related to the density of yeast present, but yeast growth is incompatible with beer making: The first requires air; the second demands its absence. Finally, the presence of alcohol inhibits yeast growth, and the alcohol needs to be separated from yeast to produce a clear product. So the stages of yeast growth and alcohol production take place one after the other, and it did not seem that such contrasting requirements could be met simultaneously in a single vessel operating continuously. The Guinness solution (and similarly that of other pioneers) was a sequence of three fermenters in which the yeast density was progressively reduced so that yeast growth and alcohol production took place in different vessels. This was a demanding process, requiring new control instrumentation such as continuous density measurement. It was hard to emulate exactly the flavour of existing beers in the new pilot plant, but it was shown that, overall, the process did produce an almost acceptable beer. Nonetheless, the need to use stainless steel construction meant that the economics were not particularly attractive.[71]

The tower fermenter

The Guinness plant with its sequence of traditional vats was pragmatic rather than revolutionary. Meanwhile, A.P.V. began to investigate a process that looked much more in tune with modern chemical engineering. The use of a single column, containing a variety of yeast densities with wort pumped in at the base and beer pumped out at the top, was made possible by making the yeast form a plug in the middle. It separated zones of high yeast concentration and fermenting at the bottom, and low yeast and separation at the top. The tower fermenter was indeed a major advance, and several pilot plants were installed. Although the first full-scale brewery at Valencia proved too difficult to commission, it appeared that Dummett was vindicated.[72] By 1970, John Hastings, one of the elder statesmen of fermentation engineering who had worked on the early penicillin plants, took pleasure in seeing brewing as once again the leader in fermentation technology.[73] Others recognized that this leadership reflected difficulties with applying continuous fermentation elsewhere.[74]

Unfortunately, it became clear that, even in breweries, the excitement over continuous fermentation had been at best premature. The economics did not prove particularly attractive. The standard of control equipment and technical competence of the labour force had to be much higher than in conventional plants. Moreover, whereas the advantages of continuous working were most apparent when a large quantity of a single product was produced and the process was allowed to run for long periods, most British breweries produced a wide variety of beers and had to be able to switch between products, according to demand. Finally, the taste was not always acceptable at a time when traditional 'real ales' were coming back into fashion. Watney's Red Barrel, the best known of the continuously brewed beers, was more a technological than a gastronomic marvel. At the same time, the tower fermenter had stimulated new interest in the cylindrical batch fermenter proposed by Nathan as long ago as 1908. There, too, because of faster yeast settling, brewing could be completed in hours.[75] Thus, continuous brewing was, by 1990, used only by New Zealand's Dominion breweries, its original pioneer.

This account is not retold to scorn the failure of the process. Indeed, many of the lessons learnt from continuous brewing were applied in the highly automated conical systems familiar to those who drive past modern breweries. The episode illustrates general issues in the history of biotechnology. Vision drove development of the new technology over many hurdles, though problems arose in applying the generalized technology to a particular industry, with specialized markets and a complex product defined in qualitative terms.

At this point, it is worth summarizing the argument. By the end of World War II, there was an international interest in integrating biology with engineering, represented variously by the words 'bioengineering', 'biotechnics', and 'biotechnology'. Promoted on both philosophical and educational grounds, this interest had a wide scope. There was a particularly dynamic commercial example of such interaction in the case of the collaboration between chemical engineers and microbiologists. Whereas in Sweden this alliance failed to wrest the use of the word 'biotechnology', in the United States the term was successfully adopted, inheriting the tremendous dynamism of chemical engineering, the special success of the antibiotic industry, and the general ambitions and philosophical depth of the attempt to integrate biology with engineering.

One other key feature characterized the thought of Elmer Gaden and other leaders of biotechnology. This was not just one more advanced technology for the developed world. The most appropriate use for continuous fermentation in 1970 seemed to be the cultivation of hundreds of thousands of tonnes of single-cell protein to feed the starving. Biotechnology seemed to be particularly benign, indeed futuristic, precisely because it seemed specially relevant to the majority of the human race who did not live in the developed world, and who could not afford to import oil to make unnecessary consumer goods, but whose agricultural base gave them plenty of biomass to process.

6
Biotechnology – the green technology

> The biologists will then have to take the responsibility for providing biotechnology with a solid foundation of biological research and for seeing that it is applied with due regard to ecology and long range consequences.
>
> (Carl-Göran Hedén, 1961)[1]

Introduction

Biotechnology in the postwar era was more than just another bundle of techniques. It was an ideal alternative to a list of earth-destroying new-ologies, associated with the 'military-industrial complex' (as Eisenhower himself dubbed the legacy of his presidency). Beer customers may have been conservative, but the world outside the public house could not afford to move so slowly in its response to poverty, overpopulation, and starvation. Here, surely, was a worthy and appropriate focus for a revolutionary technology dedicated to the transformation of agricultural produce.

During the 1960s and 1970s, biotechnology was energetically promoted as the use of the rich countries' scientific resources to solve the problems of the poor. Heralding a new industrial revolution, it was a symbol of hope, and answer to the 'Trinity of Despair' – hunger, disease, and resource depletion.[2] This vision did not apply just to the Third World, it informed too the developing idea that biotechnology would be a core technology for future development even in wealthy nations.

The dramatic end to World War II engendered the menace of future disaster, but it also drove forward those with an urgent sense of hope for a new world order. To Albert Einstein at least, even a world government seemed in prospect. The gathering pace of decolonialization and the emergence of the United Nations gave grounds for hope. An industrial revolution was prophesied, but there was little agreement as to what would lie at its heart. To some, nuclear power and the promise of limitless clean energy through science illuminated the future. Others came to dream of a human escape from this planetary prison, soaring to other worlds. Similarly, there were those who believed that modern microbiology offered humanity a way out of its history of starvation and disease. Against such optimistic uses of science was set the intensity of the cold war, which meant that space technology, nuclear physics, and microbiology could be used to make frightful weapons. The nuances of natural idealism came to be translated into hard decisions. Many scientists involved in advanced science during the 1950s found themselves presented by a Manichaean choice: Either use their skills towards the production of weapons or for the good of humanity.

Elmer Gaden and Carl-Göran Hedén would be prime movers in the movement to use biotechnology for good. Its most famous visionary, though, was Leo Szilard, a symbol of physicists' postwar revulsion of nuclear warfare and its association with their science. As a brilliant refugee, first from the disintegrating Austro-Hungarian empire of Ereky and Goldscheid and then from Germany, during the 1930s he had dire forebodings of a new form of warfare based on atomic energy. Safely in America, he was one of those who encouraged Einstein to write a letter to President Roosevelt warning of the potential danger of Germany getting a bomb first.

The Nazi menance having disappeared in the ruins of Berlin, Szilard was appalled by the prospect of an arms race with the Soviet Union and was a founder of the Pugwash conferences which brought together Soviet and U.S. scientists. He was also one of the remarkable physicists who, turning to biology, created the new discipline of molecular biology. His 1950 classic paper on continuous fermentation had grown out of an interest in a means of mass-producing cells for the study of phage.

A decade later, Szilard's efforts moved to a more social plane. In his 1960 science fiction fantasy, *The Day of the Dolphins,* Szilard imagined a better technology invented by dolphins, which was the very epitome of what was just then coming to be called 'biotechnology'.[3] Imagining a breakthrough for international relations, he relates how the Americans and Russians take, as the focus for a new collaboration, the development of a joint institute of molecular biology. Its scientists show that the brain of the dolphin is superior to the human, the animals are trained in science, and they then start winning Nobel prizes. Their greatest discovery is an alga that fixes nitrogen from the air by itself, produces antibiotics, and can be processed into a nutritious and cheap human protein source. As a result, the Indian poor no longer find it necessary to have large families, and the population explosion is prevented. Even the product's name would symbolize its world roots: 'Amruss'.

A year after publishing *The Day of the Dolphins,* Szilard created a civil rights organization, the Council for a Livable World. The council identified crucial problems in human development, and although Szilard died in 1963, its work was continued into the 1970s by his assistant and biographer, John R. Platt. In 1972, Platt and Cellarius used an article in *Science* to call for task groups to identify crucial problems of the future and to work towards their solution. Population problems, famine, the environment, and health could all be tackled, they felt, through 'Biotechnology'.[4] Such a vision expressed a hope seen to be antithetic to atomic technology. Robert Jungk, the German journalist and author of *Jahrtausend Mensch* (translated into English as the *Everyman Project*) saw Szilard as a prophet of the new age.[5]

Thus, the post–World War II generation of idealists diagnosed new possibilities in the complex of techniques of fermentation, enzyme technology, and the processing of microorganisms that, increasingly, they called 'biotechnology'. It seemed particularly well suited to otherwise disadvantaged developing countries, rich in biological raw materials and in great need of products such as fermented foods, fuel alcohol and biogas for energy, and nitrogen-fixing bacteria. The technology would be deployed by small enterprises meeting local needs, not by remote multinationals. The vision of biotechnology that emerged then had strangely close affiliations with the prewar bioaesthetic ideas of Mumford and Hogben. Mumford himself became a cult figure, and in 1971 his word 'biotechnic' adorned the title of a commune in Wales, modelled on the New Alchemy Institute in the United States.[6] The organism would be the philosopher's stone of the new era. In his introduction to a 1979 book entitled *Biological Paths to Self-Reliance,* the Swedish economist Johan Galtung, echoing Armstrong at the beginning of the century, suggested that arrogant Western man had thought he could do better than nature. Now was the time to reverse a disastrous trend.[7]

The biotechnic dream may have been, principally, the preoccupation of intellectual scientists of the Western world; but it was not therefore commercially inconsequential. It did highlight to big business the link between an emerging technology and potential solutions to an emerging world problem. The name given to that threat was couched in suitable dramatic military terms: the population explosion, with its corollary mass starvation. The secondary implications for political instability were a cause of anxiety to politicians now accustomed to proxies for the cold war – in developing countries, disputing the distribution of scarce resources over increasing numbers of mouths. The scale of the problem was not in doubt. It was an issue that seemed to make the traditional challenges to technology – how to speed machines and multiply energy efficiency – and even the manufacture of weapons, seem trivial. At a meeting entitled 'The Engineering of Unconventional Protein Production', sponsored by the normally pragmatic American Institute of Chemical Engineers in the summer of 1967, the vice-president for research at Buffalo's State University of New York conjured up the frightful image:

> Imagine, if you will, a giant sea monster rising from the depths of the ocean. The monster heaves up out of the ocean until it reaches 100 ft. above the sea, a fearsome sight indeed, a super Loch Ness monster. But the monster keeps rising until it is 1,000 ft. high, 5,000 ft. high, 20,000 ft. high, bigger than anything else in the world.... The food/population problem seems likely to reach such enormous proportions even by 1975 that it will dwarf and overshadow all the problems and anxieties which now occupy our attention, such as the threat of nuclear war, Communism, the space race, unemployment, racial problems, Vietnam, China, the Middle East, Cuba and others.[8]

In developing countries, demand for agricultural produce was rising even faster than the world's overall annual population growth of 2–3%. Before the twentieth century, agricultural production had grown at but 1% every year, and even industrialized agriculture was only giving annual growth of 1.5–2.5%. Conventional breeding, leading to the so-called Green Revolution, was one response.[9] Biotechnology offered a more direct route to the production of nutrients, that might lead to the bypassing of traditional agriculture altogether. After all, petroleum had replaced coal as the major source of chemical organic compounds within a generation. Could it now replace plants and animals in the production of protein?[10] Multinational oil companies made an immense investment in research towards the fermentation production of single-cell protein foods, as yeasts, moulds, and bacterial protein foods collectively came to be described. The product might be unfamiliar to them, but oil seemed a promising feedstock, and they were used to providing vast quantities of commodities.

Ironically, marketing research came to change focus from the impoverished people of the developing countries to the animals of Europe. Third World consumers were more conservative about their foods and had less access to oil than their richer neighbours. The very success of the Green Revolution, in commercializing the high-yielding products of conventionally bred rice and wheat, also engendered cynicism. While, in terms of food production, the consequence had been remarkable, and world famine was prevented, the new strains were resource intensive. They favoured the cash-rich over poor farmers who could not afford the increasing amounts of fertilizer. No longer could one think just of food production: Equity, rural dislocation, and dependence on complex technology came to be considered important and questionable factors.

Thus, while the biotechnology dreamed of in the 1960s would be environmentally benign, resource efficient, and benevolent, within twenty years it was widely seen as just another threat. This reversal, though due partly to anxieties over the new kind of science associated with recombinant DNA, had deeper roots. Biotechnology was seen to be 'unnatural', and the distinction with chemical technology was lost. Visions of bounty that did survive encountered a new cynical strand of thinking that could be caricatured as 'the use of poor countries' genetic resources to solve rich country problems'.[11] Biotechnology, the threats go, will lead to the privatization of public genetic assets, the remote control of local industry, and the concentration on a few money-making drugs.[12] Vanilla, an important cash crop of Mauritius, could now be made in a vat in New York. By the 1980s, biotechnology came to be widely seen as another high-technology threat to the Third World, transforming export centres into importers of virtually the same commodities, rather than a vision of a real idealism. Gradually, the memory of inspirational hopes for single-cell protein foods, nitrogen-fixing plants, and gasohol faded.

The scientific breakthroughs to which the idealists had looked took much longer than they had hoped, and the complexity of technology, even of biotechnology, was more daunting than expected. These problems were particularly challenging because, despite the impression of representing a formidable weight of opinion, the number of scientists massed to meet them was small. Men such as Hedén, known by a friend as 'a man of many hats', obtained positions and funding to promote biotechnology from impressively titled agencies such as UNESCO, the UN, the Food and Agriculture Organization, International Association of Microbiological Societies, and the World Academy of Arts and Sciences.[13] Even if success was limited, those few achieved a surprising amount both in the direction they hoped to develop a global vision and in the perhaps unexpected affirmation of the 'alternative' quality of biotechnology in the developed countries. Like Gideon's army, a few men by working tremendously hard and running very fast gave the impression of unstoppability.

International organizations

Though the 1950s was the high point of the cold war, internationally minded scientists of the time worked hard to engender a new era of worldwide collaboration and concern. The International Geophysical Year in 1957 showed what could be achieved when scientists from East and West, North and South collaborated. Inspired by this example, the ecologists obtained international support for the International Biological Programme (IBP), which lasted not just a year but a whole decade. While the IBP was concerned principally with man's relationship to the ecosphere, it did have sections dealing with practical issues.[14] If microbiology was akin to space and nuclear technologies in the magnitude of its implications, it was much cheaper. In the case of space technology, the basic requirements of research were so great that only highly funded, state-backed, enterprise was conceivable. Microbiologists could operate in the great national laboratories such as Fort Detrick, or Britain's Microbiological Research Establishment at Porton Down. Nonetheless, their capital requirements were sufficiently low that, elsewhere, they could operate much more freely of defence priorities than their contemporaries in electronics.[15]

A significant contribution could even be made through UNESCO, born in London in 1946. Its first director general was Julian Huxley, who took as his head of science another left-wing English biologist and friend of the developing countries, Joseph Needham. Though the two were in post only two years, as the biochemist Marcel Florkin points out, they coloured the future development of the organization.[16] Needham was certain that it would be important to introduce appropriate elements of Western science to developing nations. As early as the first meeting of UNESCO, it was decided to support a cell culture collection in Brazil. During the early 1950s, there was discussion in UNESCO over the support of cancer research in the Third World. Opposition from countries such as Britain, concerned about the duplication of established developed countries' work, led to the generalization of the initiative as the Cell Biology Programme.[17] Assistance to this programme was provided by UNESCO in 1962 through the establishment of a specialist body, the International Cell Research Organisation (ICRO). Though this naturally had broadly medical ambitions, when the newly assertive Japanese government suggested to UNESCO in 1963 that it should establish a microorganism programme, the ICRO, recognizing the importance of microbiology, could provide a mechanism for responding. It appointed an advisory panel on applied microbiology to coordinate international culture collections, to explore microorganism resources on land and sea for industrial, agricultural, and medical purposes, and to provide assistance to researchers in less developed lands.[18]

Pressure on UNESCO from Japan, an increasingly important member government, was matched by professional initiatives. The microbiologists had their own organizations through which enduring visions of a useful science could be transmitted and promoted. At the first postwar meeting of the International Association of Microbiological Societies in Copenhagen, the aged Orla-Jensen who had created his school of biotechnics before World War I, was still an active figure. The then young Swede, Hedén, still remembers Orla-Jensen talking to him about the current international dimension of applied microbiology.[19]

Hedén came to be closely identified with the benevolent uses of his science. He saw the compact fermenter he developed at the Karolinska Institute with Swedish government sponsorship purchased by foreign governments interested in biological warfare, and he believed that the rapidly advancing technology of microbiology could be used to more valuable purposes. Thus, by developing a nitrogen-fixing strain of wheat, expensive chemical fertilizers could be abandoned. As he wrote, 'If the aim [of producing fertilizers] cannot be reached by conventional means, such as building big factories, we must then obviously consider unconventional ones, for instance making untold numbers of self-replicating, self-repairing, minute factories work for us.'[20]

In 1962, completing his term as secretary general of the International Association of Microbiological Societies, Hedén encouraged its Montreal meeting to issue a special statement on the contribution of microbiology to coping with the world food crisis. Microbiologists could play their part to prevent the Malthusian disaster allegedly already underway, through research on new ways to improve soil fertility, on agriculturally useful substances such as plant growth factors, on the prevention of food spoilage, and on nonorthodox foods. The statement called for particular attention to organisms, such as the rhizobia bacteria which could convert atmospheric nitrogen into protein. On all these, governments were urged to sponsor research, even if the products were not yet economic.[21]

As part of his Montreal initiative, Hedén promoted the creation of a section dealing with economic and applied microbiology. Its committee included some of the chief founders of biotechnology. Two of the new editorial board of *Biotechnology and Bioengineering* had key places – Marvin Johnson of Wisconsin was president with Hedén himself as a vice-president. The other vice-president was Ivan Malék, doyen of Czech continuous fermentation. The seventeen-member international council included Ernst Chain from Britain, Kai Arima, pioneer of Japanese biochemical engineering, and the ageing German-Czech Karl Bernhauer who was about to reintroduce the word '*Biotechnologie*' to Germany.[22] Through the International Association of Microbiological Societies therefore, one can see an institutional continuity between Orla-Jensen, founder of biotechnical chemistry early in the century, and the crystallization of biotechnology fifty years later. At the Montreal meeting, a successor was 'conceived, gestated and born'.[23] It took the form of the conference entitled 'Global Impacts of Applied Microbiology' (GIAM) held in Stockholm at the end of July 1963, sponsored by Hedén's network of organizations – including Sweden's IVA (he was chairman of the biotechnics section), the International Association of Microbiological Societies (IAMS; the section on economic and applied microbiology of which he was vice-president), and a new worldwide association of idealistic intellectuals, the World Academy of Arts and Sciences. Hedén was chairman of the organizing committee and other familiar names were involved. The vice-presidents working with the meeting's president, Nobel Prize–winning Swedish microbiologist Tiselius, included Marvin Johnson, Ivan Málek, and N.E. Gibbons, the new secretary general of IAMS.

The meeting held in Stockholm was not intended to bring out contributions to fundamental knowledge but to articulate the emerging role of microbiology as a language of international cooperation. In the published proceedings, the first section was entitled 'Some Philosophies of Applied Microbiology'. Contributions came from both biologists and outsiders such as Israel's Abba Eban and the distinguished Swedish economist Gunnar Myrdal. With the human population rising to roughly 3 billion in 1960, an increase of 19% during the previous decade, food would apparently run out. Disease was rampant. World wars would be ever more destructive. In the face of such issues, the organizers proclaimed: 'A sense of unity is provided by the unquestioned belief that micro-organisms are a prime natural resource of all mankind, regardless of national boundaries. If man makes the effort to understand, control and utilize this resource, it may well affect his future most profoundly'.[24] Food and spoilage, disease and drug production were addressed by specialist speakers. There was an attempt by many to focus on the special problems of poor countries. Elmer Gaden spoke on the need to depart from the high-technology penicillin-based paradigm of stainless steel equipment. He felt many useful fermentations could employ wood or concrete equipment.[25]

The finale was an affirmation of support for the IAMS statement on the contribution of microbiology towards world food supplies and for the proposal of a UNESCO-sponsored 'microorganisms decade'. In addition, the meeting called for special attention to international culture collections, a coordinated creation of a 'Needed Projects Catalogue', and planned approaches to vaccine research, production, and administration. Educational needs were highlighted: An international network of teaching laboratories under the auspices of a new 'International Organisation for Bioengineering and Biotechnology' would be required.[26] Though the speakers in Stockholm were generally Europeans and Americans, when the next meeting was held in 1967 in Addis Ababa, scientists from the developing countries themselves were given greater prominence. The opening ceremony was performed by Emperor Haile Selassie of Ethiopia, and Kenya's Jomo Kenyatta was also present.[27]

Issues with important scientific and social implications to policy makers and scientists away from the mainstream of world science have been highlighted by the GIAM meetings held since 1967 in Bombay, Sao Paolo, Bangkok, Lagos, Hong Kong, and Malta. Yet other organizations emerged for the purposes of coordinating research, training, and culture collections. Following Hedén's suggestion in Montreal, the International Organisation of Biotechnology and Bioengineering was eventually founded in 1968, with help from UNESCO, by the now predictable Hedén and Gaden, Japan's Terui, and Finland's Gyllenberg who wished to help member laboratories in both developed and developing nations collaborate for research and advanced training.[28] Initiative piled on initiative. The problems of conserving and exploiting the microbial gene pools in developing countries led to the network of Microbiological Resource Centres (MIRCENS) established from 1975 with the backing of the United Nations Environment Program and UNESCO.[29] The first development here was a central registry for world culture collections in Brisbane, Australia. By 1991, there were twenty-four MIRCENS.

The emphasis on gene pools represented the growing interest in the industrial significance of genetics that will be discussed in the next chapter. In 1982, the movement was taken a further stage by the formation of the International Centre for Genetic Engineering and Biotechnology (ICGEB). The sponsor was the United Nations Industrial Development Organization, which had been working since

1975 to raise the developing nations' share in the world's industrial production to 25% by the end of the century. Support of biotechnology had come to be seen as the best use of its limited resources, and Hedén was commissioned to write a working paper, which he entitled 'The Potential Impact of Microbiology on Developing Nations – A Fountainhead of Hope or a List of Lost Opportunities'.[30] Clearly, there was much to be done, but though the formation of the ICGEB was accepted at an international meeting in Madrid in 1983, bickering between potential hosts delayed its creation by another year. A split site between Trieste and Delhi was finally accepted for the new organization in 1984. Trieste would deal with biomass conversion, hydrocarbon microbiology, industrial-scale fermentation, and protein engineering. Delhi would concentrate upon biological nitrogen fixation and soil microbiology. Vaccines and immunology for animals and man would also be explored there. These themes expressed enduring issues in biotechnology for developing countries. The issues of rhizobia, energy production, and single-cell protein would be particularly pressing.

Rhizobia

The mystery of the means whereby the rhizobia bacteria nesting in the root nodules of leguminous plants convert atmospheric nitrogen to ammonia was described by Sir Harold Hartley as one of 'the great vital unknowns' of the world's economy.[31] If only man could emulate this process in cereals, the need for artificial fertilizers would appear to vanish. In 1960, the U.S. company Du Pont attracted much attention with its study of this potential challenge to their chemical market. The researchers identified active cell-free extracts of a bacterium that could 'fix' nitrogen. Shell investigated Du Pont's work, seeking either a fermentation process to ammonia or a low-pressure ammonia synthesis.[32] Though the research was only secondarily directed towards bacterial fertilizers, such possibilities galvanized researchers concerned with poorer countries. In the Du Pont work, Shell's investigator found 'an air of optimism'. When he suggested to the team that it might take twenty years for their research to pay off, there was a 'spontaneous, almost agonized, cry of protest from Dr Carnahan.... I gathered they were thinking of 5 years (or less).' Despite its tone, the report failed to stimulate Shell into supporting such research, but it was passed to Britain's Agricultural Research Council, which enthusiastically set up a unit. The group under Joseph Chatt, at the new University of Sussex, was to have its own well-recognized achievements. Yet breakthrough after breakthrough, the goal of a self-fertilizing cereal has remained distant. Early successes faded, and repeatedly over the following four decades hopes that the genes which give such marvelous powers to leguminous crops could be transferred to cereals have been raised and lowered. The most ambitious hopes have still not been fulfilled, although inoculation of impoverished ground is a useful process.

The relevance of this work became starkly clear when the cost of synthetic ammonia for fertilizer rose from $30 in 1972 to $140 two years later.[33] From the beginning of the MIRCENS programme in 1975, at the time of the first oil crisis, rhizobia were a major preoccupation. Of the fifteen MIRCENS set up by 1980, five (in Brazil, Kenya, Senegal, and two in the United States) were already concerned with rhizobium technology and were combing the tropics' rich reserves for a more effective bug.

Biogas and gasohol

The search for the secret of rhizobia was partly a response to escalating oil prices. These had more direct effects on poor countries than just through rising ammonia costs. Energy became more expensive. Biotechnology could also make a contribution here through 'biogas' and 'gasohol'. The idea of producing gas commercially through the use of rotting farmyard manure has a century-long history. It has been developed principally in China and India, where its history goes back to before the Second World War. The first commercial Chinese methane digesters were developed in 1920 by Luo Guorui, who created a company to market his invention.[34] Steam-generating digesters built in 1937 were still operating half a century later. Waves of rural plant building occurred in the 1950s and again in the 1970s. More recently, the effort has moved to industrial uses. By 1985, China could boast 6–7 million family-size rural digesters, 600 biogas power plants with a capacity of 6MW, and 1,200 biogas-powered electricity-generating stations with a capacity of 16MW.

Elsewhere, biogas acquired an image as a second-rate technology. And indeed it was reported that only half of China's rural digesters were actually working by the early 1980s. Nonetheless, their attraction as a decentralized energy source was immensely powerful in the early 1980s. India committed $62m for a five-year plan between 1981 and 1986 to introduce almost half a million digesters as part of its rural development policy.

A feature of biotechnology is its use of the rich raw materials of tropical lands. And of the tropical countries, the most ambitious and rapidly growing in the 1970s was Brazil. There, since the Portuguese invasion 500 years ago, sugar cane has displaced forest. In 1974, a sugar magnate, Urbano Stumpf, persuaded his president that alcohol produced from sugar could by itself power all Brazil's cars, without a need for oil, and could replace 80% of diesel fuel. Needed would be 31 billion litres of ethanol. At first alcohol was mixed as a 10% solution with conventional fuel, but from 1979 cars that would run on 100% gasohol were produced. Production reached 4.1 billion litres the following year.[35]

Brazil committed great resources to this programme. In 1980, it was seen as a model to the world, though later it would have a dispiriting history as world oil prices dived. Other tropical countries, from Costa Rica to India, intended to emulate its example. U.S. delegations visited, for in the wake of the oil crisis their country too was interested in reviving what Hale had called 'agri-crude' and what was now coming to be called 'gasohol'. Its story will be recounted in the next chapter, for this product was a key part of the dynamics of U.S. policy.[36]

Single-cell protein

The shift of biotechnological solutions from the problems of the poor to their employment by the rich can be seen in the development of single-cell protein, which was at first earnestly seen as a key source of nutrition. Even if nitrogen-fixing bacteria would produce conventional plants more cheaply, it appeared that alternative sources of protein should also be sought. Fast-growing algae (cultivated on the surface of water), traditional fermented foods, and most ambitiously bacteria grown in petroleum seemed candidates for a nutrition revolution. In the 1950s there were great hopes for the green chlorella algae, which were grown and processed in both Japan and Czechoslovakia, though after fifteen years production was still only experimental. By 1967, the Japanese were producing about 110 tonnes.[37] In the Far East, there is a tradition of using fermented soy beans, and Indonesian tempeh became a fashion food in the United States. Soya itself could be texturized to make an artificial meat.

Nonetheless, it was the possibilities of growing microorganisms on oil that captured the imagination of scientists, policy makers, and commerce. Of course, the idea of growing yeast for food was not new; after all, Max Delbrück had called yeast an edible mushroom. Following the German effort of the First World War, both U.S. and

German scientists continued to explore the use of yeast as fodder. In Germany, there was a particular interest in using the waste sulphite liquor from paper mills as the nutrient, and work was intensified by the isolation of World War II. By the war's end, six innovative plants were in operation, producing 15,000 tonnes a year of yeast.[38]

The British too had been interested in producing food, but they did not have available either sulphite liquor from paper making or waste starch from corn. Instead, a wartime committee identified high-sugar waste from the production of colonial products like bananas, plantains, and sugar in colonies such as the West Indies. Thaysen, who had worked for Weizmann in World War I, led a project to grow yeast on the byproducts of sugar manufacture. A laboratory demonstration was successful, and in 1944 it was decided to build a pilot plant in Jamaica. This opened in 1947 and ran for a decade.[39] Unfortunately, just as Dr Johnson dismissed oats as a food for animals, so yeast's technical success was marred by the unacceptability of the product as a human food. The cosmetic entitling of the product as 'food yeast' to distinguish it from 'fodder yeast' failed to convince. Moreover, a slightly unpleasant flavour left by the defoaming agent, consumers' conservativism, and the limits on its use of 7gms a day caused the project to fold.[40]

There was also an effort to look for innovative foods that could be produced at will, with the most ambitious plans based on microorganisms that would grow on oil. It had been known since the nineteenth century that certain bacteria will live on hydrocarbons. During World War II, the Germans were already producing oil from coal, first by breaking down coal with steam to produce carbon monoxide and hydrogen (the Lurgi process), and then by building up the components under pressure, through the so-called Fischer-Tropsch Process. By growing bacteria on the product they might effectively produce food from coal.[41] In 1947, Felix Just and Willy Schnabel, based appropriately at Delbrück's Institut für Gärungsgewerbe, reported experiments showing that both yeast and bacteria could be grown on a paraffin product of the Fischer-Tropsch Process. Although this was an interesting piece of science, a Cologne University scientist, W. Hoerberger, in 1955 concluded that current knowledge could not sustain hope for a commercially viable process. The requirements for oxygen would be great, and there was a tenacious oily smell to the resulting yeasts.

Nonetheless, it took only another two years for a way forward to be found. The experience of its French pioneer, Alfred Champagnat, is worth recounting. Not only was his work and that of his employer, British Petroleum (BP) important; it also inspired most other oil companies to follow suit.[42] At the laboratories of BP's French affiliate, Champagnat was exploring the microbiology of petroleum to remove traces of hydrocarbons from refinery waste water. Needing help, he brought in Jacques Senez of the CNRS laboratory in Marseilles. After eighteen months of work, the project was incomplete and appeared to be doomed. Then, Champagnat records, in a meeting with Senez, the two had the idea of using the yeast they had managed to grow on oil as food. The amino acid range was complete, and the product would therefore meet the world's needs for more protein. Nonetheless, even here they met mass scepticism – petroleum was carcinogenic, microbiological processes were slow, prices would be too high, and there was that negative study by Hoerberger, not to mention another by Raymond. Undeterred, Champagnat reasoned that the problems were technological rather than microbiological, and the technology had not been explored. The potential of continuous fermentation had so far been neglected, and no actual toxicological studies had been conducted. Moreover, with millions of hungry people there had to be a market. The petroleum industry with its multinational scale could mass-produce fermentation products just as it could petrochemicals.

By 1962 BP was ready to build a pilot plant, at Cap de Lavèra in southern France, and to publicize its product, Toprina (a near anagram of protein). Over the next five years, interest grew enormously. BP itself took the project to heart and, in addition to supporting Champagnat, launched a variant process in Britain. As yet there was no well-accepted term to describe the new foods, but in 1966 the term 'single-cell protein' (SCP) was coined at MIT (by Scrimshaw) to provide an acceptable and exciting new title for an old product, avoiding the unpleasant connotations of microbial or bacterial, and was launched at a major conference there the following year.[43] The field's rapid growth is indicated by the number of papers on the fermentation of petroleum: 92 in 1962 and 163 the following year (excluding Soviet and Japanese publications). By 1974, Champagnat could list several dozen projects in ten countries.

The Soviet Union was particularly active with production of single-cell protein on a variety of feedstocks, increasing from 58,000 tonnes in 1963 to 1.1 million tonnes by the end of the 1970s.[44] The Japanese were also planning large plants in Italy and in Rumania. Since 1955, the United Nations had been advised on food issues by an international panel entitled the Protein Advisory Group. They were impressed by the prospects of SCP and held a large number of working parties, producing a range of papers exploring the safety of the product that was seen as a solution to world hunger.[45]

In the West, BP seemed at first the strongest contender. By the mid-1970s, 40,000 tonnes of its product had been tested and marketed, and the company was ready to go up another scale; a 100,000 tonnes a year plant in Sardinia was announced as early as 1973. Other European companies besides BP, including Shell, Hoechst, and ICI, also entered the race to grow biomass on hydrocarbons examining both bacteria and fungi. Shell dropped out before their process went beyond the laboratory. Hoechst built a pilot plant, spending DM 80m, to grow bacteria on methanol. A rather similar process was developed by ICI, Britain's largest chemical company, but this went beyond the pilot plant stage to launch the world's largest fermenter with a capacity of 50,000 tonnes a year to grow its bacterial product 'Pruteen'. Like BP, it spent as much as £100 m on its investment.[46] Its work, however, had begun in the late 1960s, a full decade after BP, and only reached a decision point in the late 1970s.

By then the cultural climate had completely changed, for the growth in interest had taken place against a shifting economic and cultural context. First, the price of oil rose catastrophically in 1974, so that its cost per barrel was five times greater than two years earlier. Second, despite continuing hunger across the world, anticipated demand also began to change. The programme had begun with the vision of growing food for Third World people and BP advertisements showed Champagnat himself eating biscuits made of Toprina. However, the product was, instead, launched as an animal food for the developed world. The rapidly rising demand for animal feed made that market appear economically attractive. The demand for animal feed was growing even faster than that for the human population because of the inefficiency of animal conversion. A chicken producing 2kg of protein consumes 8.4kg of protein. In Japan, consumption of compound animal feed tripled between 1963 and 1975. While the demands for protein to feed humans directly and to feed animals were roughly equal in 1980, in the next twenty years, the animal demands were expected to rise 2.47 times, about half as fast again as human demand, which would rise with population, 1.61 times.[47]

The emphasis on animal feed was convenient because there are low limits on the safe amount of unprocessed yeasts humans can consume, and the removal of the nucleic acid (RNA), which can cause gout, is expensive. Only one Western company has commercially produced a human food, and its orientation and history were slightly different.

During the early 1960s, as Champagnat's work was becoming known, Ranks Hovis Macdougall, the British bread-making company, was led by an idealistic Methodist, Lord Rank. He employed a brilliant research director, Arnold Spicer, who saw the potential of the company's surplus starch as an ideal medium for growing microorganisms for food. Unlike the scientists in the oil companies, Spicer, as a food technologist, started with the belief that his product needed to be 'delicious to eat' and have a pleasing texture.[48] He therefore sought an organism with a natural consistency, choosing the long-stranded microfungi on which work had already been done. As well as a commercial convenience, Lord Rank saw the speculative and highly expensive project as a personal mission for the world's poor and offered help whenever it was needed. A suitable organism was identified by 1970, but it took more than a decade to turn this into a palatable product, in which the RNA had been destroyed and the safety validated. Even then Mycoprotein would be launched first in 1985 to cater to British vegetarian and low-cholesterol tastes rather than to the developing countries. Moroever, the economics of the process meant that it could not be based on the waste product of baking, but on imported American maize used for glucose syrup. To grow the fungus, the Ranks team used an ICI fermenter on which Pruteen had been developed. And from there ICI took over the entire project.

The Ranks experience showed how difficult it was to produce a novel human food. The easier choice made by other companies to go for animal nutrition had ironic consequences. It appeared that here was a rapidly growing market with potential in developed as well as poor countries which could attract profit-seeking commercial organizations such as BP. As part of its marketing strategy, BP had purchased large animal feed companies in order to become a major supplier of conventional products. If the simplicity of animal feed attracted big companies, it also proved the undoing of the technology, for it brought the novel product in competition with cultivated competitors, particularly the rapidly growing soya bean industry. While the price of oil soared, soya bean prices sagged. Argentinian and Brazilian producers entered the market to sustain a tripling of production from the early 1960s to the late 1970s.

David Sharp, secretary of Britain's Society of Chemical Industry in the 1970s, has carried out an elaborate analysis of the economics of SCP and the soya competition. Clearly, there were major problems in competing with an agricultural product whose price could be reduced just by increasing acreage. The rising price of oil and the declining cost of agricultural protein created such economic barriers

An advertisement for Mycoprotein: Stir-fried Quorn® in Blackbean Sauce. Courtesy Marlow Foods.

to the success of SCP grown on oil that the strength of public opinion was not seriously tested. Yet even as the economic balance tipped, there was another nonfinancial factor in the case of oil-based foods – public resistance. This was particularly vocal in the two countries where production came closest to fruition – in Italy and Japan. For all their normal enthusiasm for innovation and traditional interest in microbiologically produced foods, the Japanese were the first to ban the production, and the plans of Dainippon and Kanegfuchi were arrested. Kei Arima, one of the pioneers of Japanese biotechnology, identified five questions raised by newspapers and consumer groups.[49]

1 – Why is the public forced to eat petroleum and micro-organisms, in this case yeast? (emotional opposition)
2 – Food originally came from the natural world and must be the one which has been eaten since earliest times. (rule of experience)
3 – The safety data has been presented only by the industrial side. (distrust of enterprises involving the problem of environmental pollution)

4 – Petro-protein only cannot solve the food shortage. Petroleum will be exhausted shortly. (problem of resources)
5 – Agriculture and fisheries will be destroyed by development of SCP industry. (political opposition)

Systematically, the Japanese 'petro-protein' makers had failed to separate the idea of their new 'natural' foods from the far from natural connotation of petroleum.[50]

These arguments were made against a background of suspicion of heavy industry. The companies gave the technology a name so bad that the biological roots were insufficient to validate it. In Italy the problem was the same. Anxiety about minute traces of petroleum was expressed against the background of the hideous damage caused when the chemical plant at Seveso exploded, inundating the environment with toxic PCBs. Biotechnology had been associated with all the odour, metaphorical and literal, of the chemical industry. Neither the plant built for BP and its Italian partner, the Italian holding company ENI, nor one erected according to a Japanese design, was allowed to operate.

From Arima's careful catalogue, it is clear that the offence was made more through the association with industry than through ignoring any particular scientific result. One might have expected the vegetarian movement to take the Mycoprotein developed by Ranks, approved scientifically and by the regulators as it has been, to its heart. That this has not happened is perhaps as much because the product is produced from organisms grown in a fermenter at ICI's Billingham chemical complex, as because of the use of animals to prove its safety. Anxiety about the safety of the product was therefore founded on suspicion of those who had dismissed other fears at an early stage only to find them widely accepted – the sense of the unnaturalness of the process and the economic instability caused if it were successful. A decade later, the very same issues would be raised in questions about products based on recombinant DNA techniques.

Conclusion

After half a century of marginal intellectuals pushing the idea of biotechnology as a new kind of technology meeting a new kind of need, it appeared that the right moment had come in the 1960s. The new need was urgent and agreed on: world starvation. Radical measures would be needed to meet it. Biotechnology seemed to offer the answer. However, the solutions proved to be either too expensive or unacceptable, and solving world hunger through SCP food and the engineering of nitrogen fixation by cereals seemed to be mirages. In the 1970s, the food crisis was succeeded by the energy crisis. Here too, biotechnology seemed to provide an answer for the poor, but once again, costs proved prohibitive as oil prices slumped in the 1980s. So, in practice, the implications of the technology could not then be tested.

Still, the approach has kept its adherents, particularly as a way of generating completely new organisms appeared to be available through genetic engineering. In 1982, the veteran Dutch microbiologist la Rivière summed up the enduring attractions of biotechnology that were then, and are still, valid. The low-cost, rural, and small-scale nature appealed, he said, to the 'ethical' track of Dutch development policy aimed at the poorest of the world's needy. At a national level, the potential contribution to upgrading agricultural outputs appealed to the 'rational' track of Dutch policy.[51]

The distinction is a useful indicator of the two sides of the appeal of biotechnology. It would, however, be misleading if it suggested that the ethical side could be considered secondary. It was the combination of these two characteristics that translated the early twentieth-century ambitions into the context of the 1970s. Elsewhere, the use of technology in poorer countries might have been a desirable, but still secondary aspect of its development. In the case of biotechnology, by contrast, this provided the key context for the development and preservation of the technology's international networks. The shift in focus from developing countries to the home front in the 1970s made the failure of the techniques integrating the ethical and rational tracks particularly tragic. But the techniques of recombinant DNA that appeared in the 1970s offered hope of a 'miraculous' reversal. The switch to Western needs, and the marriage of industrial microbiology with genetics, will be explored in the next three chapters.

7

From professional to policy category

> Is biotechnology really revolutionary? Can industrial products be made economically from natural or waste materials? Just what does come under the umbrella of biotechnology? A recent meeting at the Royal Society provided some of the answers.
>
> (Pearce Wright, 1979)[1]

This headline indicated a new approach to biotechnology, even if the 'answers' would be harder to find than it suggested. Previously, those feeling the call to conjure up a vision of this as the 'next stage' in human technology had often left the 'how' of the transition to secondary detail. Even in the 1960s, the advocates of change were evangelists rather than managers. However, during the decade of the 1970s, biotechnology became the focus of detailed state policy. First in Japan, then shortly after in Germany, throughout the European Economic Community (EEC), and in states throughout the world, administrations clutched at this resolution to the paradox of a need both for a stronger industrial base and for environmental enhancement.

The lesson of a century of prophecy came to be widely accepted: The existing technological system was obsolete and had to be superseded. Two escalations in the price of oil, in 1974 and 1980, increased the cost of the Western world's energy tenfold, but even beforehand, developed nations were concerned that established industries had matured, and their growth had peaked. Then, to the fear of declining momentum in the industries which had driven rapid postwar growth – chemicals, steel, shipbuilding, and automobiles – was added increasing political concern over the pollution that those industries had caused. In response, officials and politicians plucked the concept of biotechnology, together with the hopes surrounding it, out of the intellectual firmament and inserted it into their industrial development programmes. The implications (and the subsidies) were explored by small investors and big business.

Thus, biotechnology came to be defined by programmes and by a nascent industry, bounded commercially, administratively, and politically. The metaphor by which the subject was promoted, 'the next industrial revolution', resonating with other such familiar catchphrases as the 'information age', was translated into the language of urgent and ambivalent current affairs. Astounding scientific advance seemed to provide ever more reassurance that biology's century-long promise would be redeemed through what would be called 'life sciences' in Japan, 'biotechnology' in Europe, and 'bioresources' in the United States. Because most histories of biotechnology cast these advances as themselves the cause of change, it is worth reiterating that the perception of scientific techniques was refracted through the existing language, visions, and current aspirations. Emerging in the aftermath of the genetic engineering revolution, the widespread popular hopes for biotechnology continued mid-1970s interpretations driven by policy concerns.

Such concerns had responded to challenges posed by the United States, which had long dominated the global economy and whose ebullient culture stimulated responses worldwide.[2] Four examples were put to other industrialized countries: the environmental movement, the emerging philosophy of enzymes as an increasingly integrated part of bioresource development, the scale of life sciences research in general, and the model of Silicon Valley.

Historically, the problems of agriculture have stimulated thinking about biotechnology. The pattern was repeated in the 1970s, although on this occasion the issues were initially most vocally expressed outside the scientific establishment and grew out of the environmental movement beginning with Rachel Carson's attack on the use of the pesticide DDT, in her 1963 exposé *Silent Spring*. In particular, the consequences of an endless use of nonrenewable resources attracted extensive analysis in the age of the *Whole Earth Catalog*. Best known was the 1972 report of a group of MIT scientists, *Limits to Growth,* which seemed to show that the exponential use of oil and minerals was about to become nonsustainable.[3] Experiencing, almost immediately, the implications of resource scarcity in the costs and queues of the first oil crisis, readers of the report thus saw the prophecy of shortage apparently being fulfilled even sooner than expected.

In response, the Midwest offered vast amounts of agricultural produce – bioresources. In 1971, even before the oil crisis, tests of alcohol fuels by Nebraska's Agricultural Products Industrial Utilization Committee demonstrated the continuing viability of the prewar 10% blends. A year later, the term 'gasohol' was coined by William Scheller, a University of Nebraska professor, and capitalizing on his work, the state took out a trade mark on the word. Winning congressional and presidential approval, the fuel became increasingly well known. The turning point came in 1979, when the Soviet Union sent troops to Afghanistan and, in retaliation, the Carter administration cut off supplies of agricultural produce. Now, as in the 1930s, corn products had to be used industrially if farmers were to survive, and, with another oil crisis impending, gasohol would be the prime solution. On 11 January 1980 an alcohol fuels program targeted a 600% increase in ethanol production within two years. In June, Congress approved a synthetic fuels bill that set aside $1.27b for federal aid towards alcohol and other biomass fuels. Before the new direction could be taken, the wind changed again; the Reagan administration came to power on 20 January 1981 and, with the declining oil prices of the 1980s, ended support for the industry before it was born.[4]

Support for the gasohol program was anomalous in the history of U.S government industrial policy. The vigour of industrial R&D and attachment to the market economy has repeatedly inhibited national leaders from formally identifying and supporting 'winning' technologies of the future. Instead, the federal government has given funds for fundamental science research that would not be seen as upsetting the market or giving undue preference to individual companies in civilian industry. This self-denying ordinance has not extended to health, military, or space issues, and during the 1970s, under the pressure of so-called corporatist tendencies, there was a tentative approach to the promotion of other technologies.[5] In 1971, the National Science Foundation (NSF) established a small programme, Research Applied to National Needs (RANN), in which enzyme technology had its own special budget from the start. The sums awarded were small, rising from half a million dollars at first to just over $2m in 1976.[6] Enzyme technology's potential key role in the manufacture of economic gasohol and the energy efficiency of enzymatic processes meant that the enzyme programme was incorporated within the energy programme from 1977. There, its distinctive image was lost, and the vote declined. So, although the United States sponsored key research, and U.S. companies were the central players in enzyme utilization during the 1970s, the government was not a creative sponsor of biotechnology policy. Nonetheless, the existence of the RANN programme and the contemporary investments by industry promoted a vision of the multiple applications of enzymes in health, agriculture, the environment, industry, and energy. RANN's meetings and papers helped develop and publicize models of the central role of enzymes that would look very familiar a decade later.

Though agriculture and its problems continued to be an important stimulus to biotechnology, life sciences were increasingly funded on account of their medical significance. In the United States, public interest in health matters, elsewhere resulting in 'socialized medicine', underpinned a vast growth of research. In 1930, half the civilian chemists employed by the federal government were employed in the Department of Agriculture; by 1978, scarcely 13% were employed there, compared with almost twice the number in Health, Education, and Welfare, an agency that hadn't even existed in 1930.[7]

The biomedical funding body National Institutes of Health (NIH), founded in 1930, saw its budget rising from $3m in 1946 to $70m by 1953 and $1.1b in 1969.[8] Expenditure on basic life science research in academe almost doubled between 1964 and 1972, while funding for physical sciences increased by only 50% in the same period.[9] Heart disease, mental illness, and cancer remained powerful challenges to the American way which had been so successful in developing technical solutions as diverse as the tungsten light filament, the atom bomb, and landing on the moon. In 1971, two years after the last of these achievements, President Nixon launched the War against Cancer. The new Cancer Act authorized him to spend $1.59b on cancer over three years. By 1975, two-thirds of a million people were engaged in the campaign, from Nobel Prize–winning scientists distantly removed from patients to bedside caregivers.[10] Meanwhile, molecular biology, its manifold implications particularly for new drugs, and the means of regulation were being vigorously explored, and the scale of U.S. life science research, in government and commerce, impressed the world.

If biomedical research demonstrated the vitality of government supported activity in the United States, the electronics boom of the Santa Clara Valley in Northern California, known as Silicon Valley, seemed to presage a change in the form of industrial organization. Small companies based on high technology and little capital seemed to provide a base for the future.[11] In 1968, there were thirteen start-up chip makers alone. The most heroic of all stories was that of the home computer makers: Apple Computer, started in a garage in 1977 by Stephen Jobs and Stephen Wozniak, became a $100m company by 1980. The semiconductor industry, barely a couple of decades old, was worth $6b by 1979. Every country would aspire to its own 'Silicon Valley'. A popular 1982 text about the electronics boom was entitled

The New Alchemists. Similar titles would be popular for books about biotechnology, for there was a similar aspiration. Would there be a parallel with the silicon miracle, by which science was turned into money? The question was urgent, as maturing economies looked for new bases of future economic growth.

Japan

Picture a technology similar to the edifice built upon chemistry, but jointly meeting the needs of industrial renewal and environmental balance. This was the desirable vision that 'sold' biotechnology in the world's two postwar success stories, Japan and Germany. By 1970, Japan had achieved the world's third-highest GNP, after the United States and the Soviet Union. The achievement had been based on industrial growth of 14% p.a., with energy use tripling in the 1960s. Because of the small size of the country, and its even smaller habitable area, industry and its effluents were geographically concentrated. The amount of copper used per square kilometre was almost ten times that of the United States in 1969. Both air and water were, in many cases, unacceptably polluted. Almost all fresh surface water contained industrial effluents by 1970. The rivers passing through the major cities of Tokyo, Osaka, Fukuoka, and Nagoya were contaminated far beyond government standards.[12]

These chronic problems were brought to a head by local issues and by a series of national scandals. Named after the village where it was first observed, the 'Minamata' disease was caused by mercury poisoning induced by eating fish from polluted water. By 1972, with evidence accumulating from 1964, sixty deaths and almost three hundred cases had been recorded. Another disease, Itai-Itai caused by eating rice grown in cadmium polluted water, had killed thirty-four by 1972. Whereas in earlier decades the people had been willing to accept what authority had decreed, by the late 1960s this was no longer true. The issue was termed *'kōgai mondai'* by the Japanese and is explained by Bennett and Levine as having the 'same generalized and evocative meaning that "ecology" has in public American English: it is a symbol of man's inhumaneness to the environment, and it includes social dangers and discomforts derived from damage to nature.'[13] This concern informed discussions within Japan's trade ministry, MITI, the Science and Technology Agency, and in large companies. The 1977 *White Paper on Science and Technology* stated that people expected 'science and technology to give assurances of "safety, conservation and environmental integrity", instead of "speed, low cost, quantity and convenience" '.[14] The emerging consensus suggested that there would be need for the development of new nonpolluting technologies and a knowledge-intensive workforce. During the early 1960s, an overwhelming 87% of prize-winning Japanese technologies had been devoted to the pursuit of efficiency. The pursuit of environmental integrity had attracted only 3%, conservation only 4%. By the early 1970s, environmental technologies had gained 13% of awards and conservation 7%, while efficiency was down to 69%.

The early 1970s saw ambitious schemes for reform of Japanese society in general, especially as the shock of the oil crisis was particularly intense in a country without its own petroleum reserves. The sense of a need for a new kind of industry led, most dramatically, to the focus on the new information technologies whose conquest of world consumer markets are familiar. Less well known are the new philosophies of manufacture, for example, the 1976 coinage, 'mechatronics', to represent the integration of mechanical and electronic engineering.

The life sciences were another beneficiary of the Japanese reevaluation. As so often in Japan's industrial development, the view of industrial prospects was based on a reinterpretation of accumulated experience. Here the broad conception of the interface between biology and engineering was similar to Western analogues. Although different words were used, the spirit was the same, as was explained, unself-consciously, in a 1970 paper by Sakaguchi, the grand old man of Japanese applied microbiology:

> The tendency for applied microbiology to be presented here as approximately synonymous with fermentation science may appear strange to the Western reader. However, the term 'hakkō' in Japanese, although basically similar to the English word fermentation is used to include a broader range of phenomena. Its essential idea is that of the production of useful substances (and destruction of harmful substances) through microbial activity. By some, it is further extended to include non-degradative processes ... thus giving rise to a concept which in literal translation would appear as 'synthetic fermentation'.[15]

By 1970, the Japanese were already secure in their achievements and tradition in applied microbiology. The subject had a far higher profile than in the United States or Europe, as the official volume with the all-inclusive title *Profiles of Japanese Science and Scientists,* which included Sakaguchi's paper, demonstrated. It featured, among its sixteen chapters, essays entitled 'Applied Microbiology in Japan', 'Amino Acid Production by Fermentative Processes', 'Molecular Structure and Function in Living Matter', and others on natural product chemistry.[16]

Applied microbiology was therefore a ready resource for the chemical industry fighting off its reputation for pollution. In 1971, the Mitsubishi Chemical Company established an institute for the 'Life Sciences'.[17] This complex was also identified as a key area for the 1970s by the Council for Science and Technology advising the prime minister in its 1971 report 'Fundamentals of Comprehensive Science and Technology Policy for the 1970s' which stated that strategic issues should be targeted and the research infrastructure improved. Teeth were given to such aspirations when, two years later, the Committee for the Promotion of the Life Sciences was established to coordinate activities across the government and within the Science and Technology Agency. A special Office for Life Science Promotion was set up to implement the plans, within RIKEN, the Institute for Physics and Chemistry. The government's commitment was maintained, as the 1977 *White Paper on Science and Technology* pointed out: 'Life Science, in particular, is for the study of phenomena of life and biological functions that will be made useful; for industrial, medical, agricultural and environmental purposes, and so this area of science is expected to set the pace for the next round of technical progress.'[18]

The policy debates of the early part of the decade had translated specific experience into the conception of a new generic technology for the future. Then, in a characteristically Japanese manner, the 1970s saw two parallel processes. While the dynamic enzyme industry developed along commercial lines, the Office for Life Science Promotion developed a philosophy of enzyme technology and a programme for implementing a bioreactor revolution. A 1975 report in *Nature* by the Office for Life Science Promotion's head, Professor A. Wada, reflected current obsessions with robots. Entitled 'One Step from Chemical Automatons', it expressed an entire philosophy of bioreactors, which was further articulated in two Japanese reports the following year.[19] This warning of a jump into the industrial future was closely scrutinized by Europeans. The later reports predicted:

> Industrial applications of enzymatic reactions which have been known to play a central role in highly organized and efficient biological activities are recognised in recent years as one of the important and urgent tasks for the benefit of human welfare. Some of the greatest benefits which human society can expect once such application is made practical are: 1) reduced energy consumption, 2) chemical industry based on aqueous solutions under normal temperature and pressure, 3) streamlined processing of complicated chemical reactions, 4) self-controlled chemical reactions and 5) minimum disturbance to ecology.[20]

From this model emerged a complex analysis showing the manifold ramifications of bioreactor technology. Though the scheme developed by Professor Wada was impressive, it was not radically new. Instead, it brought together established thinking into a coherent framework. Japanese achievements were viewed with alarm in the West, where the thorough philosophy and planning were put together with the indicators of industrial activity and apparently impressive (possibly overestimated) patent statistics, to cast the industrial endeavour in a light that made it seem potentially world dominating.

Germany

The Japanese construction of life sciences as a policy category contrasted with the U.S. category of biotechnology as a field of applied science. But in each of these countries, the concept of biotechnology itself was weak and, in Japan, not even used at the time. It was the Germans who brought together the scientific and policy dimensions and who, for the first time, gave policy significance to the term 'biotechnology'.

For two decades after World War II, Germany had prospered by making chemicals, steel, motor cars, and electronics superbly well. But by 1967, the economic rebirth appeared to be ending at the very time when the Americans were pioneering and dominating the development of new technologies in such industries as computers and aerospace.[21] In addition to envy of the Americans, a second factor was also becoming significant in the promotion of German biotechnology during the late 1960s: *Umweltschutz* – environmental protection. In Germany, the environmental movement has a more vigourous history than in any other country. In the aftermath of the Vietnam War, nuclear power plants were the main target. *'Atomkraft – nein danke'* (Atomic power – No Thanks) read the badges. The chemical industry that had polluted the Rhine for over a century was also cast as an ogre by the generation of the student radicals of 1968. As in Japan, environmental issues were linked with questions of political organization. In response to the polluting industrial age, Germans recaptured some of the romantic idealism of the 1920s. Although the formal 'Green' movement was not established and the name not chosen until the spring of 1978, the individual elements had been in ferment for years earlier. By 1972, there were reportedly 7,000 different groups. Underlying their policies were eight emphases: decentralisation, participation, reduction of power, sparing use of natural resources, ecological behaviour, healthy technology, freedom from coercion, and multiplicity. Hence, the emerging Green tendency favoured technology, which would depend on renewable resources, was associated with low-energy processes that would produce biodegradable products and waste, and was concerned with the health and nutrition of the world.[22]

The Green movement was not a promoter of biotechnics. Nor was it, in any case, a powerful political force in its own right until 1985. Nonetheless, it did pose a challenge to the established political parties, particularly to the Social Democrats to whose ranks many of the more left-wing members would otherwise have gravitated. So when the new socialist government of Willi Brandt, elected in 1971, formed its industrial support department, BMFT, the agency took, in political scientist Sheila Jasanoff's phrase, 'a very broad definition of the public interest, encompassing health, nutrition and environmental quality'.[23]

In the aftermath of the U.S. moon-landing success, a mission-based grand project had come to be seen as the potent means of technological advance. Impressed as they were, the Germans analyzed their own requirements of the benefits sought from new technologies in a 1970 report entitled *Erster Ergebnisbericht des ad hoc Ausschusses 'Neue Technologien'*.[24] This identified three types of need: fundamental requirements such as food or raw materials, infrastructural issues such as transport, and environmental concerns. In 1972, a programme of biology and technology was established in the new BMFT. Six areas were identified as priorities: security of food and feed supplies, reduction of environmental pollution, pharmaceutical production, development of new routes to raw materials, chemicals and metal production, and development of biotechnological processes and basic research. The close association with the New Technologies Programme was clear. At a time when increasing priority was being accorded to environmental issues, biotechnology could be shown to address them all.

The infrastructure was, however, inadequate. Fermentation and biology could not match the status of chemistry, and as a discrete subject, chemical engineering was poorly developed. Unlike the United States, where, even by World War II, there had been a mature profession of chemical engineering which could then expand its horizons to take in biochemistry, in Germany the engineering of chemical works had, until recently, been the result of close collaboration between chemists and engineers. Now, considerable thought was being given to the disciplinary basis of an integration of engineering with other sciences.

In this dynamic context, the word 'biotechnology' would serve as a catchphrase for the importance of developing a new type of industry. The first enthusiast seems to have been the Stuttgart Professor Karl Bernhauer, author of the classic prewar text on fermentation chemistry.[25] During the 1930s, he had taught in Prague, becoming increasingly interested in biochemistry. After the German takeover, his teaching commitment to biochemistry, fermentation, and nutrition increased. With an understanding of submerged fermentation (covered in his book), he worked on penicillin production during the war. Three years after the German defeat, he moved to Stuttgart, working for the drug company Hoffmann-La Roche. There he concentrated on vitamin B and the production of cobalamines. From 1960, he was given the honorary title of professor of biochemistry, and four years later he retitled his group 'Biochemistry and Biotechnology'.[26] Bernhauer had close links with the penicillin industry, and it is reasonable to interpret his adoption of the word 'biotechnology' as a U.S. import. At the same time, he had been so active before the war that he may well have been familiar with the earlier German uses. Bernhauer was no modernist molecular biologist. He stood solidly in the technological line of Delbrück and Lindner (whose name is the first to be cited in his 1936 book). The history of fermentation chemistry was for him the history of yeast and its use in alcohol and then foods. Pasteur is hardly mentioned in the three-page historical summary to his prewar text.

Hanswerner Dellweg, a former colleague of Bernhauer, became director of Berlin's Institut fur Gärungsgewerbe in 1967. Feeling that the scope of the institute, from its roots in brewing, needed to be broadened to encompass such major modern industries as penicillin, he followed Bernhauer's example and renamed Delbrück's creation as the 'Institut für Gärungsgewerbe und Biotechnologie'.[27] Adorning the title of such a majestic organization, the word had 'arrived' in Germany. Ironically, for a prewar inhabitant of distinguished German dictionaries, *'Biotechnologie'* had traveled to the United States and back and had to be associated with a new industry developed abroad, antibiotics, to acquire significance.

A first Symposium on Industrial Microbiology was held in 1969. At its successor the following year, the microbiologist H.-J. Rehm called for a bringing together of biochemists who were ignorant of engineering and of engineers ignorant of biochemistry: 'A future aim should therefore be to close the gaps by suitable training, to rise above classical fermentation technology, and to build up a modern science of biochemical-microbiological engineering'.[28] About the same time, an article appeared in the news magazine *Nachrichten aus Chemie und*

Technik, entitled 'Biotechnik und Bioengineering'. Arguing that in Germany, applied microbiology and bioengineering were woefully behind the United States, Japan, and even Britain, it called for a new research and teaching organization to cover the new field whose name had clearly not yet stabilized.[29]

These scientist-inspired formulations complemented an industrial movement to respond to the 'greener' colour of national needs.[30] In 1972, the chemical company Bayer established a centre, a *'Biotechnikum'*, for research into biological engineering. The same year DECHEMA, the German association of chemical plant manufacturers, having already taken the initiative and established a working group, was formally commissioned by BMFT to carry out an enquiry into what should be done about biotechnology.[31] Beginning with the comment that the word 'biotechnology' had a variety of meanings and that here the biomedical contexts were definitely not intended, the report defined 'biotechnology' in a way that was to influence European ideas thereafter: 'Biotechnology is concerned with the use of biological activities in the context of technical processes and industrial production. It involves the application of microbiology and biochemistry in conjunction with technical chemistry and process engineering'.[32]

The report systematically analysed the opportunities of the field and complained that compared with Britain, the United States, and even Czechoslovakia, Germany had undervalued applications of microbiology. Now the range of possible products and the environmental friendliness of the methods made it an urgent priority, for it could be related closely to the objectives of the recently formulated science policy.

The impact of the report was accentuated by its timeliness, for its appearance in 1974 coincided with the final shattering of faith in cheap oil induced by the Organization of Petroleum Exporting Countries (OPEC) oil boycott. It was the basis for a series of plans by the Federal Research Ministry designed to upgrade German work in the area. Between 1974 and 1979, support for the biotechnology programme increased from DM 18.3m to DM 41.3m.[33] The availability of federal government funds further encouraged the chemical industry to take the area seriously at a time when petrochemical plants seemed to be dinosaurs, and chemistry itself seemed to be failing the industry built on it.

The enthusiasm for biotechnology also led to the take-over of a major institute. In 1965, the Volkswagen Foundation had provided DM 11m for the purchase and conversion of a laboratory complex in Braunschweig to address the research front in molecular biology, an area in which it was felt Germany was falling behind. This became known as the GMBF – Gesellschaft für Molekularbiologische Forschung. The institute was given independence in 1968. At the same time, the federal government was reflecting upon the institute's future, and at a meeting in May 1969 it was determined that a pilot plant was required. With this rather more practical orientation, the state would be willing to support and in 1975 to take over the institute.[34] Its title was changed to Gesellschaft für Biotechnologische Forschung (GBF) as the role changed from fundamental molecular biology to mediating between academe and industry. Its head, Professor Fritz Wagner, was another Bernhauer pupil. In his introduction to a celebratory booklet about the GMBF/GBF, Hans Matthöfer, the BMFT's minister, expressed sentiments that drew on a century of hopes and promised that the institute would encourage the better use of natural resources and environmental protection.[35]

Despite its 'greenish' colouration, the Germans constructed biotechnology with a strong industrial orientation. The concept was filled out during the later 1970s by H.-J. Rehm who had chaired the DECHEMA group. For him the subject had passed through four generations: There was the ancient brewing, then the more scientific approach stimulated by the generation of Pasteur, followed by the penicillin age; the fourth generation now imminent would be distinguished by the understanding of processes of biochemical engineering, of immobilized enzyme technology, and of microbial genetics. Unlike the millennial visions found earlier in the century and seen later in the same decade, this generational approach did not suggest revolutionary changes. Instead, it was describing a gradual, if cumulatively impressive, improvement in humankind's ability to control useful microbial processes.

The German influence

The German and Japanese visions of the 1970s were formulated before the full impact of new recombinant DNA techniques was felt and reflected a biotechnology which, while benefiting from many innovations and discoveries, was as much need-pulled as science-pushed. The subject was interdisciplinary; DECHEMA officials portrayed it at the interface of chemical engineering, microbiology, and biochemistry, but it was also a conventionally recognizable technology. Laid down on the template of chemistry, it would be advanced through the traditional forms of industrial research approaches. This model proved stable and influential, modifying only slowly through the

1970s, as new results emerged from the laboratories of molecular biology, and enduring to the 1980s. By then, seen from the U.S. perspective, biotechnology policies were either a terrible threat (in the Japanese case) or a strange case of corporatism (in the German). Their distinctive roots in the response to the environmental movement had been forgotten.

Nevertheless, in the 1970s, Rehm's philosophy, and the report which he had engineered, had importance to the whole of Europe. A visit by Cornell University's Robert Finn, who in the early 1950s had promoted one of the first biochemical engineering programmes, catalyzed activity at a European level. Staying at Zurich's prestigious university, the ETH, Finn noticed how divided were European biologists and chemical engineers. He proposed the formation of a European biotechnology society to bring them together. Despite initial support from scientists who were circulated with a proposal, the idea was killed at a meeting of microbiologists in Berlin. Finn describes the attitude as being 'you can't dance to two tunes at once'. Still, the seed was planted and DECHEMA was well placed to capitalize on the freelance attempt to give biotechnology an institutional home. Recruiting Britain's Society of Chemical Industry and its French equivalent, the Institut de Chemie Industrielle, it created a nonmembership organization, the European Federation of Biotechnology (EFB) which held its first meeting in 1978.[36] European though the federation is, its secretariat is based in DECHEMA, which also issues the newsletter (first edited by Klaus Buchholz, who had been the secretary to the original DECHEMA biotechnology working party). Again, the first meeting at Innsbruck was hosted by DECHEMA and organized by Rehm and by Fiechter of the ETH, with the themes of immobilized enzymes, bioreactors, and biochemical engineering. The definition of its subject, adopted by EFB in 1981, clearly descended from the German formulation a decade earlier: 'Biotechnology is the integrated use of biochemistry, microbiology and engineering sciences in order to achieve the technological application of the capacities of microorganisms, cultured tissue cells and parts thereof'.[37]

The second of the EFB's international meetings was in England's seaside resort of Eastbourne, reflecting the importance of the British to the organization. Britain provides an interesting parallel to Germany. Its scientists had closely followed the development of biotechnology; as early as 1972, the Society of Chemical Industry had retitled its publication *Journal of Applied Chemistry* to *Journal of Applied Chemistry and Biotechnology*. British companies had played an important role in the development of antibiotics: witness Beecham's place in the semisynthetic penicillin story. Moreover, British enterprises had been world leaders in the attempt to produce novel protein foods. Established institutions in both Germany and Britain, the Institut für Gärungsgewerbe and the Microbiological Research Establishment, respectively, sustained pragmatic conceptions of biotechnology as the basis of development that would build on (in the German case) or reverse (in the British case) an economic legacy. There were of course differences: In Germany the government acted early, whereas in Britain it would wait until 1979. By then, it could take account of the new genetic engineering breakthroughs, though even its programme grew out of the ideas of the previous decade. The British case therefore provides an interesting demonstration of how the German-inspired model could accommodate new discoveries.

Britain

British developments were coloured by the legacy of the country's Great Power and imperial status, which left the country with disproportionately large government research institutes, with its great oil and chemical companies now seeking to diversify, and with a national obsession over economic decline counterbalanced by hopes of technological greatness. Though, by comparison with Germany and Japan, environmental concerns were relatively minor, the strong tradition of biology and the hopes and experience of its industrial potential meant that biotechnology came to be seen as a means of revival. Industrially, British enzyme work, as report after report showed, was, by international standards, small scale. At the same time, the single-cell protein interests of BP, Shell, ICI, and Ranks Hovis Macdougall meant that the British had a key role in what seemed to be a central market. The continuous fermentation expertise permeating out from Porton provided expertise at the cutting edge of bioreactor technology. In the background, the country's excellence in molecular biology provided hope for the next generation.

Despite the military victory, on many industrial fronts the postwar world would leave Britain with the after-taste of defeat. In the case of biotechnology, failure was much less decisive. Chemical engineering developed at a speed second only to that of the United States, Britain rapidly developed an antibiotics industry, and it had a strong tradition of applied microbiology. The postwar decades saw repeated enthusiasm for bringing these ingredients together, though the opportunities presented by such large government institutes as the Microbiological Research Establishment were lost repeatedly.

The stage was set by Sir Harold Hartley, an enthusiast for biotechnology even before World War II. In a much-cited 1951 presidential address to the Institute of Chemical Engineers, he called on the rapidly growing profession to take seriously the prospects for biochemical engineering.[38] Enzymes, he argued, had tremendous potential as biological catalysts. Two years later in another presidential address, he echoed his call with mixed success.[39] During the 1950s, several departments of biochemical engineering did spring up, and though one major laboratory was killed, it seemed that an international centre was about to be born.

Manchester, the industrial centre where Weizmann had worked, experienced a biochemical renaissance under one of Weizmann's trainees, T.K. Walker, who had taken a post there after World War I. A chemist, he was somewhat out of place in biochemistry, traditionally a medically oriented subject. In 1958, with new fermentation plants springing up in the area, he relaunched his department under the title 'Biochemical Engineering'.[40] At Birmingham University, the old British School of Malting and Brewing widened its ambit, in a way that would have been familiar to Orla-Jensen, and also became the Department of Biochemical Engineering (appropriately, Orla-Jensen's present-day successor, Professor O.B. Jørgensen, is a graduate of the Birmingham department). A group at University College London was formed within the Department of Chemical Engineering. It was from there that Donald and Crook called for the establishment of Gaden's journal. The most dramatic and certainly flamboyant innovation was by Professor Ernst Chain at London's Imperial College. At times seeming a caricature of the brilliant East European scientist – often temperamental, impatient of what he considered others' incompetence, sure of their grudges against him – Chain had been irritated by the British wartime failure to patent penicillin. A Nobel Prize winner for the discovery of penicillin (with Florey and Fleming), he defected to Italy after the war to a lavish institute provided for his personal use. There he developed his programme of penicillin manufacture, latterly in conjunction with Beecham. He returned to the offer of an institute built to his specification at Imperial College. Within it was installed a major pilot plant, the largest in British academe, funded by the industrial magnate Lord Rank.[41]

Against these successes was set the failure of the government's Chemical Research Laboratory team, another part of Weizmann's legacy. The development of the World War I acetone–butanol process had trained people experienced in relating biochemistry to engineering. The navy had created a small research group around its pilot plant at Holton Heath in Dorset. This included the enthusiastic Danish microbiologist Thaysen, who had previously worked at the Lister Institute, H.J. Bunker, and L.D. Galloway. In 1932, they were moved to the Chemical Research Laboratory and there carried out many studies of microbiological deterioration for the government. After World War II, the laboratory came into its own. It led the study of sulphur metabolizing bacteria, though, as we have seen, its scientists had to turn to Porton for continuous fermentation expertise. While the enthusiasm of the moment had highlighted the potential of fermentation, the Chemical Research Laboratory was restructed after its manpower and its budget had been cut in government economies of the late 1950s.

So, by the mid-1960s the British had a variety of institutions concerned with biochemical engineering. Most had grown out of departments dedicated to specialized professional training and lacked the size and national stature to take a role equivalent to Germany's Institut für Gärungsgewerbe or the new GBF. There was, however, one candidate whose future has been raised time and time again. It is characteristic of Britain, which has spent an unusually large proportion of its research budget in military contexts, and experienced a greater continuity from World War II than other European states, that this should be the Microbiological Research Department (MRD, or Microbiological Research Establishment after 1957) at Porton Down.[42] By the late 1950s, the MRD's military role seemed to be diminishing. The successful test of a British hydrogen bomb and a defence review suggested that, in future, leadership in biological warfare would be less crucial. The abolition of its parent department, the Ministry of Supply, in 1959 seemed an opportunity to civilianize the laboratory to create a National Institute of Applied Microbiology. The societies of Chemical Industry, Applied Bacteriology, and General Microbiology together formed a committee, which after an elaborate analysis concluded that such an institute, somewhere, was required.[43] There were problems: The proposed supporter, the Department of Scientific and Industrial Research (DSIR), had a poor record in supporting the life sciences, and MRD's director, David Henderson, with a medical bent, was averse to seeing his institute swallowed up within DSIR. 'Sulphur and Sewage', he snorted.[44] Still, by 1962 it did seem that an acceptable formula had been worked out: Porton Down would become the world's best civilian laboratory for applied microbiology. At the last moment, however, the plan was defeated by the deepening of the cold war after the Cuban missile crisis. Nonetheless, Porton did provide ideas and people for the discipline in universities, both in

Britain and overseas, and acquired the responsibility for producing specialized chemicals for research when the Medical Research Council (MRC) closed its penicillin production centre at Cleveden in 1962. In that role, it produced key fermentation products, such as large quantities of the enzyme asparaginase for treatment of acute lymphoplastic anaemia. It also supplied t-RNA to the molecular biologists at Cambridge's Laboratory of Molecular Biology, famous as the birthplace of the double helix model.[45] As was typical, it was a Porton scientist, John Pirt, who suggested the title 'biotechnology' to the Society of Chemical Industry in 1972, the year he moved to Queen Elizabeth College, London, to begin a distinguished academic career.[46] Yet it is true that, in 1962, the year of the entitling of the journal *Biotechnology and Bioengineering,* the British decided not to create a national institute. Five years later, another initiative was launched, again by the now 85-year-old chemical engineer Harold Hartley.[47] He was as sure as he had been fifteen years earlier that what he was still calling 'biochemical engineering' was the way of the future, though he did admit that possibly he had overestimated the chemical engineer's role and had underplayed the part of the biologist. Though no longer well, he did have an enormous range of contacts ranging from the Queen's spouse, Prince Philip, to the energetic chemical engineer Leo Hepner, who created and edited the campaigning magazine *Process Biochemistry.*

In an era of technological *dirigisme,* Hartley pushed the National Research and Development Council, which replied by supporting the development of a national champion in commercial enzyme production, Whatman Biochemicals.[48] Once again, Porton Down was suggested as a national centre, but the idea did not survive and no professor was sufficiently dominant that his laboratory would provide an acceptable alternative.[49] Though Hartley's energies were spent, others fought on. Ernst Chain at Imperial College did persuade the Science Research Council to establish a special budget for applied microbiology. The programme was cancelled after four years, on the grounds of the inadequate quality of applications, reflecting the dominance of Britain's superb pure biochemists and molecular biologists. The small scale of the biochemical engineering enterprise was highlighted in a 1976 report: There were just four world-class departments, at University College London, Birmingham, Manchester, and Swansea.[50]

Small as the community was, it did include several highly articulate leaders, including the molecular biologist Brian Hartley, Chain's successor at Imperial College, John Bu'Lock at the University of Manchester, John Pirt at Queen Elizabeth College, London, and John

Ashworth, a biochemist who was the scientific advisor to the cabinet. The last was in a particularly good strategic position to promote an awareness of the continuing problem of what to do about Porton Down and its context of biotechnology. When the laboratory was eventually transferred to the Public Health Laboratory Service on account of its expertise in dealing with pathogenic organisms and renamed the Centre for Applied Microbiology & Research (CAMR), there was a clash of cultures and a sense of unhappiness. In 1978, the creation of an enquiry into biotechnology was announced by Secretary of State Shirley Williams. Unusually, this was jointly sponsored by the government's advisory committee, Advisory Council for Applied Research and Development (ACARD), and the scientific community through the Advisory Board for the Research Councils (ABRC) and the Royal Society. A June 1979 meeting at the Royal Society established the community consensus, and the so-called Spinks report, named after its chairman who had been research director of ICI, was completed that autumn.[51]

The report can be seen as a successor document to the DECHEMA study of seven years earlier. It shared the general German approach: Biotechnology is 'the application of biological organisms, systems or processes to manufacturing and service industries'.[52] It was also based on a very similar generational view to that of Rehm, which had already been explored in England in a study by three English microbiologists: Alan Bull (who had worked at Glaxo), Derek Ellwood (a Porton scientist), and Colin Ratledge (of Hull University and an expert on developing countries).[53] Where the English report differed was in its giving greater significance to new developments in recombinant DNA technology. As in Germany, this was not seen in isolation. Though already very important, recombinant DNA was still just one of a number of new biotechnologies, also including enzyme technology and fermentation, which had major implications in the new climate of energy and environmental awareness. The key areas identified were the general applications of genetic engineering and enzymes, monoclonal antibodies and immunoglobulins, waste treatment, plant cell culture and single-cell protein, and production of fuels (ethanol and methane) from biomass. In a more popular context, Alan Bull together with Manchester's John Bu'Lock published an article in the magazine *New Scientist,* one of whose vivid cartoons which would adorn office walls for a decade is included here.[54]

This British formulation proved to be influential in its own turn. The Organization for Economic Co-operation and Development (OECD), the international grouping of industrialized nations based

Biotechnology illustrated in 1979. This first appeared on 7 June 1979 in *New Scientist* magazine, London, the weekly review of science and technology. Courtesy *New Scientist.*

in Paris, was looking around to see what technologies would change the world. An analyst, Salomon Wald, picked on biotechnology, as so many others were doing, and having been recommended three alternative authors for a report – the key British strategists Alan Bull, Holt at the Polytechnic of Central London, and Malcolm Lilly in University College London's Subdepartment of Biochemical Engineering – he found they could work together as a team. Their report, which appeared in 1981, was to be a widely cited classic reflecting half a century of European and U.S. thought: 'Biotechnology is the application of scientific and engineering principles to the processing of materials by biological agents to provide goods and services'.

The parallel discussions in European countries, such as the Netherlands and France, shared the same overall concerns, though each was embedded in local policy.[55] So, for instance, the French, concerned to raise the status of their science and the efficiency of a large agricultural sector, issued three reports.[56] The first, by three eminent biologists, identified a new philosophy of biology, ending with a reflection on biology and society. A more focused report on top priorities for France, emphasizing such issues as gasohol and nitrogen fixation, was then issued, to be followed by a report on bureaucratic implementation. The outcome was a heavily funded national biotechnology programme.

In addition to such national-based reflections, there was also, unusually, a key Europe-wide dimension. For the European Commission itself was examining more broadly the requirements for, and economic and social implications of, radical technological change – new emphases for an organization which hitherto had been principally concerned with tariffs and with supporting an ever more expensive agricultural policy.

The European Commission and the Biosociety

During the 1970s, interest in biotechnology within the Commission came from two distinct directions: long-range planning and research support. Since the two initiatives were integrated only in the 1980s, they will be treated separately, with the genetic engineering push recorded in the next chapter. In 1975, a group entitled 'Europe Plus Thirty' created under the auspices of Commissioner Ralf Dahrendorf engendered the European Commission's own forecasting unit, FAST (Forecasting and Assessment in Science and Technology). Established in 1979, the FAST group identified three kinds of change on three time horizons: The future of work and employment was an immediate problem; an information society based on computing and telecommunications technology would create fundamental changes within ten to fifteen years; while in thirty years the new biotechnology could be expected to effect major transformations. Interestingly, with the exception of energy and agriculture, contrary to the fears at the beginning of the decade expressed by *Limits to Growth*, shortage of resources did not in general now seem to present problems for the community.

The programme was laid out in a preliminary work, *The Old World and the New Technologies: Challenges to Europe in a Hostile World* by Godet and Ruyssen.[57] Europe was falling behind Japan and the United States, and it was essential that the initiative be reclaimed particularly in the areas of the major challenges. The FAST group labelled its three concerns 'Employment and Work', the 'Information Society' (the currently fashionable phrase), and for neatness, the 'Biosociety'. The last was addressed by two people, neither of whom was a biologist: Mark Cantley, an Ulster-Scot mathematician who had worked on global systems in Vienna's International Institute of Systems Analysis, and Ken Sargeant, a British chemist who had come from the Microbiological Research Establishment at Porton Down, after it was taken over by the Health Department. Each brought a very broad view of the potential of biotechnology.

Cantley began by talking to Bu'Lock at Manchester, who recommended him to Behrens, head of DECHEMA. From the beginning then, Cantley was inducted into the tradition of European biotechnology. The strength of the link was enhanced by the contemporary formation of the EFB which proved a powerful industrial partner to Cantley's Bio-Society group. The latter's first document quotes a definition of 'biotechnology', by the British microbiologists Bull and Bu'Lock from their then recent article, 'The Living Micro Revolution'. This closely followed the DECHEMA definition and would be the basis of the OECD report which Bull coauthored two years later:

> 'The meaning that is most widely accepted is that it is the industrial processing of materials by micro-organisms and other biological agents to provide desirable products and services. It incorporates fermentation and enzyme technology, water and waste treatment, and some aspects of food technology. Its scope and potential therefore are enormous and like microelectonics before it, novel ideas and tangible benefits are being reported at an ever increasing rate.'... and with the advent of techniques of genetic engineering, the *modification* of such micro-organisms can be added to this list. Nor should 'industrial processing' exclude the application of biological science to agriculture and other non-industrial fields.[58]

The first of the topics covered by the FAST group was 'A Community Strategy for European Biotechnology'. It was coordinated by DECHEMA, from whose 1974 report it was clearly descended. Seven bottlenecks and six product and process areas, ranging from energy to fine chemicals, were identified, but the report hardly mentioned genetic engineering or indeed novel proteins such as the human insulin Eli Lilly was just bringing to market.[59] Indeed, the entire FAST report reflected the formulation of biotechnology as conceived at DECHEMA in the early 1970s.[60] Thus, the intellectual framework for European action in the early 1980s was the result of a slow evolution, rather than the radical revolution that would be experienced in the United States.

Conclusion

The history of biotechnology, conceived as a descendent of zymotechnology, can be seen to have evolved triumphantly in the 1970s. In principle, biological processing was seen as the means for converting low-value locally produced raw materials into high-value products, superseding older, cruder, more wasteful methods of manufacture. New and promising techniques had become available. Nonetheless, the vision continued the hopes and aspirations that had long been expressed by Berlin's Institut für Gärungsgewerbe and articulated by Karl Ereky, and even Julius Wiesner exactly a century earlier. From its position at the industrial margin in 1900, biotechnology had become the subject of arguments for official support in successful economies such as Germany and Japan (seeking to change priorities), in historically unsuccessful economies such as Britain (where again, a change of direction was felt to be desperately needed), and in coordinative administrations seeking a useful leadership position such as the European Commission, the OECD, and even state governments in the United States.

Underlying the scientists' promises of the bounties to be had from biotechnology in Europe were threats of being left out, highlighted by the rush towards integrating information technology throughout industry. The pattern of winners and losers, already becoming obvious, could always be repeated. So, declining economic growth rates, the possibilities of fermentation, and admiration for the model of U.S. life science research, made biotechnology a power in Europe and Japan even before the full impact of recombinant DNA technology. Urgency was given by the oil crisis and by the experience of information technology. However, 'biotechnology' came to connote much more than a latter-day zymotechnology. Julian Huxley, in 1936, had heedlessly used the word to describe his vision of human biological engineering, when introducing Hogben's much more agriculturally oriented piece. Again, in the late 1970s, the concepts of biological and genetic engineering would become intertwined with the conception of biotechnology.

8
The wedding with genetics

We are now, though we only dimly begin to realize the fact, in the opening stages of the Biological Revolution – a twentieth-century revolution which will affect human life far more profoundly than the great Mechanical Revolution of the nineteenth century or the Technological Revolution through which we are now passing.

(Gordon Rattray Taylor, 1968)[1]

Introduction

Microbiologist Arnold Demain, addressing the first International Symposium on the Genetics of Industrial Microorganisms in 1970, reflected on the 'engagement' of genetics and industrial microbiology. He suggested that 'the partners have not yet fully committed themselves to working towards the mutual benefit of the pair.... If the union is ever consummated, we can look forward to a bright future for both'.[2] So effective was the subsequent marriage at defining a new beginning that, within scarcely more than a decade, it would be frequently remarked of the late 1970s that it was a time 'when the word biotechnology was hardly known'. An important perception was communicated by that, albeit technically incorrect, statement: Biotechnology had come to be seen as an outcome of genetic manipulation techniques. Its conception as primarily concerned with genetic enhancement, itself long established, thus came to dominate the forty-year-old technological tradition which emphasized the selection, cultivation, and processing of naturally occurring organisms. Though biotechnology did acquire many of the hopes, already building up, for bacterial factories, there was, particularly at first, an apparent disjunction with its more conventionally 'technological' tradition. Despite the importance of both partners, the marriage would be frequently celebrated by the recitation of a family genealogy that emphasized solely the genetic side and a historical view of biotechnology that focused exclusively on the lineage of genetic engineering.

This chapter will follow the romance that led to this seemingly one-sided relationship.

Whereas much of the recent push for biotechnology had come from Germany and Japan, new expectations came out of a science dominated by Americans. The key events have come to be seen as two scientific breakthroughs. One was the 1953 discovery of the structure of DNA, by James Watson, now head of Long Island's prestigious Cold Spring Harbor Laboratory, together with Francis Crick, at the Laboratory of Molecular Biology in Cambridge, England. The other was the 1973 discovery, by Cohen and Boyer in California, of a recombinant DNA technique by which a section of DNA was cut from the plasmid of an *E. coli* bacterium and transferred into the DNA of another. This approach could, in principle, enable bacteria to adopt the genes, and therefore produce proteins, of other organisms, even humans, on demand. Popularly referred to as 'genetic engineering', it came to be defined as the basis of new biotechnology.

The logical leap from single tool to economy-reviving industry was improbable; nonetheless, the role of other skills was relegated to obscurity. By 1980, Elmer Gaden, pioneer of the chemical engineering interpretation, was commenting on 'a reduced pace in biotechnology itself', during the 1970s.[3] Though his concern might have seemed bizarre to most others awestruck by the promise of recombinant DNA in 1980, it reflected a fundamental shift in the dominant interpretation of biotechnology from an evolving aspect of engineering to a field defined only by biological feasibility. A study of the uses of the word 'biotechnology' in scientific papers has contrasted the steady exponential growth from 1969 to 1984 found in the articles indexed by *Chemical Abstracts* with the quite different pattern of an explosive emergence since 1983 found in a biological index (*Biosis Previews*). A sudden awareness was of course characteristic of society in general and is indicated in the *National Newspaper Index* and an investment data base, *Investest.*[4]

However novel the scientific basis, the benefits of biotechnology, as highlighted in 1980, were expressed in terms established long before recombinant DNA. The nine chapters of a popular 1981 survey, *Life for Sale,* were entitled: 'Turning DNA into Gold', 'Biobusiness Takes Off', 'A New Era of Miracle Drugs', 'The Next Industrial Revolution', 'Feeding the World', 'The Age of Human Genetics', 'Is It Safe? Revisiting the Great Debate', 'Who Owns Life? And Other Easy Questions', and 'Epilogue'.[5] Though the emphasis on genetics was new, prophesies of a new industrial revolution were already well established. Nonetheless, in the excitement and with many people suddenly becoming aware of biotechnology, this continuity was overlooked. The chapter titles of *Life for Sale* indicate the recovery of the biological emphasis of this boundary subject, which had been so dominant in Europe before World War II, but they show no sign of the earlier lyrical vision that here was a means of transcending the fundamental problems of earlier technologies.

In part, the explanation must be ignorance; the flame of the new biotechnology flared up so fast that there was no time for earlier traditions to be recalled. Instead, there were the more well known legacies of the Nazi eugenic experiments and the cynicism of the Vietnam War era. The chemical industry, once seen as a benevolent source of clean modern technology, was associated with explosions, polluting products, and worse waste. Above all there was the terrifying precedent of atomic fission. Once seen as presaging a hope of infinite energy, possession of atomic power seemed to epitomize the character of the modern individual as the 'Sorcerer's Apprentice'. Engineers designed reactors to cope with maximum credible accidents, and still the public feared the 'China Syndrome'.[6] There was a parallel between nuclear weapons and biological warfare, but the metaphor would go much further. Biologists themselves would recall the experience of the physicists who had first built the atomic bomb, without having an opportunity to reflect on its long-term consequences, and willed that they would do better. Indeed, the extent to which technical parallels could be drawn between nuclear technology and biotechnology would continue to underlie debate.

Following the success with silicon, the juxtaposition of hopes of turning DNA into gold with dramatic fears over safety (as seen even in the chapters of *Life for Sale*) even then seemed paradoxical; and even more odd, both hopes and fears had arisen at practically the same time. Explaining the coincidence of a rising crescendo of expectations for the commercial applications of genetic engineering with a period of deep anxiety over its risks, the participant-observer, historian, and science analyst J.R. Ravetz has noted that the field acquired a notoriety which made extravagant claims for its power more plausible.[7] An even stonger link can be propounded: The outer limits of genetic engineering's commercial potential were delineated and publicized as part of a process of 'technology assessment' first stimulated by the question of what controls to impose on commercial exploitation.

Scientists, industry, and, increasingly, governments met deep popular concern by hitching the power of recombinant DNA to the carefully formulated functions of biotechnology as a new technology offering manifold benefits, devised in the 1970s and before. Whereas, hitherto, the border between biology and engineering had been underdeveloped by commerce, in the face of the spectacular opportunities frequently reiterated, however punctiliously pointed out, the lines between short-term practicality and speculative benefits of linkages across that border were quickly blurred.

The stakes

For all the interest in Europe and Japan in biotechnology during the 1970s, the boundary between biology and engineering was most important in the United States. Here, life sciences had already become a major enterprise, and as Naisbitt's popular 1982 *Megatrends* indicates, it was widely believed that, in future, the biological sciences would be the basis of technology.[8] U.S. investment in life science research may have been intended primarily to lead to better health, but a whole new range of benefits could also be discerned even by the 1970s. Time after time, congressmen would reflect that U.S. industry's strategic advantage and U.S. commercial breakthroughs would be dependent on key scientific research supported by the National Institutes of Health (NIH). Biotechnology would provide the means for that translation.

This vision not only sustained and justified enormous NIH funding. It also diverted public consciousness from the frightening consequences of NIH-supported breakthroughs. To many, the outcome might be the 'biological time bomb' in which man playing God would alter other species and even humanity itself. That would be called an irrelevance by 1980, yet in the 1960s, following prewar concerns, the phrase of the moment, 'genetic engineering', was reserved for humans. Even then, an important principle had already been established: Society was on the portals of a new stage of evolution, far quicker than traditional eugenics could have offered. Conscientious scientists, fearing the precedent of atomic fission, felt an urgent responsibility to integrate these biological changes with social innovation.

Genetics

The fears and dreams of the 1970s would be drawn nearer by breakthroughs in genetics. This is an area of biology particularly well developed in the United States.[9] It has been closely associated with animal and plant breeding, but widely separate from the study of biochemistry. However, by the postwar years, the distance between the two disciplines was rapidly lessening. The most famous link is the 'double helix', the model of DNA traced by Crick and Watson at Cambridge explaining, in molecular terms, how heredity could occur. Crucially important as was that well-remembered achievement, it has to be seen in the context of a general rapid change. In the 1940s, the geneticist George Beadle and his partner, microbiologist Edward Tatum, had demonstrated that each gene controlled the production of an enzyme.

As the distinctions between biochemistry and genetics crumbled, so did the divide between the study of microorganisms, which hitherto many had seen as autonomously developing, and genetics.[10] Such eminent scientists as Julian Huxley and Oxford's Cyril Hinshelwood had even doubted whether bacteria contained genes. Although in 1948 Joshua Lederberg showed that microbes could exchange genetic information (for which he shared the 1958 Nobel Prize with Beadle and Tatum), the historian of bacterial genetics, Thomas Brock, points out, 'Until the late 1950s, textbooks of genetics contained little about bacteria, and textbooks of bacteriology had almost nothing about genetics'.[11] As Demain's comment quoted at the beginning of this chapter reminds us, even in 1970, there was only an 'engagement' between genetics and industrial microbiology, not a marriage. Thus, whereas later the link would appear 'natural', in the 1950s the relationship between the studies of genetics and of cells was fresh and exciting, with breathtaking implications. The first speculative step was to reflect on the engineering of people, not microbes. Only later did commercial and medical opportunities on the one hand, and political and cultural restraints on the other, together ensure an emphasis on microorganisms.

Edward Tatum was brought up in a family used to discussing the practical implications of science, having a father who had been a well-known pharmacologist working for industry and the University of Wisconsin. Perhaps this was why he was among the first to go from the concept of molecular biology as a science to the technological concept of biological engineering.[12] Stimulated by the challenge of a Noble Prize acceptance speech, Tatum laid out the implications of the 1950s revolution in biochemistry.[13] Raising the possibility of the use of genetics to counteract hereditary defects, he urged people to think about the prospects of biological engineering. Ten years later, Tatum was to rally his listeners, again with a ditty, at the inauguration of new laboratories of pharmaceutical giant Merck, Sharp, and Dohme:

> The time, has come, it may be said,
> To dream of many things;

> Of genes – and life – and human cells –
> Of Medicine – and Kings.[14]

This alarming song was composed amid a widespread debate growing since Tatum's Nobel speech. Despite the tragedies of the Nazi experiments, the discussions of the 1960s displayed strong continuities, perhaps surprisingly, with prewar positive eugenics. Even many of the participants were the same: Julian Huxley, J.B.S. Haldane, Theodor Dobzhansky, and H.J. Muller. The quarter-century taboo on eugenic speculation among molecular biologists had been broken in 1963 with two seminal meetings that had brought together the now aging eugenists and the younger generation of molecular biologists. The first was suggested by the venerable Gregory Pincus (a faithful disciple of that early twentieth-century rigourously biological reductionist, Jacques Loeb), who was by 1963 himself known as the father of the contraceptive pill. He suggested to London's CIBA Foundation that it hold a workshop to be entitled 'Man and His Future'.[15]

This meeting provided the occasion for the prewar pioneers of 'reform' eugenics, with Huxley giving the first paper and J.B.S. Haldane the last, to meet the younger generation. Reference was made to the 'old days', and Haldane recollected the concern over the 'twilight of parenthood' a generation earlier. That meeting can be seen to be the pivotal point in modern biotechnology in which the torch was passed on. Reacting against the eugenic dreams of the older men, Joshua Lederberg, Stanford professor and Nobel laureate, emphasized the importance of focusing on curing living people.

Lederberg's paper, 'Biological Future of Man', which he abstracted and reprinted in *Nature* for those who missed the conference and subsequent book, began: 'Darwin's theory set off the historic debate on man's past. Today, with a new biology we mirror his future'. The talk was expressly designed to alert a sleeping world: 'As biological technology dissolves the barriers around individual man and intrudes on his secret, germinal continuity, we must face the issue of a definition of man, taking full account of his psychosocial progeny [psychosocial was a recent Huxley coinage]'.[16] He suggested that while molecular biology might one day make possible change in the human genotype, 'what we have overlooked is *euphenics,* the engineering of human development.' This was not another word for eugenics: Its very construction indicates Lederberg's emphasis on changing the phenotype after conception rather than the genotype which would affect future generations. Just as earlier pioneers of biotechnology, Lederberg placed the new revolution in the context of a sequence of previous

Joshua Lederberg. Courtesy Joshua Lederberg.

revolutions: language, agriculture, political organization, the physical technologies. Embryology he explained 'is very much in the situation of atomic physics in 1900; having had an honourable and successful tradition it is about to begin!'

The theme of a new biological revolution was taken up immediately by Tatum himself, at the second meeting, held at Ohio Wesleyan University. Now he went beyond his earlier use of the phrase 'biological engineering' to warn of 'genetic engineering'. His use in this highly charged meeting seems to have been the basis of the modern tradition.[17] Tatum's vision would inspire and terrify scientists and the public for more than a quarter of a century, for all that the speech itself was forgotten:

Biological engineering seems to fall naturally into three primary categories of means to modify organisms. These are:

1. The recombination of existing genes, or eugenics.
2. The production of new genes by a process of directed mutation, or genetic engineering.
3. Modification or control of gene expression or, to adopt Lederberg's suggested terminology, euphenic engineering.[18]

The immediate impact of Tatum's words was weakened by a two-year delay in the publication of the proceedings to 1965. As a result, they appeared in print at the same time as an extended critique by an older geneticist who had himself participated in the Ohio meeting two years earlier. Rollin Hotchkiss's 'Portents for a Genetic Engineering' alerted the scientific community to their responsibility not to act too fast, in the face of ignorance.

But do not be deceived – these gifts (all but unseen at first, quite unlike eugenic measures) will come to us by that unique synthesis of altruistic purpose and enterprise that makes so much happen so quickly in America. The eligible preparations of DNA or modified DNA will be prepared by our aggressive industry and available for any medical man who has a heroic or ambitious stripe.[19]

Publishing his article two years after Rachel Carson's *Silent Spring* had warned of the consequences of DDT, Hotchkiss feared that, like pesticides, and unlike the atom bomb, genetic engineering would creep in unobserved. There was therefore a responsibility placed on the scientific community to stop this from happening.

Hotchkiss's warning would have been specially meaningful to those who had read the article by Muller with which that organ of the socially aware, the *Bulletin of Atomic Scientists,* launched the year 1964. Repeating a contribution to the recent CIBA conference, he concluded that 'man will transcend himself genetically'.[20] But in the section just before that statement, he struck a new note. It was entitled 'Unlimited Reaches of Biological Engineering'. In a sentence of which Raoul Francé would have been proud, he alerted the reader that 'lessons learned from living things can be applied in artificial devices of diverse kinds'. Still, the potential practical import of the new biology was even greater: 'The possibilities of transforming microorganisms, plants, and animals on our own behalf, with resultant enhancement of our ecological and economic circumstances, are so nearly unlimited as to defy attempts to outline them here.' The section concluded with a reference to Lederberg's euphenics. The link between this vision and the past was made clear by a quotation from Jacques Loeb, the teacher of Muller as well as Pincus.

Lederberg shared the wish to develop public awareness of the new biochemical genetics. However, as he had said in 1962, he saw the future in modifying organisms after conception. The fundamental modification of genetic lines he saw as neither desirable nor practical. In 1966, he returned to his theme, dismissing the idea of changing the genotype as genetic alchemy or 'Algeny'.[21] Lederberg's concern in voicing this warning was solely with humans. Despite his emphasis on human euphenics, he saw the possibility of eugenics with some nonhuman species to produce genetically homogenous material. Lederberg's approach was therefore still medically focused and indeed went beyond speculation. Having become friendly with the chemist Carl Djerassi, who had pioneered the synthesis of progesterone, the hormone used in contraceptives, he persuaded Djerassi of the potential of molecular biology and was given a laboratory to explore its consequences by Djerassi's company, Syntex. The company's history, published in 1966, concludes with an affirmation that it would contribute its share to 'euphenics': 'It will seek clues to the mechanism of hormones, of the nucleic acids and of host resistance to foreign organisms. It will seek to gain new understanding of reproductive physiology and to develop better methods of controlling fertility in humans, animals and plants.'[22] Others in large pharmaceutical companies were beginning to have similar thoughts. By 1967, the biochemists Brian Richards and Norman Carey at the British laboratories of G.D. Searle were also beginning to contemplate the technological implications of biological engineering. In a manner typical of the time, they envisaged the transfer of genetic material between higher organisms. Though their techniques proved impractical, the visions of what they might be used for provide an interesting foretaste of future developments. A memorandum suggested four uses:

1. Lifetime cure of inherited diseases in Man
2. Normalizing malignant tumours
3. Inducing tolerance prior to tissue or organ transplantation
4. Genetic improvement of economical importance in domestic animals and plants including those required for ocean farming and extra-terrestrial farming.[23]

Public concern

Technical developments seemed to justify scientists' concern to raise public consciousness, while at the same time, public suspicions of their motivations increased. In December 1967, the first heart transplant by Christian Barnard reminded the public that the physical identity of a person was becoming increasingly problematic. The poetic imagination had always seen the heart as the altar of the soul, and now there was the prospect of individuals being defined by other people's hearts. The same month, Arthur Kornberg announced that he had managed to biochemically replicate a viral gene.[24] Life had been synthesized, said the head of the NIH. President Johnson, otherwise engaged with the increasingly controversial Vietnam War, hailed this as one of the supreme achievements of mankind. Genetic engineering was now on the scientific agenda. Moreover, it was becoming possible to identify genetic characteristics with diseases such as beta thalassemia and sickle-cell anaemia.

Responses to scientific achievements were coloured by cultural scepticism. *New Scientist* reported Khorana's subsequent achievement in combining nucleotides under the heading, resonant of Johnson's campaign slogan, 'All the Way with DNA.'[25] Scientists, their expertise, and even expertise in general were looked on with suspicion. The mood was caught by James Watson, publishing a cynical description of his own pathbreaking discovery of DNA's structure, *The Double Helix,* in February 1968.[26] A month later, American cynicism was seen to have a more bloody side, with the My Lai massacre of Vietnamese civilians by U.S. troops. Such diverse phenomena were being seen together at that moment of peculiar tension in Western society. The month of May saw European universities erupting and the walls of Paris daubed with warnings about the threats of pseudo-objectivity. In the United States, professors' papers were being 'trashed'.

In the atmosphere of the day, the commendation of President Johnson raised as much suspicion as faith. An immensely successful popular work and Book-of-the-Month Club selection, the *Biological Time Bomb* by the British journalist Gordon Rattray Taylor, appeared just at that moment. The author's preface, describing events in the year since he had finished writing the main body of text, saw Kornberg's discovery as a route to lethal doomsday bugs. The publisher's blurb for the book warned that, within ten years,

> YOU MAY MARRY A SEMI-ARTIFICIAL MAN OR WOMAN... CHOOSE YOUR CHILDREN'S SEX... TUNE OUT PAIN... CHANGE YOUR MEMORIES... AND LIVE TO BE 150 IF THE SCIENTIFIC REVOLUTION DOESN'T DESTROY US FIRST.[27]

Although the book's text was more measured, the message had been accurately reflected. The chapter entitled 'The Genetic Engineers' took the reader from the DNA story, through the 'New Eugenics', past 'Eliminating Defects' to 'The Spectre of Gene Warfare'. It ended with a chapter called 'The Future – If Any.' Concern was not restricted

'Pumping Gas': Genetic engineering pictured by the *National Enquirer* on 1 July 1980. Reprinted by permission. Copyright 1980 National Enquirer, Inc.

to the more sensationalist end of popular culture. In 1969, Britain's staid Third Programme, the BBC radio channel for intellectuals, publishing the text of a series as a book, chose the stark title *Genetic Engineering*.[28] It is rare for current science to be represented in the movies, but in this period of 'Star Trek', science fiction and science fact seemed to be converging. 'Cloning' became a popular word. Woody Allen satirized the cloning of a person from a nose in *Sleeper*, and cloning Hitler from surviving cells was the theme of the book and later film *The Boys from Brazil*. These works were concerned with the threat of alien people. Strange organisms were also in vogue.

Lederberg himself put his interest in raising public understanding into action, by writing a column for the *Washington Post*. Thinking about the space programme, he considered the possibility of pollution from extraterrestrial organisms and promoted the term 'exobiology' he himself had previously coined. Massive investments were made to ensure against infection from polluted space men. The successful 1971 movie *The Andromeda Strain*, based on the book of the same name, portrayed the invasion of deadly alien microbes which threatened to destroy the world. Lederberg has noticed that the fictional character of Stone, the scientist, resembles his own biography.[29] California was already showing its remarkable ability to marry the fantasies of Hollywood and Stanford.

Thus, when in 1971 two Californian scientist-entrepreneurs, Ronald Cape and Peter Farley, established a microbiology research company, they chose the name of the next constellation to Andromeda, Cetus.[30] Lederberg advised the board, and, although the company's early emphasis on screening technology did not particularly interest him, it was he who later pointed the new company to the work of Cohen and Boyer, suggesting 'this is going to be the greatest thing since sliced bread'.[31] Lederberg also sought to convert public interest into congressional understanding. In 1971, he suggested the creation of a $10m genetics task force to root out diseases such as cystic fibrosis.[32] The British magazine *New Scientist* commented that his congressional testimony 'heralded the dawn of an age which many have forecast – some with hope, some with fear: an era when the spectacular research achievements of molecular biology begin to be applied to the correction of genetic abnormalities in man'.[33] This was perhaps an overestimate, particularly since a major funding commitment was not in the end made, yet it does highlight the excitement over molecular biology's potential, in serious circles, as early as 1971.

So fast were developments that, within a few years, the prime movers, such as Lederberg, would be marginalized as new scientific methods for recombining DNA were forced ahead by a younger generation. Graduate students in biology in the late 1960s were carrying out the radical new science but questioning its uses. They wished to continue this research without creating long-term devastation. The idea that genetic engineering would have major human and commercial consequences was thus established, when in 1973 a conference in Hawaii brought together Cohen's understanding of plasmids and Boyer's expertise in using enzymes to cut and rejoin DNA, leading to recombinant DNA.[34]

The achievement of Cohen and Boyer was defined at the time as pathbreaking, and its significance has been preserved. But its meaning was created by those with prepared minds. The immediate response was from scientists who felt that the consequences of the science were potentially so dire that society would ban it, unless it was properly controlled. Clusters of younger biologists had been discussing the potential implications of advances for some years. At Harvard, a graduate student, Leon Kass, had convened a discussion group questioning the benefits of biomedical advance as early as 1967.[35] As the executive secretary of the Committee on Life Sciences of the U.S. National Research Council, in 1970 he wrote to his friend Paul Berg of their professional responsibility for bringing the ideas of 'genetic engineering' to the public and suggested sending letters to magazines such as *Science*.[36] He was still thinking of genetic engineering as the modification of humans and recommended Paul Ramsay's then recent *Fabricated Man*.

In July 1974, a group of eminent molecular biologists headed by Berg wrote to *Science* suggesting that the consequences of this work were so potentially destructive that there should be a pause until its implications had been thought through. The suggestion was explored at a meeting the following February at California's Monterey Peninsula, forever immortalized by the location, Asilomar. Whatever *New Scientist* might have said three years earlier about Lederberg's congressional speech launching the new world, Asilomar was its founding convention. Its historic outcome was an unprecedented call for a pause in research until it could be regulated in such a way that the public need not be anxious, and indeed it led to a sixteen-month moratorium until NIH guidelines were available in mid-1976.

The events at Asilomar have been recorded by half a dozen books. Detailed and reflective as these are, they tend to perpetuate the image of the event as the beginning of the biotechnology age, rather than the culmination of almost two decades of reflection on the threats posed by genetic engineering. In part, this demonstrated the dominance of a younger generation.[37] The classic description of the meeting by political scientist Sheldon Krimsky in his book *Genetic Alchemy* grants Joshua Lederberg only three incidental mentions, even though he invented the very phrase 'Genetic Alchemy' (a fact not mentioned by Krimsky).[38] The contribution of Lederberg was chiefly noted as one of opposition to formal controls. He even felt moved to write to the *New York Times* to protest his interest in the social control of science.[39] In part, discussion had also moved on from the 1960s debates, though their legacy could not be forgotten, for Asilomar had explicitly excluded human genetic engineering from discussion. Although Leon Kass was to reflect that its shadow had fuelled public debate and anxiety over the use of genetic knowledge may have focused scientists' concerns on safety issues, it was at that very moment that the phrase 'genetic engineering' ceased to be associated just with the manipulation of humans and came to be associated with any organism.[40]

Ironically, the molecular biologists at Asilomar were only loosely connected with microbiologists and others with really practical interests. Whereas the scientists discussed in previous chapters had been interconnected socially as well as intellectually, this was not true of the molecular biologists.[41] Even though the Asilomar discussion had been prompted by experiments on bacteria, the molecular biologists were using microbes only as convenient experimental animals and were accused of lacking a feel for the organisms. 'Scientific slobs' was one description.[42]

Rather than extensive regulations, E.S. Anderson of Britain's Public Health Laboratory called for 'the technical re-education of the average molecular biologist whose manipulation of bacteria chills the blood of anyone accustomed to handling pathogens'.[43] This comment was made two weeks after the Paul Berg letter, when Berg led a televised discussion on the control of research in London's Royal Institution. Anderson was backed up from the audience by John Pirt, the distinguished Porton alumnus who pointed out that at the Microbiological Research Establishment, he was used to safely manipulating organisms so dangerous that a single bacterium was fatal. For him, the only problem was professional containment. This confrontation was but a reflection of an enduring tension between microbiologists and molecular biologists which would stretch to issues of curricula and the novelty of the opportunities coming to be recognized. When Ernst Chain discovered that he was to be succeeded as professor of biochemistry responsible for the fermentation pilot plant at Imperial College by a molecular biologist, Brian Hartley, he threatened to sue the college for irresponsibility.[44]

As for industrial significance, much as the Asilomar discussion did reflect a concern for the impact of science on society, in retrospect it seemed to be concerned with the self-imposition of constraints within which scientists could work without further social concerns. Participants seemed only slightly interested in the practical applications of the techniques they were developing. One of the few industrial microbiologists at Asilomar was A.M. Chakrabarty of General Electric, who, as early as 1972, had filed a patent application for a bacterium modified to digest oil. From there his interests moved; as he wrote in October 1974, he had hoped that by giving an appropriate plasmid from *Pseudomonas* to *E. coli* he might discover how an animal could obtain more calories from cellulose. Recalling his altruistic contemporaries working on single-cell protein, he was seeking a way of making better use of fodder.[45]

But Chakrabarty was an unusual member of the community whose attitudes had been expressed with unusual frankness at a 1970 London meeting to discuss the biological revolution. Organized by the Nobel Prize–winning Maurice Wilkins, this was attended by such distinguished molecular biologists as Watson, Monod, and Perutz. The radical young Harvard molecular biologist Jon Beckwith defended a lone and embattled industrial scientist with the comment, 'Dr Hale is pretty much alone as a representative of the industry and at this sort of meeting we very rarely have any contact from industry, whom we have essentially ostracized.'[46] Similarly, discussion at Asilomar dealt only incidentally with commercially significant applications.

Lederberg was the leading exception in emphasizing, as he had for years, the potential benefits. As early as the mid-1960s, he had directed a Syntex Laboratory dealing with molecular biology. Now, at Asilomar, in an atmosphere favouring control and regulation, he circulated a paper countering the pessimism and fears of misuse with the benefits conferred by successful use, including:

> an early chance for a technology of untold importance for diagnostic and therapeutic medicine: the ready production of an unlimited variety of human proteins. Analogous applications may be foreseen in fermentation processes for cheaply manufacturing essential nutrients, and in the improvement of microbes for the production of antibiotics and of special industrial chemicals.[47]

Atypical as Lederberg was at Asilomar, this was the vision that would soon inform the development of the biotechnology industry. Over the next two years, as public concern over the dangers of recombinant DNA research grew, so too did interest in its technical applications. Curing genetic diseases remained in the realms of science fiction, but it appeared that producing human proteins would be good business.[48] Insulin, one of the smaller and best understood proteins, had been used in treating diabetes for half a century. Hitherto it had been extracted from animals in a chemically slightly different form from the human product. Here again was a challenge. If one could produce human insulin, one could meet an existing enormous demand with an apparently superior product whose approval would be relatively easy.

In the period 1975 to 1977, insulin represented the aspirations for new products that could be made with 'New Biology'. The number of other practical products was low, and the means of manufacture not yet fixed. A group at Harvard attempted to grow human insulin-producing cells not in *E. coli,* as would become standard, but in tumours on mice. A competing team at the University of California, San Francisco, associated with Herbert Boyer, who in April 1976 was a founder of the small company Genentech (*Gen*etic *En*gineering *Tech*nology), did use *E. coli* and claimed success in September 1978.[49]

These practical developments have often been considered separately from the simultaneous rise in concern. However, a few studies have highlighted the considerable efforts made about 1977 to reduce media anxiety and to emphasize the potential practical benefits.[50] Even earlier, in the face of pressure on innovators to defend work from its very inception, influential formulations of the potential were expressed and generalized to defend the field from regulation. Uses of recombinant DNA technology were portrayed long before most were practical, to ensure that industrial development wouldn't be impeded. Lederberg's presentation at Asilomar was one example. Before that, as early as September 1974, a few months after his famous letter to *Science,* Berg had reflected during the debate in London's Royal Institution: 'Probably because of exaggerated and misleading claims by scientists and the press, the words genetic engineering evoke terror as well as excitement at its prospects.' He then explained the new process of recombinant DNA, citing its scientific fruits and also its 'great practical significance'. Berg's list of potential products was to become familiar, even banal. In recalling the variety of its descendents, one should remember its place in his argument, justifying research in the area, and not as a business plan promising short-term gains:

> Why can't these simple organisms become the factory for producing some of society's most needed supplies — anti-biotics — hormones — and even a source of food. And for those who enjoy real speculation there are of course the dramatic possibilities for introducing new genes into human cells and thereby trying to cure certain genetic diseases.[51]

Berg went on to balance these distant benefits with equally uncertain risks, developing antibiotic resistant bacteria or converting harmless into toxic organisms. So, as early as 1974, a list of speculative gains was being set against the speculative risks.

The first debate in public was held by Edward Kennedy in the Senate Health Subcommittee in April 1975. A year later, in June 1976, the sixteen-month moratorium on research expired, with the publication, at last, of NIH guidelines on good practice. They defined the risks of certain kinds of experiment and the appropriate physical conditions for their pursuit with a list of those too dangerous to do at all. Moreover, modified organisms were not to be tested outside the confines of a laboratory or allowed into the environment.[52] Technically, the guidelines applied only to recipients of NIH funds, though industry abided voluntarily. In the autumn, the city of Cambridge, Massachusetts, held public hearings to decide whether Harvard University should be allowed to build a laboratory in which recombinant DNA experiments would be permitted. Discussions around the country were followed by debates in Congress. Sixteen bills for the control of genetic engineering research were discussed on Capitol Hill in 1976–77.[53] Public displays reached a crescendo in March 1976 with a debate at the National Academy of Sciences.

The meeting was to be the last great set-piece confrontation on the very nature of genetic engineering. The old meanings were remembered when students sang, 'We shall not be cloned'. An attack was launched by the new leader of the opposition, the populist fast-speaking Jeremy Rifkin. He argued that what was at stake was the nature of life itself, and appealed to the need to keep inhuman science in its place. Playing on the 1960s meaning of genetic engineering, he was for instance to use Lederberg's otherwise forgotten 1966 epithet 'Algeny' as the title of a book. J.R. Ravetz points out how Rifkin, by drawing on U.S. religious feeling, touched deep emotional tensions – conservative creationism and dislike of biology versus the 'primal fears' of the biologists, 'liberal, intellectual, and many of them Jewish'.[54] Later the same month, the Subcommittee on Science, Research, and Technology of the U.S. House of Representatives held hearings to clarify the general science policy implications of the new science. The mood was summed up by a liberal California Senator, George Brown, in his cross-examination of the research director from the pharmaceutical company Eli Lilly. 'There are a lot of people who think that we have a gene that gives us knowledge of good and evil, and if we get the wrong kind of research going on, it will transmit that gene so that it does not recognize good but only does evil'.[55] The response by Irving Johnson of Eli Lilly was to proclaim his adherence to a very different agenda: 'I think you have to examine the incentives and the reasons for doing some of this research. In industry our reasons and our incentives are pretty well defined. It is to make a useful product and I just do not think that some of these projects are particularly pertinent to this subject.' In an interview for the Eli Lilly internal magazine *Lilly News,* submitted in full to the congressional hearing, Johnson had already worked out a rhetorical strategy to deal with the threat of evil. The interview was explicitly targeted towards answering the anxieties, 'misunderstandings', surrounding the current congressional hearings.

As in his answer to George Brown, Johnson emphasized the practical commercial applications of science. The following exchange develops an argument which would become familiar, and is indeed essentially replicated in the chapter order of *Life for Sale* cited at the beginning of this chapter. Here it is interesting because it was being laid out so explicitly in response to the threat of legislation. The company interviewer asked a series of leading questions helping Johnson go from the science, 'How is foreign DNA inserted into a bacteria?' to the benefits to humanity, 'What do you gain from this?' gently back to scientific implications: 'The practical benefits sound fantastic. Does the technique have any applications in basic research?' Safety and all other ethical issues which had once loomed so large, now became just a matter of adequate safeguards: 'Great. But how about safety? Do the NIH guidelines really give adequate protection?' Further reassurance came with the detailed response to the question 'Could you give us a quick summary of the guidelines?' Back again to practice: 'The safety procedures sound elaborate. But what are Lilly's current research interests with recombinant DNA?' The interview ended with Johnson describing some of the 'more tangible benefits':

The theoretical possibilities – and, remember, that's all they are at present – sound like pure science fiction. Some of the things most frequently mentioned are: tailor-made microorganisms for energy production and pollution control; plants that are resistant to diseases, pests and drought; a whole range of hybrid plants such as a 'pomato' with tomatoes above ground and potatoes on its roots; beef cattle, swine and poultry designed for taste and efficient production; completely new species of plants and animals; and, finally, cures for genetic diseases through replacement of defective DNA.[56]

Again, this list was not cited to raise money or to show the financial implications of the new tools. Its intention was to dissuade Congress from legislating against research that Lilly wanted to do. This sort of argument proved compelling. The mood of Congress changed, slowly. Edward Kennedy's suggestion of regulation by means of a national commission made up of presidential appointees was dropped. Not that this move indicated great enthusiasm for the long-established dream of bringing life within the realm of engineering: Rather, it represented the victory of the metaphor of the silicon chip over the parallel with nuclear power.

The movement to counter fears of human biological engineering with the prospects of enhanced microbial technology was given official support by a study conducted by the Office of Technology

Assessment (OTA). Established as a neutral source of expert advice to Congress, during the early 1970s when environmental debates had become particularly heated, the ambit of OTA had slowly widened. As early as 1976, thirty representatives had called for an assessment of recombinant DNA.[57] Cautiously, the new body resisted this pressure until 1979 when it began to address such concerns, as its 1981 report entitled *Impacts of Applied Genetics* confirms. In his book *The New Politics of Science,* David Dickson, later editor of *New Scientist,* has pointed out how this report confined itself to practical industrial benefits and eschewed the complexities of the application to humans.[58] Just as the critics of safety had looked to the most extreme possible dangers, so the study, providing balance, looked also to conceivable commercial benefits. Information poured into the project office, which equally was looked to for advice. The director of the study, Zsolt Harsanyi, has described the role of the OTA in those days as that of 'Mount Olympus', and the balance it provided was correspondingly authoritative.[59]

Thus, the category of benefits was developed as part of the debate over regulation just before the media and Wall Street discovered biotechnology late in 1979. In the process, expectations were raised. When the microbial production of human insulin was finally announced in September 1978, the implications could be seen not merely as launching a new product, but also as beginning a new world. In December, London's influential *Economist* magazine announced its discovery of the new revelation under the title 'Industry Starts to Do Biology with Its Eyes Open'.[60]

It was neither just to a few emerging products nor even to the limited interest of large companies that the *Economist* was referring. It was beginning to notice the emergence of a new kind of innovative biology-based company, exemplified by the success of Genentech in developing insulin. Cetus, founded in 1971 in the shadow of existing biotechnology, had been the first, inspired by that same faith in microbes DECHEMA would espouse a few years later. Genetic engineering would enhance, not replace, existing skills. However, increasingly the rhetoric of the company and its philosophical president, Ronald Cape, integrated the now standard hopes for the use of microorganisms with the developing vision of genetic engineering. Thus, in the late 1970s, working with National Distillers, Cetus put considerable effort into developing an improved alcohol fermentation organism which could live at high enough temperatures for the product to be distilled as it was being produced. This was important because Cetus would influence three new companies, Genentech, Biogen, and Genex, each drawing on academic biology and exploiting it with entrepreneurial business skills.

For all the claims that molecular biology would have revolutionary effects across industry, the main foci of attention after insulin were the potential money spinners in the pharmaceutical industry: human growth hormone and what promised to be a miraculous cure for viral diseases, interferon. Cancer, the Big C of the 1970s, was a central target, because increasingly the disease was linked to viruses. Hitherto, viral diseases had proved hard to cure, and while vaccines had been known since the eighteenth century, there was still no obvious cancer vaccine.[61] Unproven products, such as laetril, had acquired wonder-drug status, and hopeless sufferers proved willing to pay remarkable prices in Mexico for a product banned in the United States.

In more reputable circles, attention was moving to interferon, a protein that the body itself produces to fight viruses, and discovered in 1957 by the British scientist Alick Isaacs. He was particularly interested in the influenza viruses, and the discovery did seem interesting to a number of companies. But interferon is present in natural blood only in very small quantities. By the early 1970s, the main supplier, the State Serum Institute in Helsinki, needed the blood from 90,000 donors to produce a tenth of a single gram. Across the world, such costs and the ambiguity of the results achieved with the precious material discouraged further work. However, in the hothouse of American cancer research even such problems could be overcome. In April 1975, the American Cancer Society sponsored a meeting on the potential of interferon. The society was committed and that meant the race was on to find an economical source. By 1980, a new company, Biogen, founded by Harvard's Walter Gilbert and Charles Weissmann of Zurich, had produced interferon through recombinant DNA.

The emergence of interferon and the possibility of curing cancer raised money in the community and the enthusiasm of otherwise penurious scientists. A financial analyst specializing in the pharmaceutical industry, Nelson Schneider, drew the moral that here was an outlet for risk capital now being used more courageously because of the new tax law which had halved the maximum rate of tax on capital gains to 25%.[62] Unusually, for a leading analyst, Schneider was based in Washington, D.C., rather than New York, and was therefore familiar with the work then being conducted at OTA. Although not a scientist, he was impressed by the hopes vested in insulin and in the impressive array of technologies which he was being told might be revolutionized by recombinant DNA. After all, here was the acknowledged leader of the insulin world, Eli Lilly, taking out a licence on the production of its key product from a company less than three years old.

In June 1979, on the recommendation of Cetus's Peter Farley, Schneider attended London's Royal Society's conference entitled 'New Horizons in Industrial Microbiology', which brought together a wide range of promoters of what in Europe was already being called 'biotechnology'. Genetic engineering was indeed mentioned. It was the topic of the last paper and the subject of the most emotional discussion, as speakers reflected on the lack of government support for their pathbreaking science.[63] Schneider returned confident in his belief that there was something new. A whole range of potential technologies could be transformed through recombinant DNA, and the concept could be marketed by his company, E.F. Hutton, a high-technology investment house.

This experience formed the background to Schneider's individual contribution to the formation of modern biotechnology. He brought the idea to Wall Street and the word to American lips. On 1 August 1979, Schneider wrote a paper to investors depicting his vision: 'DNA — The Genetic Revolution'. There he described the potential of insulin, the long list of other industries that might be affected according to the OTA, the hopes for smaller companies that were already dependent on the new technology — and the ambitions of larger companies such as Eli Lilly. The significance of the paper became clear on 17 September 1979, when Schneider rented a room in New York's Plaza Hotel to introduce his concept to a few institutional investors. The speakers would include a Hutton executive, followed by an industrial scientist (Herbert Weissbach of Roche), Zsolt Harsanyi of OTA, and finally the director of the government's regulatory body, the Office of Recombinant DNA Activities. Advertised in the afternoon were talks by leaders of the four major new biotechnology companies, Cetus, Genentech, Biogen, and Genex. Instead of the 30 participants expected, the announcement drew more than 500. All the excitement revolved around interferon.[64] Investment analysts who did not know a gene from a protein knew that here was a technique to make a cancer drug that otherwise needed thousands of litres of blood, and it would be but the first of many products over the next thirty years in areas which, Schneider predicted, would make up 70% of the GNP.[65]

The success was astounding but not unprecedented. Amplification was the idea of the moment, amid the thrill of information technology.

To Hutton, the idea of a key enabling technology was not new. It was already running seminars on the implications of the new 'microprocessor technology'. Biological technology seemed to be going the same way. But there was not even a language to describe this marketing concept; biological technology hardly rings. So, Schneider recoined the word 'biotechnology' to market the concept of an industry based on a new technology. Wanting to use the word to title a newsletter to investors, following a legal search in September, Hutton applied for a trademark on the use of the term 'biotechnology' within any mass-produced journal in December 1979.[66] At a 1982 meeting, Zsolt Harsanyi, by that time working with Hutton himself, reflected general opinion in the United States when he declared flatly, 'Biotechnology is a neologism'.[67] Unaware even of the recent uses of the word as he was, Schneider had recreated the subject in a manner suitable for the moment.

Schneider's parallel with information technology was occurring to futurologists and policy makers around the world. However, the United States was unusual because of the speed with which enthusiasm spread. As late as February 1978, Cetus's Ronald Cape was complaining to the American Association for the Advancement of Science of what he considered the incredible conservativism of U.S. industry, which had been reluctant to become involved.[68] He also warned that profits would not come until the 1990s. Within two years, with insulin production solved and interferon on the horizon, the world was transformed. When, just two years after Cape's speech, in 1980, Genentech went to the market, its shares, inflating as they were launched from $35 to $89, marked the fastest rise of any stock in the history of the New York capital market.

The OTA report, *Impacts of Applied Genetics*, was published in 1981 in a context far removed from the anxieties over regulation that had surrounded its conception in the late 1970s. Still, more than expressing immediate hopes, the report laid out many possible long-term uses, as well as the issues that might confront congressmen. Quite explicitly, it drew on the older European definition of 'biotechnology' ('the use of living organisms or their components in industrial processes') to conjure up images of an impending revolution ('a number of different industries have learned to use micro-organisms as natural factories') and to suggest that applied genetics would facilitate improved design of such factories.[69] Engineered microorganisms might be used to extract metals from low-grade ores more efficiently or to speed up the degradation of waste. Benefits in terms of fermentation technologies, pharmaceuticals, the chemical industry, food processing, environmental issues, and agriculture were looked to in both the short term and longer term to set against the question of risk. Thus, this familiar model of biotechnology was officially expressed, developed, and published explicitly in response to the questions in Congress, and beyond, about the legitimacy of recombinant DNA research.

Beyond the United States

The intense argument between the potential benefits of biotechnology and the countering calls for industrial regulation came first to the United States. In part, this was because it was the acknowledged leader in the area, with recombinant DNA research sustained by an overwhelmingly large research enterprise. Academic biomedical research support exceeded $2b in 1979, and the competence of American scientists in recombinant DNA techniques was far greater than any elsewhere, with the possible exception of Britain.[70] There were, of course, local factors elsewhere which made it inappropriate to build up the potential benefits of recombinant DNA to beat down the threats of regulation. In Japan, for instance, research based on recombinant DNA techniques was not permitted until 1979.[71] Moreover, U.S. academics were far more willing to engage in commercial enterprise than their counterparts in Europe or Japan. In Britain, the failure to patent the discovery of monoclonal antibodies in 1975 had become a national scandal, with the scientists asserting that they had acted to make the authorities aware of the potential significance. British academe was demoralized and small compared to the ebullience of America. The official report of the 1979 Royal Society meeting entitled 'New Horizons in Industrial Microbiology' ends on the bloodless note: 'The discussion ended on a note of cautious optimism, with the hope that some of the successes that had been presented during the meeting would encourage more support for research into industrial microbiology'.[72]

Even in 1984, the analyst Edward Yoxen, noting the American enthusiasm indicated by Herbert Boyer's 1981 appearance on the front cover of *Time* magazine, could lament, 'But that kind of fascination with the emergence of a new industry has been noticeably lacking in Britain.... Technology assessment has been absent here from the process of building a consensus behind a set of policies for technological innovation'.[73] With opinion about genetic research almost equally balanced in 1979 (32% believing it worthwhile, and 36% fearing unacceptable risks), regulations were recommended by a secretive committee which did not open itself to the type of public debate found in the United States. Ravetz, a member of the British Genetic Manipulation Advisory Group (GMAG), recounts the experience of returning from the hothouse of the U.S. controversy to a meeting of GMAG. On arriving at their places, members found copies of the Official Secrets Act awaiting their signature, by which they would be forbidden from making any disclosure whatsoever of proceedings.[74]

While the U.S. model of regulation by expert committee and detailed rules of containment were widely accepted in Europe, scientists there were less eager to seek out industrial opportunities than their counterparts across the Atlantic. In Germany, molecular biologists preferred to keep working in fundamental research.[75] Moreover, the feared horrors of genetic engineering were so great that no amount of income would have counterbalanced the nightmare of human eugenics. The Europe-wide 1979 poll showed that only 22% of those polled in Germany felt genetic research was worthwhile, compared to 45% who felt the risks unacceptable.[76] Not until 1981 was there significant industrial pressure to conduct more research, and even in 1983 there were only twenty-nine industrially funded gene technology research projects.[77]

Again, the European Commission reacted in a more measured way than enthusiasts would have wanted. Despite the alert provided by the FAST analysts, it took six years for the European Commission to decide to support even a small programme on genetics. Hitherto, science in the commission had been deployed solely towards nuclear power with molecular biologists employed to understand the genetic mutations caused by radioactivity. Of course, their attention was drawn to the new discoveries, and in 1975 Dreux de Nettancourt, a French botanist at the commission, had begun to drum up official interest. He drafted a report, finalized in 1977, entitled 'Applied Molecular and Cellular Biology: Background Note on a Possible Action of the European Communities for the Optimal Exploitation of the Fundamentals of the New Biology'.[78] This highlighted the exciting scientific prospects in three areas: enzyme technology and bioreactors, transfer of genetic information between organisms (genetic engineering), and the molecular understanding of pathology. He related these to the four acute problems of the world: population growth, food shortage, pollution, and energy shortage.

The immediate outcome was the commissioning of three major detailed reports by the very people much cited in the original proposal: Thomas of the University of Compiègne reported on enzyme technology, Rörsch on genetic engineering, and de Duve on molecular pathology.[79] This work led to the formation of the Biomolecular Engineering Programme (BEP), so small that it was largely symbolic, dispensing 15m ecu (an ecu is roughly $1US), during 1982–5, across the community with the brief to remove the bottlenecks to the application of modern biochemistry and molecular genetics in agriculture and agrifood industries.

Though this first programme was small scale, it did portend larger things, and its genesis does indicate that the roots of enthusiasm for recombinant DNA were built on the German and Japanese interest in enzyme technology. This was to be seen in the orientation of de Nettancourt's proposal and in the subsequent books. Rörsch discussed Japanese conceptions of life science research and Wada's model of enzyme use at considerable length. Thomas reproduced a U.S. model of the potential of enzymes that had been presented to a 1975 RANN meeting. Thus, here too, the concepts of the multiple dimensions of the biotechnic revolution formulated before the widespread appreciation of recombinant DNA technology were deployed to provide a context for interpreting the new development.

Conclusion

In the United States of the 1970s, the new frontier for engineering as a whole was clearly in the direction of electronics. For the engineers, biology was in fact less important than it had been in the 1930s. In any case, engineering as a whole was at a low cultural ebb in the wake of the Vietnam War. Meanwhile, health care was a rapidly growing industry. Perhaps this is why the 1970s saw a reversal of the pre–World War II situation: Now biotechnology represented a technological development in Europe, but in the United States it was a response to issues within the biological professions. The 1970s expression of genetic engineering grew out of the movement primarily concerned with human modification, albeit taking a very different direction from earlier years.

The radical shift in the connotation of 'genetic engineering' from an emphasis on the inherited characteristics of people to the commercial production of proteins was nurtured by Joshua Lederberg. His broad concerns since the 1960s had been stimulated by enthusiasm for science and its potential medical benefits, by the zymotechnic zeal of his friend Djerassi at Syntex, but also by irritation with the eugenic proposals of an earlier generation. Countering calls for strict regulation, he expressed a vision of potential utility that fitted the technology assessment approach of the 1970s. Against a belief that new techniques would entail unmentionable consequences for humanity and the environment was ranged a growing consensus on the economic value of recombinant DNA. Of course, this drew on the parallel visions of the benevolent uses of bacteria developed by the existing biotechnology movement. Indeed, by 1980 molecular biologists in the United States had coopted the technological promise of that century-old tradition.

In Europe, the relationship between microbiologists and molecular biology might have seemed inverted, with the established biotechnology coopting the new techniques. Even in the early 1980s, there was a considerable difference in emphasis between the signification of biotechnology on either side of the Atlantic.[80] There would be continuing policy implications in this difference; but whatever the emphasis, on both sides of the Atlantic the 1970 'engagement' of genetics and industrial microbiology had been consummated, and many saw great commercial potential in the marriage. It seemed to many investors and statesmen alike as if here was another member of a very select class of high-technology industries. However, in both the United States and Europe, such common perceptions as eugenic uses of the new biology preserved the memory not just of atavistic anxieties, but also of concerns expressed by leading scientists in the 1960s. In the public's failure to follow, or to believe in, the scientists' shift of emphasis lay the seeds of what would often seem a dialogue of the deaf, between proponents and opponents of biotechnology.

9

The 1980s: between life and commerce

> The impact of biotechnology will be pervasive. Public perception, and governmental response will be of paramount importance in setting a regulatory framework and determining the rate and direction of the diffusion of the technology. The power of public feeling must not be underestimated; consumer resistance and fears for safety and pollution for example, can seriously encumber commercial prospects.
>
> (Advisory Council on Science and Technology, UK, 1990)[1]

Practically the entire period covered by this book so far could be considered as the 'prehistory' of biotechnology, for only in the 1980s did biotechnology acquire internationally recognized significance. Certainly, it was then that the subject became an economic category, with careful measurements of national investments and outputs. By 1990, *Books in Print* was listing seventy titles including the term.[2] Whereas this book began by discussing concepts communicated among a few individuals, by the 1980s opinions about biotechnology were being polled as if they were political preferences.

However, as its significance was accepted, so the clarity of local visions was replaced by a cacophony of dissonant interests, speaking each in their own tongue. This, rather than any single philosophy, became the striking characteristic of 'biotechnology'. The term acquired different implications for commerce, for the promoters of new technology, for the opponents of the 'perversion', and for the regulators. However confusing, those multiple meanings are not dispensable. To regard the subject as if it were either a real thing, which has particular technical properties, or else as just a dispensable word is to neglect the lesson of the analyst Gregory Daneke: The U.S. economic system, he has stated, 'is impregnated with certain myths and metaphors which rarely correspond with reality, yet they have a profound impact upon the parameters of policy change'.[3]

The extent of the debate, on a scale that would have amazed Lancelot Hogben, dead for barely a decade, justifies its description as an industry in itself (a 'metaindustry' perhaps). For all the scientific breakthroughs of the previous decade, and despite the sense of novelty, this debate was not historically isolated, but was the sequel to a century of discussion. Out of still unresolved problems came two issues that urgently required resolution. The first was raised in the search for research policies: To what extent is biotechnology a branch of engineering, subject to the managerial thrust so characteristic of technology, and to what extent is it an applied science? This question for civil servants behind closed doors was matched by a more public dilemma: Is biotechnology merely the latter-day descendent of a pragmatic zymotechnology, or is it also the legatee of the concerns and ambitions of those other ancestors, the early proponents of a human-oriented biological engineering?

Industry

Before exploring the process of negotiation, it is necessary to appreciate biotechnology's most obvious characteristic in the early 1980s: It seemed to characterize a nascent 'real' industry, providing titles for the emerging trade organizations such as the Industrial Biotechnology Association and the Association of Biotechnology Companies.[4] Although the coherence and prospects were often exaggerated, the dynamism and pressure of industrial development played the crucial role in forcing policy makers and the public to contemplate biotechnology.

Whereas, originally, the interest of some major companies had preceded breakthroughs in genetics, early hopes for stimulus from conventional products of biochemical engineering were now fading. In the wake of declining oil prices, concern with gasohol and biogas became less pressing. Single-cell protein foods slipped from industrial consciousness as it became clear that markets would be hard to find. Even industrial enzymes, which were the pioneering products of the field in the 1960s, commanded a total world market of only $500m by 1985.[5] Certainly, these applications were supplemented by new ones, such as the 'biosensors' whose potential came to be among the most exciting aspects of 1980s biotechnology, but it was from the impetus of scientists, rather than the engineers, that hopes sprang for a major new industry.

Recombinant DNA technology and, in the pharmaceutical industry, hybridoma techniques stimulated hopes for both therapeutic proteins and biological organisms themselves, such as seeds, biological pesticides, engineered yeasts, and modified human cells for treating genetic diseases. From the perspective of its commercial promoters, scientific breakthroughs, industrial commitment, and official support were finally coming together, and biotechnology became a normal part of business life. Early in the decade, a triumphant future was predicted in such market analyses as T.A. Sheets's widely quoted 1981 estimate of the scale of biotechnology by 1990. This suggested that though the market for biotechnology products was currently $25m a year, it would grow 2,592-fold by the year 2000, reaching $20b even by 1990. The calculation had allowed for every sector of fermentation products. The implication was that each sector would provide enormous scope for genetically engineered organisms. How this was calculated has been examined in the case of antibiotics. Sheets had assumed a 7.5% annual growth in the world market to the year 2000, giving a total of $33.5b. Of that, it was assumed that 'biotechnology', by which genetically engineered products were implied, would capture 10%, giving over $3b in that market alone. The antibiotics specialist Ralph Batchelor of Beecham is reputed to have described the analysis in dismissive terms; however it did show the power of combining the scale of zymotechnic processes with hopes for recombinant DNA.[6]

Sceptics might point to an apparent disparity between expectation and performance. In 1990, the *Boston Globe* was lamenting: 'When you strip away the hype, the biotechnology industry is an intriguing little flyspeck on the national economic picture. Biotech is about as significant to the national and regional economies as, say, horse breeding.'[7] Such scepticism, though, was now a radical position. No longer were the proponents for the economic and technological significance of biotechnology the iconoclasts. Their message had become accepted and incorporated into the policies of governments and the chemical industry. Whether one considers the United States or other countries, 10% of the 500 largest companies reported that they were involved in biotechnology by 1986.[8] Certainly, the importance lay in future prospects, and the expectations of wealth had been delayed.[9] Although, in its perceived economic importance, biotechnology ranked near information technology, this status still came from promises (or threats) in the future rather than from the currently commercial applications.

The first area in which the potential value of biotechnology products helped companies overcome the costs of innovation, was pharmacology. The rewards of success in this traditionally chemically dominated industry are enormous. The single (chemically derived) drug Zantac, used to combat ulcers, commanded sales of over $2b in 1990 and enabled its manufacturer, Glaxo, to become one of the world's major corporations.[10] Moreover, to the 1970s scourge of cancer, was added AIDS in the 1980s, offering an enormous potential market for a successful therapy, more immediately, a market for diagnostic tests based on monoclonal antibodies, and stimulating the already nascent market for proteins hitherto extracted from blood. In 1984, shortly before leaving the by-then struggling Biogen, Walter Gilbert reflected that, six years earlier, he had had great expectations of a great variety of applications, but Biogen had had to focus on pharmaceuticals: 'We now concentrate our efforts there because we see it as the new field in which the technology will be the most commercially rewarding over the next ten to fifteen years. We do not see the technology being equally effective in the chemical, and the food and beverage industries in that period'.[11]

Even in pharmaceuticals, developments were slower and much more costly than originally envisaged, though their promise was still enormous. By 1988, only five proteins from genetically engineered cells had been approved as drugs by the FDA: insulin, human growth hormone, hepatitis B vaccine, alpha-interferon, and tissue plasminogen activator (tPA) for lysis of blood clots.[12] Interferon, the early favorite, proved a disappointment in clinical trials, and many companies changed their interests, though eventually it did prove a commercial product as antiviral and anticancer drug (despite its occasional side-effects). According to the *Economist,* the total value of engineered products in 1988 was $400m.[13] By 1991, the prediction for two years hence was now $4b.[14] The original simple vehicle for recombinant DNA had been supplemented by more complex organisms and even animals, which were modified to express otherwise rare proteins. In addition, monoclonal antibodies, produced from a hybridization of cells rather than genetic manipulation, came to be useful diagnostic tools, if not yet of therapeutic importance. Their value had reached $300m by 1988.[15]

A new generation of biotechnological processes was coming to be contemplated. The shift was described by a 1990 report of Britain's science advisory committee ACOST. 'Whereas efforts have so far been concentrated on the generation of biologically important molecules, the next phase is likely to lead to the engineering of higher levels of biological organisation: the engineering of pathways and cells'.[16] There were even prospects for new 'engineered' proteins that could be expressed by modifying genes, but the separation between such biologically originated drugs and chemically produced analogues was increasingly nominal. The invention of a biochemical method for replicating DNA, PCR (Polymerase Chain Reaction), meant that, by the late 1980s, cloning was no longer always necessary to reproduce significant bits of genetic material. Industrially, the distinction between chemical and biological techniques, which had been so marked in the early 1970s, was once again becoming blurred. Moreover, traditional screening techniques were still a primary source of new drugs. Of fourteen new biological remedies approved by the FDA in 1990, only one was genetically engineered.[17]

The integration of biotechnology with chemical approaches can be seen also in industrial restructuring. After an initial burst in which generic biotechnology companies were founded, firms came to be defined by the industries within which they operated. Microelectronics had already alerted the public to the genre of the small high-tech firm based on academic entrepreneurs. During most of the 1970s, only four companies, Cetus, Genentech, Biogen, and Genex, had been specifically associated with the subject. Then in 1980, the year of the Cohen and Boyer patent and of the Genentech flotation, twenty-six biotechnology companies were founded, and more than forty the following year. This burst was followed by disillusionment, with only twenty-two new companies in 1982.[18]

Although, by 1988, over a thousand U.S. companies were in some way connected to biotechnology, no longer would the genre of the small scientist-driven company define the technology.[19] Disillusionment with this model became public, when Biogen parted from Nobel Prize–winning Gilbert in 1984. Increasingly, companies were led by experienced executives rather than scientists, even if close relations between university and industrial laboratories did endure and develop. A study of the founders of U.S. biotechnology companies has shown that whereas in the period 1971–80 almost twice as many had academic as had business origins, by the mid-1980s, two-thirds were from business.[20] At a congressional hearing in 1981, the chemist Carl Djerassi scorned the estimated sizes of the market being attributed to biotechnology conceived as genetic engineering. He caricatured current thinking by analogy to the car industry.[21] Just because welding is important in car manufacture, one would not define the financial importance of welding by the annual production of cars. Djerassi was proved to be right. Whereas the balance of skills varies across specialties, nowhere was the industry simply reducible to the techniques of recombinant DNA. The scientific techniques of genetic modification were integrated into complex development, production, and marketing processes over the next decade.

Increasingly, biotechnology companies either became 'research boutiques' or tried to grow into pharmaceutical companies.[22] The emerging pattern of strategic alliances between small U.S. firms and larger pharmaceutical companies, often foreign, symbolized biotechnology's integration within the pharmaceutical industry. When Genentech, the largest of the new biotechnology companies, was taken over by the Swiss pharmaceutical firm of Hoffmann-La Roche in 1989, the vision of the new biotechnology companies growing from acorn to oak, nutrified by science alone, was ceremonially interred. Two years later, Cetus, the first of the new biotechnology companies, announced that it intended to amalgamate with its competitor, Chiron.

The small companies achieved particular prominence because the interest of established pharmaceutical companies had been slower to grow. The head of ICI's seeds business has said: 'Almost from the outset, most large companies saw the "new biotechnology" not as a new business but as a set of very powerful techniques that, in time, would radically improve the understanding of biological systems. This new knowledge was generally seen by them as enhancing the process of invention and not as a substitute for tried and tested ways of meeting clearly identified targets'.[23] At first the pharmaceutical industry's response was muted by the dominance of chemists within their organizations. As late as 1987, Arthur Kornberg, who had pioneered polynucleotide synthesis, was regretting, 'Yet chemistry and biology are two distinctive cultures and the rift between them is serious, generally unappreciated, and counterproductive'.[24] However, the need for a change in tack was pressing. A 1984 Department of Commerce analysis cited 40% overcapacity in the U.S. chemical industry.[25] Through the 1970s, Germany's large companies made low profits. In 1981, Britain's ICI declared a loss. America's Du Pont, which had promised 'Better Living Through Chemistry' since the 1930s, gave up the struggle to find a successor to nylon by chemistry alone. Its British-born president, Edward Jefferson, read in the *Economist* that biology was the way of the future, and focused his company's future research there.[26]

Pharmaceuticals have been a significant area for expansion, and so, more slowly, has agriculture. Agrochemicals have become a major if controversial aspect of the chemical industry and would potentially be complemented by herbicide-resistant plants and seeds engineered to contain and produce their own nutrients. Oil companies had been trying to diversify in many directions since the early 1970s. Seeds had been an early candidate, since their markets resembled that for the specialist chemicals with which such companies as Shell were already familiar, and in the United States, the Plant Varieties Protection Act of 1970 gave increased powers to inventors against farmers. Monsanto too had already committed itself in that direction. ICI, a late-comer, spent £150m on acquisitions of seed companies in the mid-1980s.[27] Nonetheless, at a time of surpluses, agriculture could not provide the expectation of high short-term profits (however unjustified) that sustained the first-generation venture-capital-supported companies. The growing importance of prospective agricultural markets to biotechnology have therefore raised problems of identity. As a 1990 editorial in the industry's magazine *Bio/technology* suggested:

> One astute and oft-cited Wall Street analyst is fond of saying that, while health care is revolutionary, agriculture is evolutionary. New agricultural products should offer only incremental improvements, building on a traditional base in order to be championed by industry and accepted by the consumers. (By extension, the same would hold true for other products to be introduced into the environment.) Put another way, biotech may have to shift gears and become more reactionary and less revolutionary.[28]

Reflecting on the growing importance of biotechnology to agriculture, the OECD, unknowingly evoking the spirit of Ereky, reported, 'Its direct and indirect effects will be felt in an *increasing convergence of agriculture and industrial practice*'.[29] Ironically then, as biotechnology became increasingly sophisticated and significant, it was absorbed by user industries such as pharmaceuticals and agriculture and has been losing a distinctive industrial shape of its own. However, profoundly affecting such crucial areas of individual and national life as medicine and agriculture, biotechnology has become a vital bureaucratic and cultural category. Funds on the one hand, and restrictive legislation on the other, have been sustained by hopes for industrial dynamism and by the anxieties of a suspicious public. Rather than any overarching philosophy, it was the resolution of those contrary forces that was now providing a definition.

A metaindustry

In the half century since biotechnology was last seen as humanity's best hope, biology has developed much faster than the social engineering with which it had been associated before World War II. The Harvard professor Bernard Davis reflected on the contrast:

> Those of us who were entering biology in the 1930's were very much encouraged by the essays of J.D. Bernal and J.B.S. Haldane who predicted that the age of biology would soon emerge. Equally confidently, these authors

predicted a similar success of scientific planning in solving the problems of economic and political organization. Little could the students of my generation foresee that biology would mature so rapidly, while the predicted social utopia would become more distant than ever.[30]

Unsuccessful as centrally planned economies might have been, even in the West planning has become a major activity. The Scandinavian analysts Baark and Jamison have contended that 'attitudes to new technologies are transformed into policy through a "mediating filter" of confrontations, institutional innovations, and conceptual debates'.[31] Thus, discussion about biotechnology during the 1980s produced countless books and reports concerned with promotion, regulation, commercial exploitation, and such indirect consequences as the industrial relationships of academia. A triumphant preface introduces the report of a 1985 meeting, organized by the U.S. National Academy of Sciences to celebrate the decade since Asilomar. Still, while nobody was singing 'We shall not be cloned', the meeting attended by scientists, lawyers, journalists, corporate chairmen and vice-presidents, economists, university administrators, government regulators, a U.S. senator, and Capitol Hill staffers highlighted the unique processes by which biotechnology was coming to be interpreted in the United States. No longer could proponents alone define the subject.[32]

Policy makers were being asked to create two environments in one. It would suit the subject construed as the application of unforeseen breakthroughs in fundamental biology. Rather differently, it would also have to provide for biotechnology seen as the programmable response to technological evolution. Whether parliamentarians making laws, officials implementing them, or lawyers adjudicating patent rights, regulators have mediated between the proponents enthusiastically foretelling a new biology-based industry, on the model of information technology, and worried sceptics more impressed by the cost of the Faustian bargain. They have decided who should control the release of genetically engineered organisms, and by what criteria, and whether certain sorts of work should be banned altogether. This role has entailed not just defining a legislative context for the 'safe' prosecution of an industry, but also persuading society that this has been done. In identifying the balance between engineering feat and biological hazard that is acceptable to the public, science, and industry, regulators have therefore been the driving force behind the search for a definition of biotechnology in modern society.

When the 1980s began, though biotechnology was widely applauded, there was little international agreement as to its content. Some understood it as the industrial application of the scientific process of recombinant DNA, others as a technology-integrating microbiology, chemical engineering, and biochemistry. By 1985, with growing commercial experience, visions were drawing together within industrial circles: The importance of genetic engineering in changing the boundaries of possibility was recognized by the biochemical engineers. Engineering's critical importance in controlling the implementation was becoming better understood.

Problems of research policy were set amidst broader questions of economic survival. The 1980s began with an oil crisis created by anxiety. 'Would the Iran–Iraq war lead to the terrible destruction of facilities?' asked suppliers as they stockpiled reserves. The price of oil reached $35 a barrel, ten times the price a decade earlier, and some countries saw hyperinflation. As the oil crisis diminished and prices fell back to $15 a barrel, a new threat appeared: global economic competition. For thirty years since World War II, the United States had been top nation. The election of Reagan in 1980 marked a reaffirmation of the voters' conviction that this could still be true. U.S. success in high technology, it was widely felt, would be the key. In Europe, thousands of miles from the Pacific Rim, companies and states were disturbed as they considered their industrial future. For Americans, Japan increasingly replaced the Soviet Union 'as most feared country'. Worldwide, governments identified the industrial bases of the future – including information technology, materials technology, and biotechnology. Now the integration of biology with engineering would become a challenge to programme managers.

Japan

Biotechnology seems to have received roughly equivalent amounts of support from the governments of industrialized countries, each of which shares an interest in high technology. Major European countries and Japan each spent, in the mid-1980s, about $100m a year, and the United States six times that amount.[33] The ways they disbursed such funds had to reflect wider economic policy, yet they were united by the Japanese-inspired concept of a 'core technology'.

In 1981, Japan's Industry Ministry, MITI, identified biotechnology as one of the key industries of the future – a cutting-edge technology. Such technologies, which also included new materials and information technology, were the beneficiaries of government money (biotechnology obtained about 25b yen equalling £60m) and, perhaps more important, were also highlighted by industry. Ten-year plans were established. 'Biotechnology' was a new name in Japan and it benefited from the U.S. enthusiasm for genetic engineering; indeed the MITI initiative is said to have been stimulated by the decision to grant a patent for Cohen and Boyer's invention of recombinant DNA techniques.[34] Yet the initiative also built on a decade of work sponsored by the the STA's life science project. Michael Rogers, Britain's scientific attaché in Tokyo, described the MITI interests as 'fine chemicals, alternative routes to petrochemicals, fertilisers, reduction and oxidation reactions using enzyme technology, recycling co-enzymes and so on, and in general applications of biotechnology to the chemical industry, including mass production problems.'[35] In the context of the oil price escalation of the early 1980s, these plans were far more oriented towards making up for expensive oil than the medically concerned STA projects. So the subject was defined as 'bioreactors for industrial use', 'large scale cell cultivation', and 'recombinant DNA application techniques'.

For each programme, a variety of industrial targets were defined, which in turn were translated into subjects for research. Thus, in bioreactor technology, one target was the 'realisation of reforming conventional high temperature, high pressure, energy consuming chemical industry into normal temperature, normal pressure source- and energy-saving chemical industry'. This was translated into the 'development of bioreactors for synthesis'. In turn, each of these subjects were broken down into various items for research, such as 'investigation of microorganisms or enzymes'. Research on each unit was then planned by time frame over the next decade. Microorganisms capable of producing target enzymes would be investigated in an initial phase, 1981–4. After that, the breeding of the selected organisms would be explored.

In a separate initiative, MITI set up a Biomass Policy Office (May 1980) with a seven-year programme in advanced-fermentation-alcohol technology. This became the Bioindustry Office in 1982. Not to be outdone, the Ministry of Agriculture, Forestry, and Fisheries announced its own biotechnology program in 1981. As well as stimulating competing governmental activities, the MITI lead encouraged many companies to follow. Five major companies had established the roundtable conference on biotechnology in 1980. Within a year, the Research Association for Biotechnology, with fourteen members, was founded.[36] These included traditional fermentation and chemical companies. However, other organizations were also attracted to biotechnology, wishing to diversify into a profitable new area. One of the early 1980s recruits to the STA's Life Science Project Group was Daini Seikosha, better known as the manufacturer of Seiko watches, which was suffering competition from lower cost Hong Kong products and now sought to make advanced biochemical equipment.[37]

MITI has seen this foundation period of industrialization as lasting until 1985. Then came a period of building industrial infrastructure with the implementation of guidelines for industrial application of recombinant DNA technology, and increasing internationalization. By the late 1980s, corporate biotechnology research was reportedly 'more than doubling every year', and the trend was expected to continue.[38] After 1989 would come a greater emphasis on basic research and studies, for, in fact, the rigidly programmed and MITI-controlled New Generation Basic Technology Programme was not successful in marshalling industrial effort.[39] At the same time, it was recognized that bioindustry represented 'technology, products and industrial system really favorable to human being and environment'.[40] This approach to biotechnology has emphasized the breadth of the subject. Indeed, the term 'life sciences' has often been preferred. Its scope was widened further when, in 1987, as an answer to the military-, and, of course, physics-dominated Strategic Defence Initiative aimed at preventing nuclear destruction, the Japanese announced their Human Frontier Programme, which would fund foreign researchers to conduct research with a view to ending all human disease.

In the event, the programme was not large and has been principally concerned with neurosciences. However, when first announced it appeared to be concerned with gene sequencing. Wada's chemical automata programme of the 1970s had led to the development of a first generation of automatic gene sequencers. Granted, these were apparently not competitive with U.S. models. But the same had been true of their early cars. Would the Japanese now use their programming skills and technological approach to coopt Western scientific expertise to what seemed to be the key frontier of biotechnology, understanding the human genetic code?[41] From there it would be a short step to dealing with the 4,000 single-gene disorders that affect 1% of the population and, even more remuneratively, cancer itself.

United States

As in the early 1970s, Japanese industrial policy was watched with envy and admiration elsewhere. Emulation, however, was coloured by local circumstances. In the United States, an interpretation of biotechnology that emphasized its basic science roots harmonized with the federal government's restricted concept of the proper domain of industrial policy. The logic was clear in the influential OTA report published in 1984: *Commercial Biotechnology: An International Analysis.*[42] Typically, it concluded that Japan would be the most serious competitor of the United States in the commercialization of biotechnology, but the report took an approach very different from that of the Japanese.

The U.S. emphasis was on a science-pushed industry, rather than the balance between industry-pull and science-push of the Japanese. Following the lead of the previous OTA report, the emphasis of that basic research was on recombinant DNA. The index of the 612-page report highlights the difference. The words 'oil' and 'petroleum' do not appear at all. Energy does not merit a separate entry and is subsumed in the dozen pages devoted to 'commodity chemicals and energy'. By contrast, fourteen universities, from British Columbia to Wisconsin, merit their own separate entries. Even if pharmaceuticals did seem to be more immediately profitable than energy sources, the contrast with the Japanese approach was striking. As an icon, an electron micrograph of an *E. coli* DNA appears on the first page of the first part.

The emphasis on basic science was explained by a historical model according to which biotechnology was moving along a 'trajectory of innovation' from basic research through generic applied research to applied research.[43] The report did argue that better bioprocessing technology would be required. However, the thrust of the argument was that this was symptomatic of the higher reaches of the innovation trajectory rather than a key part of the invention process from its beginning, as had been assumed in Japan. The difference can be seen in the report's very definition of biotechnology, on which the organizers took three sessions to agree.[44] The writers articulated an evolutionary classification, destined to be widely influential. Having defined the subject in very general terms, OTA decided to distinguish between the 'old' biotechnology, whose origins lay back in brewing, and the 'new' biotechnology, which was seen to include two favourites of the investment community, recombinant DNA and cell fusion (which includes monoclonal antibodies). It included, too, bioprocess technologies, and indeed the treatment of this topic (based on a subcontract report written by Elmer Gaden) was granted the most space. But the perspective was destined to be neglected in later renditions of the new biotechnology which, thereafter, came to be defined as including just genetic engineering and monoclonal antibodies. Within a few months of the publication of the report, a meeting in Germany to consider the implications of biotechnology was told by its chairman, Professor Silver of Washington University, St Louis, that these made up the new biotechnology and that the technology was secondary. Basic research was where new opportunities would emerge, he said, explicitly arguing with an analysis proffered by the industrially oriented GBF.[45] Superficially, there may have been a resemblance between the four-generation model expounded by DECHEMA's H.-J. Rehm and the two-generational image of the old and new biotechnologies. Nonetheless, Rehm's view was of an evolution in the partnership between science and technology, whereas the U.S. vision was of a decisive shift towards a science-driven biotechnology.

The OTA report did argue for the importance of assuring U.S. bioprocess technology and criticized the lack of support given to generic microbiology research and training. It showed that the U.S. government, principally through NIH, was investing $510m a year in biotechnology-related basic research. However, in the categories of generic applied research and applied research, the number was but 1% of that, closer to $5m. The distinction between European and mainstream U.S. approaches was highlighted by an independently authored comparison between the biotechnology and semiconductor industries appended to the OTA report. It identified the science-push origins of biotechnology as opposed to the need-pull of semiconductors and condemned the lack of applied research and demonstration projects in the United States.

The OTA analysis reflected an emerging consensus about the support of technology during the Reagan administration. It provided a route for the government to support biotechnology in the way it favoured for all industry, by enhancing research. However, the United States is not without its more direct industrial policy, one that came to be exercised at the state level.[46] In 1988, forty state biotechnology centres counted a total funding of $83m, 60% of which came from the states themselves.[47] A state such as North Carolina, heavily dependent on tobacco, and wishing to find new economic bases, drew on its agricultural tradition and the scholarly centre evolving in Research Triangle to specifically target biotechnology. In Massachusetts, the *Boston Globe* reported: 'We are flat on our backs and grasping at each silver bullet that comes along. Everyone has a solution – the third harbor tunnel, the candidacy of John Silber, the brave new world of biotech – for the current doldrums, each one more chimerical than the last'.[48]

State efforts may have local and industrial significance, but their scale is small compared to the more than $1b dispensed by the U.S. government for biotechnology-related work. In 1988, it was pointed out that 75% of funding for biotechnology came from public rather than private sources, and these were dominated by the NIH.[49] Moreover, there has been one attempt to integrate the U.S. focus on basic biomedical research with 'big science', pioneered in space and military areas by the United States itself and translated with such commercial effect by the Japanese. That has been the Human Genome Project, the product of a collaboration between medicine's traditional patron, the NIH, and the original sponsor of big science, the Department of Energy (DOE, which now includes the Atomic Energy Commission). Long used to large schemes, the DOE, having been deprived of its mission to find U.S. energy independence, was pushed by Robert Sinsheimer, a pioneering thinker about genetic engineering during the 1960s and by then, head of the University of California in Santa Cruz. He was seeking a way of redeploying funds originally allocated by DOE for a telescope that was cancelled. Mapping the functions of areas of the human genome and identifying the detailed order of its 3 billion base pairs has seemed a basic science project that would sustain U.S. biotechnology into the next century.[50] At the same time, though scientific in its objectives, it has taken an organized style of research more characteristic of industry than of academe.

Europe

The 1984 OTA report dismissed the Europeans as secondary competitors compared to the Japanese. Each nation formulated its own development programme, but in addition the precedent of cooperation on information technology in which Europe had manifestly fallen behind in the 1970s provided an opportunity for the European Commission to take an unusually prominent and proactive role.[51] Here was an occasion for the formation of European culture in action. With the model of a healthy biotechnology industry emerging as a key community objective, once again the term and the policy were defined together.

The dismissive U.S. attitude compounded anxiety over European competitiveness that had been building for a decade. Mark Cantley, responsible for the Bio-Society initiative in FAST, was working to get its message understood. A provisional report by Cantley, his colleague Ken Sargeant, and the head of the programme, Riccardo Petrella, entitled 'Biotechnology and the European Community: The Strategic Challenges' was discussed in the community in January 1982. It suggested that biotechnology was now important for three reasons: the multisectoral applications, the easing of resource constraints it implied, and the global challenges of the United States and Japan, with the potential to help the Third World. These were attributed to the biological revolution of the past twenty years, which implied the domestication of microorganisms and plant and animal cells. A joint steering group was set up between the Directorate General dealing with the Internal Market and Industrial Affairs (DG III) and FAST, with Cantley as secretary. The initiative was accelerated a year later. On 8 February 1983, just a couple of months after the FAST final report, Gaston Thorn, the president of the commission, gave a speech which included the one-line comment that the commission would take the same approach for biotechnology as it had done in the so-called ESPRIT programme for stimulating information technology.

This gave the go-ahead officials needed.[52] Within two days, a meeting was held between DGIII and the science directorate, DGXII, to negotiate a process for putting Thorn's proposal 'to music'. Since it was argued that agricultural applications were important and DGIII was involved only with industry, DGXII with a background in the FAST process and its neutral position with respect to the big directorates of industry and agriculture was in an ideal position to take a lead. By June, a 'Community Strategy for Biotechnology' had been published, and in October, the key document 'Biotechnology in the Community'.[53] The following spring, the Concertation Unit for Biotechnology in Europe (CUBE) including Cantley and Sargeant from FAST was established.

Meanwhile, the first of the scientific sponsorship programmes, the Biomolecular Engineering Programme, was approaching its end. In 1985, it was replaced by the larger Biotechnology Action Programme (BAP) with 50m ecu. Here, there was an interaction between the broader conceptions of the FAST team and the specific research concerns of the biologists. The BAP, unlike its predecessor, included provision for the development of an infrastructure – bioinformatics concerned with data capture, data handling modeling and software, and culture collections. The programme also emphasized the collaboration between national laboratories, a policy of international networking, 'laboratories without walls', and the participation of industry. The sequel to this programme was the yet larger-scale initiative entitled BRIDGE (Biotechnology Research for Innovation, Development, and Growth in Europe) in which a careful balance was preserved between industrial relevance and the need to concentrate on 'precompetitive research'. This time the profile of risk analysis and gene mapping had been greatly raised while biotransformations had been reduced in scale.

Beyond the science programmes, the community explored agricultural conversion. Since the 1970s, it had been struggling to evolve away from its dependence on the role of agricultural controller to which every other responsibility would be subordinate. At the Stuttgart Summit of 1983, it was decided that the cost of the Common Agricultural Policy could not be allowed to rise indefinitely. Biotechnology seemed to provide an answer. A 1984 study conducted for the European Commission, starting with the prediction of 58 million tonnes of cereals in the community by the year 2000, reported that the price of a tonne of maize had already fallen, from the equivalent of about 40 barrels of crude oil in the early 1960s, to the equivalent of less than 5 barrels. Such materials could therefore be seen as potentially competitive industrial raw materials, and the report suggested, U.S. manufacturers would benefit from their even lower prices.[54] Even though oil prices fell back and the more extreme ambitions proved impractical, as CUBE's Ken Sargeant testified to a British parliamentary hearing, 'Agriculture has been a separate world from industry and we feel that the two worlds will have to be brought closer together'.[55]

In 1986, the commission issued a report arguing that there would be specific high-technology uses of agricultural produce as biotechnology facilitated the integration of the agricultural process within industry.[56] This move led, in 1987, to the development of programmes known as ECLAIR (European Collaborative Linkage of Agriculture and Industry through Research) and FLAIR (Food Linked Agro-Industrial Research). Offering officials in Brussels a marvellous opportunity to make a contribution to the community's world position (as well as to produce clever acronyms), new technologies were providing a way of organically connecting present acute problems to future opportunities.

Germany and Britain

The efforts of the European Community demonstrate a convergence between two rather different policy initiatives: of Cantley in the FAST group and of Dreux de Nettancourt, the molecular biologist. The negotiations between the enthusiasts for molecular biology and the established protagonists of biotechnology had their analogies in each country. In Germany, the crucial event was the 1981 decision, amid national consternation, by the multinational chemical company Hoechst to enter a relationship with Harvard University Medical School. In return for $67m, Hoechst would get the first rights to commercial exploitation of research at Harvard's Massachusetts General Hospital for a decade. During the 1980s, the official German vision of biotechnology came closer to that allocated across the Atlantic. Eschewing the previous emphasis on need-pull, the preface to the German government's plan entitled 'Applied Biology and Biotechnology, 1985–1988' began, 'Basic Research in molecular biology, genetics and microbiology and resulting applied research in genetic engineering and biotechnology is currently in an extremely productive and dynamic state of development'.[57] By 1989, more than 400 German companies were involved in biotechnology.

Britain, whose culture is pulled towards both Europe and the United States, produced a typically pluralistic approach to biotechnology. Following the Spinks report of 1980, three separate initiatives were made by the government. The Department of Trade and Industry, which established a biotechnology unit to support the 'near-market or competitive end of R&D', and the two major research councils, the Science and Engineering Research Council (SERC) and the Medical Research Council (MRC), each took major steps – in diverse directions. The SERC, rather like DECHEMA before it, has treated biotechnology as a technology. In 1981, it established a biotechnology directorate under Geoffrey Potter, the head of the engineering board. His directorate took an approach traditional to the engineers: They identified a need and worked out what research was required to fill it. At first, this new central group funded areas that reflected the 1970s technological view of biotechnology. The November 1981 priorities were feedstocks for chemical processes, fermentation, enzyme and immobilized-cell technology, separation and concentration technology, product processing, waste treatment and by-product utilization, automated handling, and only finally 'molecular biology and enabling technology'.[58]

MRC took a very different, much more American turn. It ran the Laboratory of Molecular Biology in Cambridge, one of the holiest places of science. With its seven Nobel Prize–winners over thirty years, including Crick and Watson, discoverers of the double helix itself, MRC could claim that its style was successful. It emphasized free enquiry, giving first-class scientists the tools and letting them get on with the job. The research councils' job was then to generate the knowledge which industry would be allowed to apply. There may, MRC admitted, be a problem of an information gap between academe and industry as demonstrated by the 1975 failure to patent its discovery of monoclonal antibodies. So, in 1981, Celltech was launched with the intention that, with special opportunities of exploiting MRC expertise, it would be a British equivalent to the American 'new biotechnology companies'.[59]

Though the MRC and the SERC have therefore had very different philosophies, over time there has been a convergence of agendas. The government's Department of Trade and Industry Biotechnology Unit has become more and more interested in supporting basic research. The SERC in turn became more interested in genetics than in fermentation technology. In 1985, the directorate applied its approach to the new scientific area of protein engineering. This area had been largely developed by the Laboratory of Molecular Biology. Remarkably, for a country where such quarrels have normally been settled in the men's room of an appropriate club, public conflict broke out between the MRC and the SERC, both over territory – who owned protein engineering – and over management styles. In a real sense the distinction lay both between the German and the U.S. view of biotechnology and between an engineering and a scientific view of innovation.[60] After considering a neater solution, which would in its own turn have created considerable upheavals and tensions within a large single research body, the government decided to allow both organizations to develop their own styles.

In principle then, biotechnology seemed to promise the solution to two of the outstanding problems of the decade: an inflexible agricultural system and the maturity of traditional industries. Worldwide, the promoters of biotechnology have struggled, as much as the more isolated philosophical and polemical thinkers before them, to make sense of the boundary role between biology as a science, whose surprising findings make it ungovernable, and engineering with its managerial traditions. Increasingly, administrations of every hue have found it necessary to permit a variety of resolutions. If biotechnology has been defined by the relationship between biology and engineering, the demarcation and nurturing of that relationship has required a catholic interpretation of good practice.

Opposition

Meanwhile, a more bitter divide arose between official and public conceptions of biotechnology. Three strands of thought had been inherited. First, the practicality of microbial technology was well understood, but second, its significance as a fundamentally different technology was now seen as threatening, leading to unpredictable ethical or ecological disasters. The third legacy was an approach to resolving the tension: countering anxieties about the concept with prospects of medical and technological wonders. Despite the many exceptions, this pragmatic approach, particularly valuable when making arguments for resources within organizations, tended to replace earlier considerations of the fundamental attractions of a biological approach.

However, the public was not entirely convinced by this strategy. Although the subtlety of the earlier arguments was often missing, decades of scientific and industrial achievement had not yet wiped out the distinction between the specifics of zymotechnology and the unbounded potential of Tornier's *Biontotechnik* or the *Biotechnik* of Francé, Goldscheid, and Geddes. It seems that Western cultures are still deeply concerned by the interpenetration of the categories of technological and biological creation. So on the one hand, biotechnology had been conceived as a 'core technology', a specific weapon to be used by all participants in the international economic contest, but on the other, it still represented the rather vague interface between life and engineering. Even in Japan, as the magazine *Bio/technology* has pointed out, '77% of the readership of the science magazine *Newton* predicted that biotechnology will develop into the same sort of "major social problem as atomic energy" – which is, in Japan, saying a lot.'[61]

The new problem of the public understanding of biotechnology is as much concerned with the boundary character of biotechnology as the policy debates of the previous decades. In *Daedalus,* seventy years ago, Haldane warned that biological inventions were a 'blasphemy'. Today, biologists are accused of playing God. Specifically they seem to be accused of reducing to industrial and commercial practice that which ought to be exempt. The debate over the milk-producing hormone bST illustrates the confusion between the various issues. Typically, an article in *The Independent,* a respected English newspaper, amalgamated the ideas of a chemical hormone (which bST is), a genetically manipulated bacteria (through which it is made), and the spectre of genetically engineered cattle, and concluded its analysis of the question of bST use: 'In other words, do you want your milk to come from Daisy or a genetically engineered and manipulated machine?'[62] As the worlds of agriculture and industry came together, the particular cultural challenge of biotechnology was being highlighted.

That challenge has been interpreted differently across the world. Some objections have been practical, responding to the quality of controls; elsewhere the ethics of modern biotechnology have been rejected, such opposition being tagged 'fundamentalist'.[63] This discord reflects the variety of intellectual inheritances, pressure groups, and ways confrontational policies are carried out. National attitudes are even harder to summarize and compare than administrative arrangements; they are even more volatile, and the various indicators, be they polls or analysis of the media, are dangerously victim to the methods used to collect data. Nonetheless, the approaches taken, particularly in the United States and Germany, seem to have often diverged. These cases are worth highlighting as they best indicate biotechnology's continuing reinterpretation.

In the United States, particular issues were fought through the courts by relatively small pressure groups, drawing particularly on new environmentalist legislation. Jeremy Rifkin, who had argued against experiments in the 1970s, headed the Foundation for Economic Trends, which fought vigorously in the next decade against their commercial exploitation, and particularly against the release of genetically engineered organisms into the environment. A 1986 survey showed that 44% of the U.S. people believed either strongly, or somewhat, that 'we should not meddle with nature'.[64] On the health side, anxiety about our right to meddle with nature was reflected by concerns over the ethics of genetically changing people and, in agriculture, by practical concerns about the release of genetically engineered organisms in the environment. The introduction of new organisms has been taken as a threat to the natural environment, a continuation of the pollution begun by man's invasion of the wilderness. Thus, although the poll showed that a new cancer drug would be welcomed, such characteristic products of the biotechnology industry as disease resistant crops and more productive farm animals evinced only mild enthusiasm with 53% and 37% of the public, respectively, being strongly in favour.

The issue of releasing engineered organisms was most prominently debated over a bug given the appropriately science fiction title 'Ice Minus'. In 1982, Stephen Lindow and Nickolas Panopoulos of the University of California at Berkeley requested permission to test a new bacterium which would impede frost production on strawberry plants by preventing the formation of ice crystals. Seven months later, a similar request was made by a commercial biotechnology company, Advanced Genetic Systems, for its own strain of the bacterium. Debates with Rifkin's Foundation for Economic Trends and local community activists delayed field testing until 1987. Here, as so often, there was a fundamental disparity between the conceptions of proponents and opposition. The affair's historians, Krimsky and Plough, refer to the disjuncture between a 'technical' and a 'cultural' rationality.[65] Many people felt that scientists were recklessly threatening the environment. An earlier concern about capitalists controlling 'life' was widely replaced by such environmentalist concerns and highlighted in an influential 1988 National Wildlife Federation report, 'Biotechnology and the Environment'.[66]

There have also been more general anxieties about the effect of a new high technology, manipulated by large companies, on agriculture, traditionally – though decreasingly – the work of small family businesses. Jack Doyle, frequently lobbying Congress on behalf of the Environmental Policy Institute, has cited estimates that, by the year 2000, 1% of U.S. farms would account for half the nation's food production.[67] Private companies seemed to be replacing the historically locally accountable agricultural experiment stations and land grant colleges as the basic source of exploitable knowledge. Seeds would become private property at the expense of the self-reliant farmer in the United States and the Third World. Thus, economically as well as technologically, the incorporation of agriculture within the structure of U.S. capitalism would be accelerated.

Despite this consciousness of the economics of agriculture, it may be that in the United States popular anxiety reflects the unpredictable response of a complex natural environment, whereas in Germany, where the memory of eugenics is stronger, there is a greater suspicion of human malice, particularly associated with genetics. While 'biotechnology' was a word hardly known in the United States before 1980 and could not be distinguished from genetic engineering, the German word '*Gentechnologie*' is often used as a concept separate from *Biotechnologie*.[68]

Latent concern was also more decisively marshalled in Germany and Denmark than across the Atlantic. *Gentechnologie* had emerged as an issue in the early 1980s, later than in the United States, just at the time that the nuclear issue was being won.[69] It became the focus of alliances that had already been forged in the debate over nuclear power and which continued to fight over biotechnology in Parliament, where the Greens were becoming a powerful force. Public opinion expressed through elected representatives was reflected in strong parliamentary pressure for general laws. The boundary articulated by Goldscheid, the quality production of people, is still a significant factor in Germany's appreciation of the word.[70] Interestingly, in the course of modern discussions of genetic engineering, Goldscheid's work has been 'rediscovered' in Germany.[71] Even plant biotechnology has been interpreted as the thin end of a wedge ending in the engineering of inheritance.[72] The founding document of the German pressure group

Gen-ethic Netzwerk proclaimed: 'We must develop a new ethic for dealing with our knowledge and cannot entrust this task only to scientists, politicians and so-called experts, nor can we leave it to the mechanisms of the free market and international competition'.[73]

A 1991 survey shows that Germans are systematically more sceptical about the benefits, and worried about the risks, of genetic engineering than the average European.[74] Thus, to the suggestion about applying new methods of biotechnology and genetic engineering to human beings, 46% of West Germans definitely agreed that 'such research may involve risks to human health or to the environment', whereas only 20.8% of British respondents agreed definitely (European Community average 30.7%). A significant proportion of Germans even frown on the development of medicines, against the European trend. In both France and Germany, a Gallup poll revealed that misplaced eugenic applications are a significant concern to more than one-third of respondents.[75] There has been little trust that the new techniques would be used for more worthy ends. Benedikt Härlin, a German Green, reported on behalf of the European Parliament as it discussed the human genome programme in 1989:

> The stress laid on the 'international challenges', on the contribution 'to the development of 'Europe's biotechnology industry' and increasing European industry's ability to compete is completely out of place in this context, for if ever there are universal tasks and challenges calling on moral, safety and financial grounds for international cooperation and a renunciation of competition and exploitation, then research into the genetic basis of mankind is one of them.[76]

Danes expressed even more hostility than Germans to 'genetic engineering', although their attitudes were markedly more positive to 'biotechnology' which was seen as an environmentally benign, green technology. In Denmark, a country in which brewing and enzyme manufacture have long been major industries, 69.1% of respondents thought biotechnology would improve life over the next twenty years, whereas only 44.5% thought the same of genetic engineering. However, elsewhere, biotechnology too was challenged, as the use of knowledge, particularly in agriculture, raised concern. Reflecting on European attitudes in general, Yoxen and DiMartino have diagnosed antagonism to commercial biotechnology as manifestations of, variously, concern over the economic impact of existing agricultural policies and a general 'antimodernism'.[77]

Attitudes in Europe are interestingly indicated by the debates in the European Parliament. Moreover, the interest of the parliament has been a critical factor in the development of biotechnology at the European level, providing a cause for the Continent's powerful and old established socialist tradition and a new and ascendant Green grouping. In reply to the commission's plans to use biotechnology to solve the agricultural crisis, the parliament expressed the anxiety that agricultural life would thereby be transformed. Thus, the power of biotechnology as a social agent, so promoted by enthusiasts in the 1970s, has been used against them in the subsequent decade, as such questions as bST were discussed. It was not just a cultural threat. By making the milk production of cattle even more subject to high technology, bST would reduce the number of cattle needed and threaten further the European family farm. Reacting to the threatening noose of lower subsidies and higher input costs, a 1986 European Parliament discussion emphasized that the parliament 'wishes to maintain family farming as the basis to our agriculture'.[78] From that it concluded that biotechnology had to be integrated into existing measures – 'a purely biotechnological approach must be rejected'.

As the debates over bST and Ice Minus showed, the dialogue between proponents and opposition has been confused by the confluence of pragmatic and fundamentalist arguments. Pollsters have found distrust to be a common result, both in Europe and the United States. According to a recent European Commission poll, industry is trusted as a source of information about biotechnology and genetic engineering by only 6.4% of the European population, about half the number crediting religious authorities on the matter. Nor is the credibility of trade unions or political organizations greater than that of industry. Similarly in the United States, the strongest suspicions were about the authenticity of the sources of information. Small-scale testing is accepted, but not large-scale release, with considerable scepticism about both government and companies as sources of information or assurance of safety.[79]

Thus, the public has been impressed by the significance of the life sciences (even if sometimes negatively) but is dubious about the integrity of companies claiming the right to manipulate them. It is this atmosphere, in which whom to trust has become a central problem, that has emphasized the importance of the regulator.[80]

The regulators

Is there any common ground between the biotechnology portrayed by promoters as just another technology, and the biotechnology identified by a fundamentalist opposition? Krimsky and Plough's diagnosis of the Ice Minus affair highlights the fundamental dilemma to which the strategies of the 1970s led; for the regulators' attempts to adjudicate between anxieties and expectations using what Krimsky and Plough call 'technical' rationality have largely been pragmatic and have often failed to address fundamentalist critiques. In a discussion over a 1986 House hearing, Congresswoman Schneider complained: 'So I am feeling terribly impotent in terms of solving this problem. And I think it needs to be solved – we cannot continue to have the trade association which says everything is fine, and you [Jeremy Rifkin] say everything is horrendous.'[81]

The control of biotechnology has not only been fought over by public and industry. Regulatory conflicts have been exacerbated by conflicts between agencies. Just as departments within national governments have squabbled for the right to promote biotechnology, so the question of whose regulations should apply has been contested to the point of absurdity. In the United States, during the early 1980s, biotechnology had become the victim of bureaucratic battles between existing agencies, that were described as Kafkaesque, in which the right office to stamp the papers was forever elusive. The journal *Bio/technology* dedicated a long quotation from Lewis Carroll to Stephen Lindow and 'all those who have pursued the glories of biotechnology, only to find themselves still standing in the same place after five years of hard running'. – ' "Tut, tut child", said the Duchess. "Everything's got a moral, if only you can find it" '.[82] The moral of course was that traditional categories were being breached as much for corporate culture as for the individual. In an attempt to reduce such problems, the Biotechnology Science Coordinating Committee (BSCC) was established within the U.S. government's Office of Science and Technology Policy in 1985. However, it was described almost immediately as a toothless discussion group by Albert Gore, a leading congressional Democrat, and for this he has been called a 'raging optimist'.[83]

Such tension has resulted in the polarization and politicization of the very word 'biotechnology'. Once governments constructed large organizations around the concept, ideas about it had to be defined as truth, through their assimilation and then dissemination by bureaucracies. Officials and analysts needed to match expenditures across nations and, in creating neat tabular comparisons, became concerned with definitions. But the stakes here have been higher than table building. To civil servants, the definition of biotechnology has become a matter of money and power. The single tag, 'biotechnology', has been used to unify issues as varied as the development of individual companies and of broad industrial sectors, hazard regulation, science policy, and cultural reflection. In discussing departmental treatments of biotechnology, the OTA had to preface each analysis with the department's definition: 'In the current political environment, where promotion of high technology is strongly favored, the definitions used for biotechnology have important ramifications. The terms used to describe biotechnology can affect research funding and the regulatory treatment of potential commercial products'.[84] In Europe, Ken Sargeant has noted, more in sorrow than in anger, 'Most definitions of biotechnology are too restrictive, reflecting narrow perceptions and interests, sometimes as trivial as the desire to influence research funds'.[85] An example of a threatened career lends poignancy to this generalization. A researcher, Professor Gary Strobel, used to working to USDA rules and definitions of genetically engineered organisms was prosecuted when, following those conventions, he inadvertently flouted the newly adopted broad definition of genetic engineering as anything beyond traditional breeding practices.[86]

One solution has been to try to disentangle the issues, dismantling the very concept of biotechnology. 'This word has now become a significant millstone around the neck of both the industry and the government', argued David Kingsbury of the NSF. With Henry Miller and Frank Young of the FDA, he urged the removal of the word from the regulatory vocabulary.[87] On the other hand, the scope of benefits that might outweigh costs continues to make the open-ended concept appealing. At a 1987 OECD-sponsored workshop, 'several participants (France, EEC) refused to have the term abandoned completely; policy makers had to learn to cope with its negative – as well as its positive – implications, and a very broad definition could also be useful as it could lead to the "banalisation" of biotechnology in the eyes of politicians and the public'.[88] This attempt to undermine fundamentalist opposition to biotechnology is a rare sign in public, of a process widespread in private, through which the political and commercial significance of particular definitions are debated before propagation.

The implication that biotechnology presented fundamentally new problems was rejected by the Republican administration of President Reagan, which had been committed to minimizing new legislation and organization. So, ironically, while for venture capitalists recombinant DNA technology has transformed the world, as far as regulation is concerned the new methods have been construed within the context of established practice. Laboratory modification of genes can be seen as merely an alternative method to traditional breeding or drug design. The approach has been successful, and from the mid-1980s the regulation of U.S. biotechnology became more routine.[89]

Despite a lack of popular enthusiasm for uncontrolled release, regulators in the United States therefore seemed to have been successful in finding a consensus that biotechnology, properly controlled by responsible scientists, could be seen as a latter-day information technology. Jeremy Rifkin's interests moved on, and by 1990 the BSCC was seen to be superfluous. It was replaced by the Biological Research Subcommittee whose purpose had less to do with the by now accepted patterns of regulation than with exploring nonbiomedical dimensions of biotechnology research. Rather than a regulatory role, the need for stimulus was seen.[90] Concern over the need to control runaway science was being replaced by the urgency of finding new scope for technological skills.

In Europe, the promoters have been less successful in suggesting that biotechnology is just one more high technology. Genetic engineering is the target of very specific rules in which the process and not just the products are spotlighted. During the early 1980s, interest was focused by the socialist member of the European Parliament Phili Viehoff. She promoted the concept of biotechnology as a socially steerable technology, capable of vast benefits as long as applications were guided towards socially useful applications, such as the so-called orphan drugs including malaria vaccines. Even though a report she issued in early 1986, through the European Parliament, broadly supported the commission's approach to biotechnology, it emphasized the need for analysis of the social consequences of biotechnology.[91]

Legislation in the early 1980s was even more restrictive than in the United States with controls at both national and local levels. In several countries, the power of the Socialists and the Greens has been sufficient to establish tight regulation of the production of recombinant DNA products. Denmark, for instance, in its 1986 Environment and Gene Technology Act prevented the deliberate release of organisms modified through recombinant methods, gene deletion, or hybridization in all but special cases.

In Germany, the U.S. technology assessment process was repeated, and between 1984 and 1987 the costs and benefits were weighed by the Bundestag-sponsored Commission of Inquiry on the Opportunities and Risks of Genetic Technology.[92] As in the United States, this extensive process seems to have quieted media interest in the dangers of biotechnology.[93] Nonetheless, Germany made experimental release subject to local vote as well as to investigative scrutiny. When, in 1990,

European Community–wide frameworks for the control of work involving contained and deliberate release of genetically engineered organisms were agreed on, the control of release was considerably more rigourous than U.S. practice, if less restrictive than Danish legislation.[94]

The social threat perceived to be caused by the use of bST has encouraged the parliament to suggest a fourth requirement to the three criteria traditionally required for the release of new materials. Hitherto, new products have had to meet the requirements of safety, quality, and efficacy. However, at the time of this writing in 1991, the extra requirement of 'social need' is still being debated. Its enactment would indicate, in a way unexpected by early enthusiasts, recognition of the very special power seen to be possessed by biotechnology, albeit as a destabilizing force.

The distinction between European and U.S. decisions can be seen in the area of patenting, a major component of regulation. Such patents are valuable and the opportunity to protect inventions is an important economic inducement. At the same time, the question of whether we can patent life directly confronts our willingness to integrate biology with engineering. Patent offices, required to mediate between such conflicting interests, have behaved in a variety of ways. The 1980 take-off can be directly attributed to the U.S. decision to allow a patent on Cohen and Boyer's technique of gene splicing and on Chakrabarty's new organism. In the case of *Diamond* v. *Chakrabarty* (16 June 1980), the U.S. Supreme Court agreed that since a new bacterium containing two energy-producing plasmids did not occur in nature, it could be patented. Seven years later, it agreed that mammals could be patented.

This liberalism is not new: U.S. rules have traditionally been the most open to such possibilities. As early as 1873, Pasteur himself patented a sterile yeast in the United States. By contrast, before World War I, German patent office law explicitly excluded the treatment of living beings and those processes 'whose results are primarily based upon the independent functioning of animate nature, e.g. so-called agriculture cultivation processes, process for the breeding of plants or animals or for the grooming and training of animals.' This was weakened in the early 1920s as German rulings allowed microbiological innovations to be patented. Protection for plant varieties has steadily been strengthened. However, animals were long seen as beyond this ruling.

In the United States, this changed when Harvard University won a 1988 patent for a transgenic mouse. For several years the patent

Transgenic mice (patented 1988) from the collections of the Science Museum. Courtesy Trustees of the Science Museum.

was refused in Europe. Article 53(b) of European patent law states that 'European patents shall not be granted in respect of . . . plant or animal varieties or essentially biological processes for the production of plants or animals; this provision does not apply to microbiological processes of the products thereof'.[95] Goldscheid's vision of people or even animals as quality products raises too many anxieties. Although in 1992 the 'Harvard' mouse was granted a European patent, this was a special exemption. Patent examiners were quoted as wishing that a transgenic cat would eat the transgenic mouse, for in European patent law, biotechnology is still a technology unlike any other.[96] In general, life forms are seen as too special to become the property of a single organization or individual.

To enthusiasts, such public caution in the face of technological progress has been hard to fathom. However, in the case of the human genome programme, since 1986 the fundamental significance of the issues involved has been confronted in Europe and the United States much more directly than in biotechnology generally. Understanding of what programmes the development of the human being has grown. It is now beginning to offer the practical promise of solving a new range of human diseases and the range of inherited disabilities, providing outstanding new vistas with potent significance to each member of the public. At the same time, biotechnology's methods, genetic analysis of the individual, and the modification of human cells are perilously close to the perceived life–technology frontier. Questions of what is legitimately patentable and who could benefit from genetic knowledge, be it prospective parents, employers, employees, or insurance companies, have become pressing. Significant proportions of the budgets for U.S. and European human genetics have been assigned to questions of ethics.

Conclusion

It is too soon to record the outcome of the debates of the 1980s. Yet as the participants negotiated the problems of a technology of biology, it is now clear they were reflecting old and profound cultural issues. A fascination with the parallels between animals and machines has characterized biotechnology throughout the century. What is the safe boundary between humanity and industry? The specificity of current issues has of course provided a sharper focus today, and the involvement of many more people than in the past has changed the character of discussion. Yet it is still true that technological and scientific breakthroughs have occurred within contexts rich with preexisting interpretations and anticipations. Biotechnology, it could be said, describes a pattern of expectations. National differences, here, have meant that even in the era of instant communications, it still has different meanings across the world.

Regulators have had to contend with differing emphases within individual societies. To promoters, the distinctiveness of this link between biology and commerce is evaporating.[97] As expectations are framed by those with more business acumen than in the past, practical possibilities that will help the resolution of urgent problems in wealth creation, agriculture, and medicine open up. Biotechnology connotes a large range of products and genres of expertise linked variously by the technical processes of fermentation, by the biological processes of gene change, or by marketing and industrial styles. The threats are science fiction. Hopefully the profits are imminent, and the companies subject to regulation.

To many members of the public, biotechnology is different from any other technology, because of its distinctive threats to the environment and to culture. That challenge to traditional agriculture which is such an attraction to promoters can be peculiarly threatening. Nonetheless, biotechnology has also come to seem, in its distrusted sources of control, akin to the work of other unpopular branches of manufacturing such as the chemical industry. The distinction between chemical technology and biotechnology, carefully nurtured in the 1970s, has been lost.

How typical is this subject, then, of modern high technologies in general? The model of a technology that sits between engineering and science, and whose lineage can be seen in the shifting balance of forces between the two, may be applicable elsewhere, certainly. Responding to urgent problems, governments encouraged a range of new technologies during the 1980s, with corresponding interrelationships between industry, research policy, and regulation. The control of information and of atomic power, to take two other examples of controversial technologies, also have long ancestries with a cultural significance beyond technology.[98] Both have stimulated repeated proclamations that they would change the world. However, the suggestion here has been that we must be cautious about generalizing too readily about high technologies. Modern they may be, but they each have their own individual patterns of development and cultural significance. Biotechnology has been distinguished by the wealth of sophisticated expectations that have, time and again, preceded technical change. Medicine and agriculture, the two sectors likely to be most immediately affected, have special places in our culture. So, while by the 1980s, it might have seemed that, finally, bureaucracies had made their peace and that biotechnology was ceasing to signify the interface between two professions, the frontier between biology and engineering had not lost its importance. Most recently, the debate over the regulation of biotechnology has highlighted the juxtaposition of life and commerce. A 1989 Danish discussion of human genome mapping concluded:

> Some of the experts hope and believe that the knowledge thus produced will be used only for good purposes, such as a better understanding of diseases and thereby better possibilities for treatment and prevention. But others are more sceptical and believe that it will always be more difficult to manage and control.[99]

Epilogue

Today, at a time when technology enables the realization of old dreams, ironically their past is rarely recalled. Nonetheless, it is important to remember that the broader meanings of biotechnology have been profoundly considered over the past century. Though many of the hopes raised in the past have, so far, been disappointed and have faded away, other visionary interpretations, products, and processes have appeared. Even now, there is widespread discussion of the prospects of understanding the map of the human genome. Ever more urgently, the public has been urged to relinquish its scepticism and caution over a new technology. Whether it can be convinced that this power will be handled responsibly has become a major issue for the 1990s.

The survival of deep suspicions should not be surprising. Public anxiety about the correctness of engineering organisms, plants, and people has too often been answered only in terms of the demands of international competition and demonstrations of the 'objectively' low risks. When parallels have been drawn with fears of disaster caused by the chemical industry, enthusiasts have been ready to answer the public by emphasizing the economic attractions of drawing the comparison with information technology. In the debate between fundamentalist opposition concerned with the status of nature, and pragmatic proponents of a new industry, the resource provided by three generations of thinkers considering the nature of technology has too often been neglected.

Hopes for the more attractive technology which was such an aspiration for Goldscheid, Francé, Geddes, and more recently, Hedén now sometimes seem to have receded. Those visions appear as fossils of a remote age, curious more for their chronological proximity than for any continuous lineage to the dynamic modern industry. But the careers and influence of such pioneering thinkers as Julian Huxley spanned the generations from the beginning of the century to our own time. This book has shown that there are intellectual and institutional links between the prewar era, that time in the 1950s when biological engineering was accepted as covering all aspects of the relationship between biology and engineering, and the current moment in modern biotechnology.

Naturally, arguments of the past cannot, and should not, determine how disputes should be resolved, or biotechnology regulated. Their example could, though, provide a resource for discussions about the fundamental virtues and dangers of biotechnology and add a dimension to those discussions of short-term benefits which may command action but leave the taste of distrust. Biotechnology need not be explained just by its 'normality' or defended by 'banalization'. It has been an unusual field in its anticipated relevance to key issues in the twentieth century. Among the key participants identified by this book were two active politicians, Israel's Chaim Weizmann and Hungary's Karl Ereky. They were looking forward to new kinds of industry, suited to their societies. More recently, in the 1960s, biotechnology was proposed as an appropriate technology for developing nations, congruent with their cultures, resources, and needs. Perhaps those arguments were optimistic, even naïve, but they addressed issues fundamental to the place of this science in the world.

Past discussions are still embodied in the diversity of definitions of biotechnology we have inherited. Embarrassing as the apparent vagueness is when it comes to tabular comparisons, the breadth can enable a gaze on to strange, but possibly exciting vistas. However distant reflections on engineering, on human biology, or on alternative technologies may appear from the current emphasis on recombinant DNA techniques and pharmaceutical production, they each relate to conceptions intended to help humanity forward. The distinctive responses evoked by biotechnology in the United States, or in Germany, Denmark, Japan, Britain, and elsewhere, are not mere bureaucratic anomalies. In those various interpretations, we have seen reflections on that contested and mobile frontier which, for a century, has lain between life and engineering.

Notes

INTRODUCTION

1 Anneke M. Hamstra and Marijke H. Feenstra, *Consument en Biotechnologie: Kennis en meninvorming van consumenten over biotechnologie,* Report no. 85 (The Hague: Instituut voor consumentenonderzoek, 1989).
2 These scenes are on a wall relief from an Old Kingdom tomb, now in Leiden. They are discussed by Pierre Montet, *Scènes de la vie privée dans les tombeaux égyptiens* (Strasbourg: University of Strasbourg, 1925), pp. 245–52. I am grateful to Carol Andrews of the British Museum for her elucidation of the Egyptology.
3 For the history of the concept of 'biology', see Joseph A. Caron, ' "Biology" in the Life Sciences', *History of Science* 26 (1989):223–68. He suggests that T.G.A. Roose was the first to use the word in 1797, and there were several other possibly independent coinages, by Treviranus and Lamarck, within a few years. However, it only acquired its modern connotations in T.H. Huxley's Fullerian Lectures, 'The Principles of Biology', at London's Royal Institution in 1858.
4 For reflections on the fascination of the concept of a cyborg, see Donna Haraway, *Simians, Cyborgs and Women: The Reinvention of Women* (London: Free Association Books, 1991), particularly pp. 149–91, 'The Cyborg Manifesto'. For a general background of evolving thought about organisms and machines, see David F. Channell, *The Vital Machine: A Study of Technology and Organic Life* (Oxford: Oxford University Press, 1991).
5 Spencer R. Weart, *Nuclear Fear: A History of Images* (Cambridge, Mass.: Harvard University Press, 1988).
6 U.S. Congress, Office of Technology Assessment, *New Developments in Biotechnology: U.S. Investment in Biotechnology – Special Report,* OTA-BA-360 (Washington D.C.: U.S. Government Printing Office, 1988), pp. 28–29.

CHAPTER 1. THE ORIGINS OF ZYMOTECHNOLOGY

1 *The Zymotechnic Institute* (Chicago: Zymotechnic Institute, 1891), p. 1.
2 Alcohol distillation had been known in the West in the twelfth century, but had only become widely disseminated at the end of the thirteenth century. See Robert P. Multhauf, *The Origins of Chemistry* (New York:

Franklin Watts, 1967), pp. 110–11; Lynn Thorndike, *History of Magic and Experimental Science,* vol. 3 (New York: Columbia University Press, 1934), pp. 347–69. This era is recalled by our modern description of 'spirits'.

3 Emil Christian Hansen, *Practical Studies in Fermentation,* trans. Alex K. Miller (London: Spon, 1896), p. 272.

4 See Gerald L. Geison, 'Pasteur, Roux and Rabies: Scientific *versus* Clinical Mentalities', *Journal of the History of Medicine and Allied Sciences* 45 (1990):341–65.

5 Bruno Latour, *Microbes: Guerre et paix; suivie de irreduction* (Paris: A.-M. Métilié, 1984).

6 E. W. Hulme, *Statistical Bibliography in Relation to the Growth of Modern Civilization* (London: Butler & Tanner, 1923), p. 45. This shows that, out of eighteen sciences, chemistry with its 108,982 author entries was the third largest in the first thirteen years of the twentieth century, exceeded only by physiology and zoology.

7 The idea of biology as an ensemble is developed by Joseph Caron in his article, ' "Biology" in the Life Sciences', *History of Science* 26 (1989):223–68.

8 I am grateful to Dr L. Siorvanes of Kings College London for the opportunity to discuss the classical roots of *zymotechnia.*

9 The classic work on Stahl's chemistry is still Hélène Metzger's *Newton, Stahl, Boerhaave et la doctrine chimique* (Paris: Alcan, 1930). See Gad Freudenthal, ed., 'Etudes sur Hélène Metzger', *Corpus des oeuvres de philosophie en langue francaise* 8/9 (1988). Also see Wilhelm Strube, *Die Chemie und Ihre Geschichte* (Berlin: Akademie Verlag, 1974).

10 For a good treatment of the Stahlians, see Michael Engel, *Chemie im achtzehnten Jahrhundert: Auf dem Weg zu einer internationalen Wissenschaft – Georg Ernst Stahl (1659–1734) zum 250 Todestag,* Exhibition in the Staatsbibliothek Preußischer Kulturbesitz, 29 May to 7 July 1984, Ausstellungskataloge 23 (Berlin: Staatsbibliothek Preußischer Kulturbesitz, 1984). For Neumann as protégé to Stahl, see Carl Hufbauer, *The Formation of the German Chemical Community (1720–1795)* (Berkeley: University of California Press, 1982), pp. 10–11 etc. Also see A. Schrohe, *Aus der Vergangenheit der Gärungstechnik und verwandter Gebiete* (Berlin: Parey, 1917), pp. 105–19 and 164–65.

11 Mary Shelley, *Frankenstein (Or, the Modern Prometheus)* (First published 1818; Clinton, Mass: Airmont, 1963), p. 50. On Frankenstein and science, see S.H. Vaspinder, *The Scientific Attitudes in Mary Shelley's Frankenstein* (Ann Arbor, Mich.: UMI Press, 1976), pp. 69–73, in which he discusses this quotation and the possible links to Davy. I am grateful to Dr Judith Field for this reference.

12 N.D. Jewson, 'The Disappearance of the Sick-Man from Medical Cosmology, 1770–1870', *Sociology* 10(1976): 225–44.

13 The precise significance of this experiment has long been contentious. The debate is summarized by J.H. Brooke, 'Organic Chemistry' in *Recent Developments in the History of Chemistry,* ed. C.A. Russell (London: Royal Society of Chemistry, 1985), pp. 107–9.

14 W.H. Brock, 'Liebigiana: Old and New Perspectives', *History of Science* 19 (1981): 201–18.

15 E. Glas, *Chemistry and Physiology in Their Historical and Philosophical Relations* (Delft: Delft University Press, 1979); F.L. Holmes, 'Introduction', reprint edition of Liebig's, *Animal Chemistry* (New York: Johnson Reprint, 1964), pp. vii–cxvi, for Liebig's physiological ideas; also Timothy O. Lipman, 'Vitalism and Reductionism in Liebig's Physiological Thought', *Isis* 58 (1967): 167–84; and Otto Sonntag, 'Religion and Science in the Thought of Liebig', *Ambix* 24 (1977): 159–69. For Liebig's vision of chemistry as the philosopher's stone, see his *Familiar Letters on Chemistry in Its Relations to Physiology, Dietetics, Agriculture, Commerce and Political Economy,* 3d ed. (London: Taylor, Walton & Mabberly, 1851), p. 46.

16 See A.K. Balls, 'Liebig and the Chemistry of Enzymes and Fermentation', in *Liebig and After Liebig,* ed. F.R. Moulton, AAAS pub. 16 (Washington D.C.: AAAS, 1942), pp. 30–39.

17 J.S. Muspratt and A.W. Hoffman, 'On Toluidine', *Memoirs and Proceedings of the Chemical Society of London* 2 (1843–45), p. 368.

18 See Emil Fischer, 'Synthetical Chemistry in its Relation to Biology', *Journal of the Chemical Society* 91ii (1907):1749–65.

19 'Die synthetische organische Chemie ist im wesentlichen erschöpft. [D]ie Gelehrten wenden sich der Erforschung der Naturprodukte zu;' This is quoted from Eric Elliott, 'The IG Farbenindustrie: Is There Science *Here* for the Historian of Science?' IG Farben Study Group, *Newsletter* no. 2. Its source is given as: Abschrift of letter to the Ministerium der geistl. u. Unterrichts-Angelegenheiten from I G Farbenindustrie AG (gez Bosch, Kurt H Meyer). This is in the former Zentrales Staatsarchiv (DDR), Dienstelle Merseburg, Ministerium der geistl. Unterrichts- und- Medizinal-Angelegenheiten. 'Generalia: Wissenschaften. Wissenschaftlichen Sachen. Abt. XI I. Teil 20. Der Verein zur Wahrung der Interessen der chemischen Industrie Deutschlands. Bd. 1 1886–1927' 427^{v}–473. The letter continues 'auch in der Industrie spielt das synthetische Auffindung neuer Farbstoffe und der gleichen nicht mehr die Rolle wie früher.' The letter argues that in industry more physical chemists are required.

20 Synthetisch der Kaffee, Synthetisch der Wein,
Die Milch und die Butter, das Bier obendrein
Natürliche Nahrung, die find't man fast nie
Der Teufel, der hol' die synthetische Chemie.

Th. Curtius, 'Wilhelm Koenig', *Berichte der Deutschen Chemischen Gesellschaft* 45iii (1912), p. 3792. I am very grateful to Mr. Martin Bud for the translation.

21 Robert E. Kohler, *From Medical Chemistry to Biochemistry: The Making of a Biomedical Discipline* (Cambridge University Press, 1982). See also W. Coleman, 'The Cognitive Basis of the Discipline: Claude Bernard on Physiology', *ISIS* 76 (1985):49–70; T. Lenoir, 'Science for the Clinic: Science Policy and the Formation of Carl Ludwig's Institute in Leipzig', in *The Investigative Enterprise: Experimental Physiology in Nineteenth-Century Medicine,* ed. W. Coleman and F.L. Holmes (Berkeley: University of California Press, 1988), pp. 139–78.

22 Gottfried Reinhold Treviranus, *Biologie oder Philosophie der lebenden Natur für Naturforscher und Aertzte* (Göttingen: Rower, 1802).

23 Hans Querner, 'Probleme der Biologie um 1900 auf den Versammlungen der Deutschen Naturforscher und Aertzte', in *Wege der Naturforschung,*

1822–1972 im Spiegel der Versammlungen Deutscher Naturforscher und Aertzte, ed. Hans Querner and Heinrich Shipperges (Berlin: Springer, 1972), pp. 186–201.

24 Julius Wiesner, *Die Rohstoffe des Pflanzenreiches,* 2d ed., 2 vols. (Leipzig: Engelmann, 1900), vol 1., p. 1. – 'Wissen allein ist nicht der Zweck des Menschen auf der Erde. Das Wissen muß sich im Leben auch betätigen'. For a modern treatment of late-nineteenth-century German botany, see Eugene Cittadino, *Nature as the Laboratory: Darwinian Plant Ecology in the German Empire, 1880–1900* (Cambridge University Press, 1990).

25 Julius Wiesner, 'Ihrem rohen Wesen nach ist die Rohstofflehre die Vermittlerin zwischen der organischer Naturgeschichte und der Technik, wie etwa die chemische Technologie die Vermittlerin zwischen der Chemie und den Gewerben ist.' In 'Bedeutung der technischen Rohstofflehre (techn. Waarenkunde) als selbstständige Disciplin und über deren Behandlung als Lehrgegenstand an techn. Hochschulen', *Dinglers Polytechnisches Journal* 237 (1880), p. 403

26 For this observation, I am indebted to Dr Bernadette Bensaude.

27 Cited in René Vallery-Radot, *The Life of Pasteur,* trans. R.L. Devonshire (London: Constable, 1919), p. 374.

28 Michael Tracy, *Agriculture in Western Europe: Crisis and Adaptation since 1880* (London: Jonathan Cape, 1964).

29 On Danish agriculture, see Einar Jensen, *Danish Agriculture – Its Economic Development: A Description and Economic Analysis Centering on the Free Trade Epoch, 1870–1930* (Copenhagen: Schultz 1937).

30 Tracy, *Agriculture in Western Europe,* p. 104.

31 The figures are reproduced from the 1900 U.S. census by John Just, 'The Commercial Utilization of Milk Waste and the More Recent Products of Milk in a Dry Form', in 5th International Congress for Applied Chemistry, Berlin, 2–8 June 1903, *Bericht* (Berlin, Deutsche Verlag, 1904), vol 3, pp. 870–91. The classic work on pig feeding is George Rommel's, 'The Hog Industry: Selection, Feeding and Management – Recent American Experimental Work, Statistics of Production and Trade', Bureau of Animal Husbandry, *Bulletin 47* (Washington, D.C.: Government Printing Office, 1904).

32 In general my treatment of the German agricultural colleges and stations follows the recent extensive analysis of Ursula Schling-Brodersen, *Entwicklung und Institutionalisierung der Agrikulturchemie im 19. Jahrhundert: Liebig und die landwirtschaftlichen Versuchsstationen,* Braunschweiger Veröffentlichungen zur Geschichte der Pharmazie und der Naturwissenschaften, vol. 31 (Braunschweig: University of Braunschweig, 1989).

33 Herbert Pfisterer, *Der Polytechnische Verein und sein Wirken im vorindustriellen Bayern (1815–1830),* Miscellanea Bavarica Monacensia, vol. 45 (Munich: Stadtarchivs München, 1973). I am grateful to Dr Ernst Homburg for pointing out this book to me.

34 Mark R. Finlay, 'The German Agricultural Experiment Stations and the Beginnings of American Agricultural Research', *Agricultural History* 62 (1988): 41–50.

35 On agricultural chemistry in America, see Margaret Rossiter, *The Emergence of Agricultural Science in America: Justus Liebig and the Americans, 1840–1880* (New Haven, Conn.: Yale University Press, 1975).

36 Sir John E. Russell, *A History of Agricultural Science in Great Britain* (London: Allen & Unwin, 1966), p. 221.
37 Charles Rosenberg, 'Science, Technology and Economic Growth: The Case of the Agricultural Experiment Station Scientist, 1875–1914' in *Nineteenth-Century American Science: A Reappraisal,* ed. George F. Daniels (Evanston, Ill.: Northwestern University Press, 1974), pp. 181–209.
38 'Bierproduktion in den verschiedenen Ländern des Kontinents und in Nord Amerika.' *Bayerische Bierbrauer* 19(1884) vol. 7 of new series, p. 342.
39 Mikuláš Teich, 'Science and the Industrialisation of Brewing: The German Case', presented to the conference entitled 'Biotechnology: Long Term Development', The Science Museum, London, February 1984.
40 Peter Mathias, *The Brewing Industry in England 1700–1830* (Cambridge University Press, 1959), p. 557.
41 On the shifting leadership within world brewing and the role of science in German brewing see Kristoff Glamann, 'The Scientific Brewer: Founders and Successors During the Rise of the Modern Brewing Industry', in *Enterprise and History: Essays in Honour of Charles Wilson,* ed. D.C. Coleman and Peter Mathias (Cambridge University Press, 1984), pp. 186–98.
42 See Robert Bud and Gerrylynn K. Roberts, *Science Versus Practice: Chemistry in Victorian Britain* (Manchester: Manchester University Press, 1984), pp. 47–51.
43 For the development of brewing technology in Bohemia, I am indebted to Soně Strbáňová, 'On the Beginnings of Biochemistry in Bohemia', *Acta Historiae rerum naturalium nec non technicarum* Special issue 9 (1977): 149–221.
44 L. Aubry, 'Hofrat Dr Carl Lintner', *Zeitschrift für das gesamte Brauwesen* 23 (1900): 93–96.
45 Carl Lintner, 'C.J.H. Balling's Leben und Wirken', *Bayerische Bierbrauer* 1 (1866): 29–35, 47–49, 62–66.
46 On Delbrück and his institute, see F. Hayduck, 'Max Delbrück', *Berichte der Deutschen Chemischen Gesellschaft* 53i (1920): 48A–62A.
47 F. Hayduck, 'Das Institut für Gärungsgewerbe in Vergangenheit und Zukunft', in *Das Institut für Gärungsgewerbe und Stärkefabrikation zu Berlin* (Berlin: Paul Parey, 1925), pp. 1–15 especially p. 4.
48 This famous quotation is cited in René Dubos, *Louis Pasteur: Freelance of Science* (New York: Scribner, 1976), pp. 67–68.
49 ' "Mit dem Schwerte der Wissenschaft, mit dem Panzer der Praxis," so wird Deutsche Bier die Welt erringen', in Max Delbrück, 'Ueber Hefe und Gärung in der Bierbrauerei', *Bayerische Bierbrauer* 19 (1884), p. 312. For Delbrück's respect for the '*Technologen*', see H. Dellweg, 'Die Geschichte der Fermentation – Ein Beitrag zur Hundertjahrfeier des Instituts für Gärungsgewerbe und Biotechnologie zu Berlin', in *100 Jahre Institut für Gärungsgewerbe und Biotechnologie zu Berlin, 1874–1974: Festschrift* (Berlin: Institut für Gärungsgewerbe und Biotechnologie, 1974), pp. 17–41.
50 See e.g., M. Delbrück and A. Schrohe, eds., *Hefe, Gärung und Fäulnis* (Berlin, Paul Parey, 1904); Schrohe, *Aus der Vergangenheit der Gärungstechnik und verwandter Gebiete* (Berlin: Paul Parey, 1917); and E. Borkenhagen, 'Gesellschaft für die Geschichte und Bibliographie des Brauwesens' in *100 Jahre Institut für Gärungsgewerbe und Biotechnologie zu Berlin,* pp. 245–52.

51 See Thomas D. Brock, *Robert Koch: A Life in Medicine and Bacteriology* (Madison: Science-Tech Publishers, 1988).
52 See Robert E. Kohler, 'The Reception of Eduard Buchner's Discovery of Cell-Free Fermentation', *Journal of the History of Biology* 5 (1972), 327–53.
53 In 1889, *Zymotechnisk Tidende* was retitled *Zymotechnisk Tidsskrift*. For Jørgensen, see his obituary by J. Blom-Björner, *Journal of the Institute of Brewing* 32 (1926): 198.
54 E. C. Hansen, *Practical Studies in Fermentation*, trans. A.K. Miller (London: Spon, 1896), p. 65.
55 See [Alfred Jørgensen] 'Alfred Jørgensens Gjaeringsfysiologiske Laboratorium', *Zymotechnisk Tiddsskrift* 19 (1903): 80 and his 'Ansættelser fra Laboratoriet i sidste Semester', *Zymotechnisk Tidsskrift* 23 (1907): 90.
56 Hansen, *Practical Studies in Fermentation*, p. 272.
57 See Max Henius Memoir Committee, *Max Henius: A Biography* (Chicago: Privately printed, 1936). Hansen, *Practical Studies on Fermentation*, pp. 18, 71.
58 John P. Arnold and Frank Penman, *History of the Brewing Industry and Brewing Science in America: Prepared as a Memorial to the Pioneers of American Brewing Science, Dr John E. Siebel and Anton Schwartz* (Chicago: Privately printed, 1933), pp. 15–22.
59 [John E. Siebel], 'The Zymotechnic College. A School for Brewers, Distillers, Maltsters, Wine and Vinegar-makers', *American Chemical Review and Zymotechnic Magazine* 4 (1884), p. 193.
60 The advertising brochure, 'The Zymotechnic Institute', begins with a comparison between agricultural experiment stations and fermentation research centres, subsidized by foreign governments, quoted in part at the head of this chapter.
61 Schwartz averaged 250 members and about 30 students per course. From an untitled manuscript, undated but c. 1901, in support of a petition for incorporation in that year. In the possession of Mr W. Siebel, I am grateful for his assistance.

CHAPTER 2. FROM ZYMOTECHNOLOGY TO BIOTECHNOLOGY

1 '*Auf Grund des gleichen Gedankenganges weist der Verfasser alle die Arbeitsvorgänge, bei denen aus den Rohstoffen mit Unterstützung lebender Organismen Konsumartikel erzeugt werden, dem Gebiete der Biotechnologie zu.*' Karl Ereky, *Biotechnologie der Fleisch-, Fett- und Milcherzeugung im landwirtschaftlichen Großbetriebe* (Berlin: Paul Parey, 1919), p. 5.
2 This distinction is laid out quite explicitly in the classic text of chemical engineering, George E. Davis, *Handbook of Chemical Engineering* (Manchester: Davis Bros., 1901–2), vol. 1, p. 4.
3 Max Delbrück, 'Ueber Hefe und Gärung in der Bierbrauerei', *Bayerische Bierbrauer* 19(1884): p. 304, 'Die Hefe ist eine Arbeitsmaschine, wenn ich mich so ausdrücken darf'.
4 G.C. Ainsworth, *Introduction to the History of Mycology* (Cambridge University Press: 1976), pp. 289–92. Also [F. Král], *Král's Bacteriologisches Laboratorium: Der gegenwärtige Bestand der Král'schen Sammlung von Mikroorganismen, März 1904* (Prague: Privately printed, 1904).

5 Paul Lindner, 'Die botanische und chemische Charackerisierung der Gärungsmikroben und die Notwendigkeit der Errichtung einer biologischen Zentrale', *Seventh International Congress of Applied Chemistry*, Section Vib, 'Fermentation' (London: Partridge & Cooper 1910), pp. 169–72.
6 [Paul Lindner], 'Förderung eines Institutes für Erforschung technisch-wichtiger Mikroben in England', *Zeitschrift für technische Biologie* 8(1920): 64–67.
7 W. Chaston Chapman, 'The Employment of Micro-organisms in the Service of Industrial Chemistry: A Plea for a National Institute of Industrial Microbiology', *Journal of the Society of Chemical Industry* 38 (1919): 282T–86T.
8 David L. Hawksworth, 'The Commonwealth Mycological Institute (CMI)', *Biologist* 32 (1985): 7–12.
9 Otto Rahn, 'Theoretische Bakteriologie', *Naturwissenschaften* 9(1921): 374–76.
10 Ib Gejl and Povl Vinding, 'Gustav Hagemann', *Dansk Biografisk Lexicon* 5 472–76.
11 H. Munch-Petersen, ed., *Aarbog for Københavns Universiteter Kommunitet og den polytekniske Læreanstalt, indeholdende Meddelelser for det akademiske Aar 1907–1908* (Copenhagen: Universitetebogytrykkeriat, 1912), p. 367, 'Oprettelse af en fast Lærerstilling i Gæringsysiologi og landbotanisk Kemi'. I am grateful to Professor O.B. Jørgensen of the Danish Technical University for pointing out this passage to me.
12 O. Rode, *Optegnelser efter Prof. Dr. Phil Orla-Jensens: Efteraars Forelæsninger over Bioteknisk Kemi* (Copenhagen: Det Private Ingeniørsfonds forlag, 1915).
13 S. Orla-Jensen, *Lidt anvendt Filosofi* (Copenhagen: Det Schønbergske Forlag, 1934), p. 6.
14 'E. A. Siebel Co.', *Western Brewer and Journal of the Barley, Malt and Hop Trades* (January 1918): 25.
15 On Chicago's Bureau of Bio-Technology, see John P. Arnold and Frank Penman, *History of the Brewing Industry and Brewing Science in America: Prepared as a Memorial to the Pioneers of American Brewing Science, Dr John E. Siebel and Anton Schwartz* (Chicago: Privately printed, 1933), pp. 15–22, though it is incorrect in implying that the Bureau of Biotechnology was founded in 1917. There was no reference to the Bureau of Biotechnology in the article on Siebel, 'E.A. Siebel Co.' The Chicago bureau was only one of several similar enterprises under one roof promoted by E.A. Siebel and was seemingly the least important; it had no entry in the 1930s Chicago telephone directory and had a low, though visible, profile in the company's publicity, for instance in E.A. Siebel and Company and Siebel Laboratories, Inc. 'Achievement: Yesterday. Today. Tomorrow', n.d. On Emil Siebel, see *Who's Who in Chicago* (Chicago: A.N. Marquis, 1939). I am grateful to Mr W. Siebel of J.E. Siebel's Sons & Co. Inc. for assistance in tracing his uncle's firm. I am grateful too for the help of Leslie Ann Schuster of History Works Inc. for her help.
16 E.A. Siebel and Company and Siebel Laboratories, Inc., 'Achievement: Yesterday. Today. Tomorrow', n.d. Chicago.
17 The review, from the *Brewers Journal* of 15 December 1920, was reprinted

in 'Some Press Comments', *Bulletin of the Bureau of Biotechnology* 1 (1921): 83.

18 See for example F.A. Mason, 'Microscopy and Biology in Industry', *Bulletin of the Bureau of Biotechnology* 1 (1920): 3–15.

19 See E. Andreis, 'Il "Bureau" per le ricerche biologiche e l'industria delle pelli', *La Conceria* 29(1921): 164. I am grateful to Professor Luigi Cerruti for this reference and idem, 'La biotecnologia e l'industrii dei Cuoi', *Rivista italiana del Cuoio dei Pellami,* 1921, which I have not seen but is cited in 'Some Press Comments', *Bulletin of the Bureau of Biotechnology* (1921): 84.

20 John Lukacs, *Budapest 1900: A Historical Portrait of a City and Its Culture* (New York: Weidenfeld & Nicolson, 1988), p. 63.

21 On Budapest, see Lukacs, *Budapest 1900;* Péter Gunst and Lászlo Gaál, *Animal Husbandry in Hungary in the 19th–20th Centuries* (Budapest: Akadamiai Kiado, 1977), p. 46; Antal Voros, 'The Age of Preparation: Hungarian Agrarian Conditions between 1848–1914', in *The Modernization of Agriculture: Rural Transformation in Hungary, 1848–1975,* ed. Joseph Held (Boulder, Colo.: East European Quarterly Press, 1975), p. 112.

22 Ereky, *Biotechnologie der Fleisch-, Fett- und Milcherzeugung im landwirtschaftlichen Großbetriebe.*

23 See György Ranki et al. (eds.), *Magyarorszag története,* vol. 8 (Budapest: Akademai Kiado, 1976). I am grateful to Ms Judit Brody, formerly of the Science Museum Library, and Dr Ferenc Szabadvary, of the Hungarian Museum of Technology, for their examination of Hungarian sources.

24 I am indebted to Mr N. W. Pirie for a collection of correspondence (now in the Science Museum Library) concerning a paper submitted by Ereky to the *Transactions of the Bath and Western Agricultural Society* (though apparently never published) relating to his leaf protein.

25 Karl Ereky, 'Die großbetriebsmäßige Entwicklung der Schweinemast in Ungarn' *Mitteilungen der Deutschen Landwirtschafts-Gesellschaft* 34 (25 August 1917):541–50.

26 Ereky to John Hammond, 23 February 1924, Ereky Correspondence, Science Museum Library.

27 M. Herter and G. Wilsdorf, *Die Bedeutung des Schweines für die Fleischversorgung,* Arbeiten der Deutschen Landwirtschafts-Gesellschaft, vol. 270 (Berlin: Paul Parey, 1914), p. 205.

28 Herter and Wilsdorf, *Die Bedeutung des Schweines für die Fleischversorgung,* 202. See also the explication and illustrations provided by Dr Oermann-Seeste, 'Schweinemastgroßbetriebe, ihre Technik und wirtschaftliche Bedeutung', *Jahrbuch der Deutschen Landwirtschafts-Gesellschaft* 25 (1911):956–68.

29 For a reflection on modern piggeries, see David Goodman, Bernardo Sorj, and John Wilkinson, *From Farming to Biotechnology: A Theory of Agro-industrial Development* (Oxford: Blackwell Publishers, 1987), p. 179.

30 Karl Ereky, *Nahrungsmittelproduktion und Landwirtschaft* (Budapest: Friedrich Kilians Nachfolger, 1917).

31 The crucial role of converting raw materials is dealt with in Theodor Brinkmann, 'Die Dänische Landwirtschaft: Die Entwicklung ihrer Produktion seit dem Auftreten der internationalen Konkurrenz und ihre Anpassung an den Weltmarkt Vermittels genossenschaftlicher Organisation', *Abhandlungen des Staatswissenschaftlichen Seminars zu Jena* 6, pt. 1

(1908), pp. 41–42. The giant Danish slaughterhouses are described on pp. 153–54. Brinkmann's ideas on the conversion of agricultural raw materials are also discussed in English translation in Elizabeth T. Benedict, Heinrich Hermann Stippler, and Mary Reed Benedict, eds., *Theodor Brinkmann's Economics of the Farm Business* (Berkeley: University of California Press, 1935), pp. 120–62. This book was a translation of Brinkmann's *Die Oekonomik des landwirtschaftlichen Betriebes* published in 1922 but whose manuscript is dated 1912. On Brinkmann's place in the history of agricultural economics, see Joseph Nou, *Studies in the Development of Agricultural Economics in Europe* (Uppsala, Sweden: Lantbrukshögskolan, 1967), where Brinkmann's work is compared to that of Aerobee.

32 See Heinz Haushofer, *Die Agrarreformen der Oesterreich-ungarischen Nachfolgestaaten* (Munich: Dresler, 1929), pp. 20–21.

33 Ereky, 'Die großbetriebsmäßige Entwicklung der Schweinemast in Ungarn'.

34 Ereky, *Biotechnologie der Fleisch-, Fett- und Milcherzeugung im landwirtschaftlichen Großbetriebe,* p. 84.

35 'Das Ziel seiner Bestrebungen ist, einen neuen Wissenszweig zu begründen, den er "Biotechnologie" nennt und der derauf hinwirken soll, die Produktion dieser wichtigen Nährmittel auf wissenschaftlicher Grundlage zu erhohen', H. Pringsheim, Review of Karl Ereky, *Biotechnologie . . . , Die Naturwissenschaften* 7(1919): 112.

36 Paul Lindner, 'Allgemeines aus dem Bereich der Biotechnologie', *Zeitschrift fur Technische Biologie* 8(1920): 54–56.

37 'Diejeniger Gewerbestande, bei der Lebewesen als Rohprodukte oder auch für die Umwandlung von Naturprodukten eine Rolle spielen. z.b. Gärungsindustrie.' *Meyers Lexikon,* 7th ed. of 1925, vol. 2, p. 403; 'Untersuchung und gewerbl. Verwendung der Lebenstätigkeit von Kleinlebewesen (Hefe, Gärungsorganismen)'. *Der Große Brockhaus,* 15th ed. vol.2 (1929), p. 747.

38 Albrecht Hase, 'Ueber technische Biologie: Ihre Aufgaben und Ziele, ihre prinzipielle und wirtschaftliche Bedeutung', *Zeitschrift fur Technische Biologie* 8(1920): 23–45.

39 E.C. Hansen, 'Introduction' to Franz Lafar, *Technical Mycology: The Utilization of Micro-organisms in the Arts and Manufactures,* trans. T.C. Salter (London: Griffin, 1898), p. vii.

40 Raphael Meldola, *The Chemical Synthesis of Vital Products and the Interrelations Between Organic Compounds* (London: Arnold, 1904), vol 1., pp. 1–19; on Meldola, see J.V. Eyre and E.H. Rodd, 'Raphael Meldola', in *British Chemists,* ed. Alexander Findlay and W.H. Mills (London: Chemical Society, 1947), pp. 96–125.

41 On activated sludge, see H.H. Stanbridge, *History of Sewage Treatment in Britain.* Pt. 7. *Activated Sludge* (Maidstone: IWPC, 1977). On the emergence of biological sewage treatment more generally indicating its empiricist roots, see Christopher Hamlin, 'William Dibdin and the Idea of Biological Sewage Treatment', *Technology and Culture* 29 (1988): 189–218.

42 Max Delbrück, 'Hefe ein Edelpilz', *Wochenschrift für Brauerei* 27 (30 July 1910): 373–76.

43 R. Braude, 'Dried Yeast as Fodder for Livestock', *Journal of the Institute of Brewing* 48 (October 1942): 206–12.

44 A.J. Kluyver, 'Microbiology and Industry' (Translated from 'Microbiologie en Industrie'), Inaugural lecture at Delft, 18 January 1922, reprinted in *Albert Jan Kluyver: His Life and Work,* ed. A.F. Kemp, J.W. la Rivière, and W. Verhoeven (Amsterdam: North Holland, 1959), p. 175.
45 H. Benninga, *A History of Lactic Acid Making: A Chapter in the History of Biotechnology* (Dordrecht: Kluwer, 1990).
46 O.E. May and H.T. Herrick, 'Some Minor Industrial Fermentations', *Industrial and Engineering Chemistry* 22 (November 1930): 1172–76; W. Connstein and K. Lüdecke, 'Ueber Glycerin-Gewinnung durch Gärung', *Berichte der Deutschen Chemischen Gesellschaft* 52ii (1919): 1385–91.
47 See Jehuda Reinharz, *Chaim Weizmann: The Making of Zionist Leader* (Oxford University Press, 1985).
48 Austin Coates, *The Commerce of Rubber: The First 250 Years* (Oxford University Press, 1987), pp. 146–53, deals with the 1905–1910 rise in prices in response to a Brazilian cartel. Before April 1910, prices rose to 12/4 a lb but they fell to 6/- a lb by October. Malayan costs were 1/- a lb and Brazilian 4/- a lb.
49 W.H. Perkin Jr., 'The Production and Polymerisation of Butadiene, Isoprene and Their Homologues', *Journal of the Society of Chemical Industry* 31 (1912): 616–25, p. 620. Perkin's calculations as to the target price of synthetic rubber is confirmed by Austin Coates's estimate of Malayan production costs for natural rubber of 1/- a pound.
50 Quoted in Reinharz, *Chaim Weizmann,* p. 302.
51 Weizmann to Fernbach, 8 August 1910, Court Collection f.35, Weizmann Archives, Rehovot, Israel (hereafter Weizmann Archives).
52 Rowland Whincop to Weizmann, 4 April 1910,. f.25, Weizmann Archives.
53 Schoen to Weizmann, 8 December 1910, f.51, Weizmann Archives.
54 Strange to Perkin, 29 March 1912, f.232, Weizmann Archives.
55 H.E. Armstrong, 'The Production of Rubber: With or Against Nature?' *Times Engineering Supplement,* 17 July 1912, p. 21.
56 'Nature and Art', *Times Engineering Supplement,* 17 July 1912.
57 The scene is evoked by Edwin E. Slosson in his *Creative Chemistry* (New York: Century, 1921), pp. 148–53.
58 Weizmann to Strange, 1 January 1911, f.67, Weizmann Archives.
59 I am grateful to the biographer of Weizmann, Professor Jehuda Reinharz, for the opportunity to discuss this hypothesis.
60 Strange to Mathews, 6 September 1911, f.131, Weizmann Archives.
61 'Commercial Solvents, Corporation v. Synthetic Products Company Ltd.' *Reports of Patent, Design and Trade Mark and Other Cases,* vol. 43, ed. F.G. Underlay (London HMSO, 1951), pp. 218–19.
62 The episode was graphically described by J.H. Hastings in *The Pasteur Fermentation Centennial, 1857–1957,* ed. Charles Pfizer & Co. Inc (New York: Charles Pfizer & Co. Inc, 1958), pp. 100–101.
63 G.A. Dummett, *From Little Acorns: A History of the A.P.V. Company Limited* (London: Hutchinson Benham, 1981), p. 30.
64 Keith Vernon, 'Pus, Sewage, Beer and Milk: Microbiology in Britain, 1870–1940', *History of Science* 28 (1990): 289–325.
65 Sir William Pope, 'Address by the President', *Journal of the Society of Chemical Industry* 40 (1921): 179T–82T.
66 Kemp, la Rivière, and Verhoeren, *Albert Jan Kluyver,* 165–85.

67 Ellis I. Fulmer, 'The Chemical Approach to Problems of Fermentation', *Industrial and Engineering Chemistry* 22 (November 1930): 1148–50.
68 Robert E. Kohler, *From Medical Chemistry to Biochemistry: The Making of a Biomedical Discipline* (Cambridge University Press, 1982), p. 210.
69 K. Bernhauer, *Gärungschemisches Praktikum,* 2d ed. (Berlin: Springer, 1939), p. iii; Henry Field Smyth and Walter Lord Obold, *Industrial Microbiology: The Utilization of Bacteria, Yeasts and Molds in Industrial Processes* (Baltimore: Williams & Wilkins, 1930).
70 J.F. Garrett, 'Lactic Acid', *Industrial and Engineering Chemistry* 22 (November 1930): 1153–54.
71 O.E. May and H.T. Herrick, 'Some Minor Industrial Fermentations', *Industrial and Engineering Chemistry* 22 (November 1930): 1172–76.
72 J.F. Richardson, *A Digest of Farm Chemurgy: Industrialisation of Farm Products. The Way Out for American Agriculture.* Sec. 2, p. 20. The price of corn is taken from 'Agriculture, General, 9, August 1938, Research Division Republican National Committee. Bound in *Digest of Farm Chemurgy,* New York Public Library.
73 The story of the coinage of the term 'chemurgy' is told by Wheeler McMillen, *New Riches from the Soil: The Progress of Chemurgy* (New York: Van Nostrand, 1946), pp. 17–31.
74 Quoted in Christy Borth, *Pioneers of Plenty: The Story of Chemurgy* (Indianapolis; Ind.: Bobbs-Merrill, 1939), p. 84.
75 Carroll Pursell, 'The Farm Chemurgic Council and the United States Department of Agriculture, 1935–1939', *Isis* 60 (1969): 307–17.

CHAPTER 3. THE ENGINEERING OF NATURE

1 Julian Huxley, 'Chairman's Introductory Address', in Lancelot Hogben, *The Retreat from Reason: Conway Memorial Lecture Delivered at Conway Hall ... May 20, 1936* (London: Watts & Co, 1936), p. vii.
2 Cited in Henry Shefter, 'Readers Supplement' to his edited edition of K. Čapek's *R.U.R.* (New York: Simon & Schuster, 1973), p. 10. Of course, even Rossum's robots prove to be far more complex than just servile machines, themselves developing a range of unexpected humanlike traits.
3 J.H. Randall, *Our Changing Civilization: How Science and the Machine are Reconstructing Modern Life* (London: Allen & Unwin, 1929), pp. 7–8.
4 Gustav Tornier, 'Ueberzählige Bildungen und die Bedeutung der Pathologie für die Biontotechnik (mit demonstrationen)', in *Verhandlungen des V. Internationalen Zoologen-Congresses zu Berlin, 12–16 August 1901,* ed. Paul Matschie (Jena: Gustav Fischer, 1902), pp. 467–500.
5 Wilhelm Roux et al., eds., *Terminologie der Entwicklungsmechanik der Tiere und Pflanzen* (Leipzig: Wilhelm Engelmann 1912), p. 66.
6 Ernst Kapp, *Grundlinien einer Philosophie der Technik* (Braunschweig: Westermann, 1877); see also Vitus Graber, *Die äuseren mechanischen Werkzeuge der Wirbeltiere* (Leipzig: Frentag, 1886). Also see Brigitte Hoppe, 'Biologische und technische Bewegungslehre in 19. Jahrhundert'. In *Geschichte der Naturwissenschaften und der Technik in 19. Jahrhundert.* Ed. by B. Hoppe et al. (Dusseldorf: VDI, 1969), p. 9–35.
7 See Friedrich Rapp, 'Philosophy of Technology: A Review', *Interdisciplinary Science Reviews* 10 (1985): 126–39.

8 On mechanism and its competitors, see Donna Jean Haraway, *Crystals, Fabrics and Fields: Metaphors of Organicism in Twentieth-Century Developmental Biology* (New Haven, Conn.: Yale University Press, 1976). See L.R. Grote, 'Wilhelm Roux in Halle a.S.', in *Die Medizin der Gegenwart in Selbstdarstellungen,* 2 vols. (Leipzig: F. Meiner, 1924), vol 1., pp. 173–74.

9 Loeb's philosophy is thoroughly discussed by Philip Pauly, *Controlling Life: Jacques Loeb and the Engineering Ideal in Biology* (Oxford University Press, 1987).

10 Owsei Temkin, 'Materialism in French and German Physiology of the Early Nineteenth Century', *Bulletin of the History of Medicine* 20 (1946): 322–27.

11 See J. Herf, *Reactionary Modernism: Technology, Culture and Politics in Weimar and the Third Reich* (Cambridge University Press, 1984).

12 'L'absence ou le silence des instincts conservateurs exige la recherche des lois speciales de l'hygiène. Il nous faut un art, à defaut de nature; ou plutôt à cause de notre nature multiple, il y a pour elle seul une *biotechnie* anthropologique', J.J. Virey, *Hygiène philosophique ou de la santé dans le régime physique, moral et politique de la Civilisation moderne* (Paris: Crochard, 1828); this is the first use cited in *Trésor de la langue française* 4, 522. On Virey, see Alex Berman, 'Romantic Hygeia: J.J. Virey (1775–1846), Pharmacist and Philosopher of Nature', *Bulletin of the History of Medicine* 39(1965): 134–42; Claude Benichou and Claude Blanckaert, *Julien-Joseph Virey: Naturaliste et anthropologue* (Paris: Vrin, 1988).

13 R.C. Grogin, *The Bergsonian Controversy in France* (Calgary: University of Calgary Press, 1988), pp. 81–82.

14 Amidst the mass of Bergson scholarship, see Jean-Pierre Séris, 'Bergson et la technique', in *Bergson: Naissance d'une philosophie. Actes du Colloque de Clermond-Ferrand 17 et 18 novembre 1987* (Paris: Presses Universitaires de France, 1990), pp. 121–38. Bergson's concern with work was of course based on a century of increasing interest, since Smith and Ricardo, the new physiology of Milne Edwards, the sociology of Durkheim, the socialism of Marx and of Bergson's friend Jaurès, and the moral outrage of Zola. See also Henri Gouhier, *Bergson dans l'histoire de la pensée occidentale* (Paris: Vrin, 1989). I am grateful to Mme Annie Petit for the opportunity to discuss Bergson's philosophy.

15 Henri Bergson, *Creative Evolution,* trans. Arthur Mitchell (London: Macmillan Press, 1954), 146.

16 See Peter J. Bowler, 'Theodor Eimer and Orthogenesis: Evolution by "Definitely Directed Variation" ', *Journal of the History of Medicine and Allied Sciences* 34 (1979): 40–73 and idem, *The Eclipse of Darwinism: Anti-Darwinian Evolution Theories in the Decades around 1900* (Baltimore: Johns Hopkins University Press, 1983).

17 A. Pauly, *Darwinismus und Lamarckismus* (Munich: Reinhardt, 1905). On Pauly, see Hans Spemann, *Forschung und Leben* (Stuttgart: J. Engelhorns Nachfolger, 1943), pp. 147–50, 157–68 R.G. Rinard, 'Neo-Lamarckism and Technique: Hans Spemann and the Development of Experimental Embryology', *Journal of the History of Biology* 21 (1988): 95–118. Freud's debt to Pauly is discussed by Frank J. Sulloway, *Freud, Biologist of the Mind: Beyond the Psychoanalytic Legend* (New York: Basic, 1979), pp. 274–76.

18 Adolf Wagner, *Geschichte des Lamarckismus: Als Einführung in die psychobiologische Bewegung der Gegenwart* (Stuttgart: Franck, 1908).

19 For the attribution, see R. Eisler, *Wörterbuch der Philosophischen Begriffe* (Berlin: Mittler, 1927), vol. 1, p. 226; It is interesting to note that Eisler's friendship is acknowledged in the introduction to Goldscheid's *Menschenökonomie.*

20 Goldscheid himself was a complex character. He began his career as a poet, becoming a monist social reformer, founder of the German Sociological Association, yet has been neglected by historians. For one of the few extended treatments, see Doris Byer, *Rassenhygiene und Wohlfahrtspflege: Zur Entstehung eines sozialdemokratischen Machtdispositivs in Oesterreich bis 1934* (Frankfurt am Main: Campus Verlag, 1988), pp. 86–101; also August M. Knoll, 'Rudolf Goldscheid', *Neue Deutsche Biographie,* vol. 6, 607–8. For the context of Goldscheid's word, see Peter Weingart, Jürgen Kroll, and Kurt Bayertz, *Rasse, Blut und Gene: Geschichte der Eugenik und Rassenhygiene in Deutschland* (Frankfurt: Suhrkamp, 1988), pp. 255–59, and Paul Weindling, *Health, Politics and German Politics between National Unification and Nazism, 1870–1945* (Cambridge University Press, 1989). I am grateful to Jürgen Kroll for sharing with me his understanding of Goldscheid's concept of *Menschenökonomie.*

21 See Horst Zimmerman, 'Fiscal Pressure on the "Tax State" ', in *Evolutionary Economics: Application of Schumpeter's Ideas,* ed. Horst Hanusch (Cambridge University Press, 1988), 255–73. Goldscheid's work is reprinted as 'A Sociological Approach to Problems of Public Finance', in *Classics in the Theory of Public Finance,* ed. R.A. Musgrave and A.T. Peacock (London: Macmillan Press, 1958), pp. 202–13.

22 See Arnold Heertje 'Schumpeter and Technical Change', in Hanusch, *Evolutionary Economics,* pp. 71–89.

23 Edouard März, *Joseph Alois Schumpeter: Forscher, Lehrer und Politiker* (Vienna: Verlag für Geschichte und Politik, 1983).

24 Henry Bergen, 'Rudolf Goldscheid, "Höherentwicklung u. Menschenökonomie" ', *Eugenic Review* 3(1911–12): 236–41.

25 Gertrud Kroeger, *The Concept of Social Medicine: As Presented by Physicians and Other Writers in Germany, 1779–1932* (Chicago: Julius Rosenwald Fund, 1937).

26 See Paul Weindling, 'Degeneration und öffentliches Gesundheitswesen, 1900–1930: Wohnverhältnisse', in *Stadt und Gesundheit: Zum Wandel von 'Volksgesundheit' und kommunaler Gesundheitspolitik im 19. und frühen 20. Jahrhundert,* ed. Jürgen Reulecke and Adelheid Gräfin zu Castell Rüdenhausen (Stuttgart: Frans Steiner Verlag, 1991), pp. 105–13; see also Rudolf Thissen, *Die Entwicklung der Terminologie auf dem Gebiet der Sozialhygiene und Sozialmedizin im deutschen Sprachgebiet bis 1930* (Cologne: Westdeutscherverlag, 1969). For a view of the historiography from a British perspective, see Dorothy Porter and Roy Porter, 'What was Social Medicine? An Historiographical Essay', *Journal of Historical Sociology* 1 (1988): 90–106.

27 Alfred Grotjahn, *Soziale Pathologie: Versuch einer Lehre von den sozialen Beziehungen der menschlichen Krankheiten als Grundlage der sozialen Medizin und sozialen Hygiene,* 2d ed. (Berlin: Hirschwald, 1915), pp. 8–9.

28 Grotjahn, *Soziale Pathologie,* p. 326.

29 Goldscheid expressed his vision of a better technology in his contribution to the Ostwald festschrift. See Rudolf Goldscheid, 'Ostwald als Persönlichkeit und Kulturfaktor', in *Wilhelm Ostwald: Festschrift ans Anlaß seines 60. Geburtstages 2 September 1913,* ed. Monistenbund in Oesterreich (Vienna: Suschitzky, 1913), pp. 57–82, especially p. 67.

30 In many ways Koestler (actually born in Budapest) with his interests in Lamarckism and parapsychology continued to develop the interests of his old compatriots. He treated Kammerer in his investigation of the apparent fraud in his *The Case of the Midwife Toad* (London: Hutchinson, 1971). Kammerer's brief period of popularity in the Soviet Union is explored by A.E. Gaissinovitch, 'The Origins of Soviet Genetics and the Struggle with Lamarckism, 1922–1929', *Journal of the History of Biology* 13 (1980): 1–51.

31 Paul Kammerer, 'Höherentwicklung und Biologie', *Archiv für Rassen und Gesellschaftsbiologie* 11(1914): 222–33; followed by W. Schallmayer, 'Antwort auf P. Kammerers Plaidoyer für R. Goldscheid', pp. 233–40. Kammerer's 1910 speech to the Berlin Scientific Society was translated and published as 'Adaptation and Inheritance in the Light of Modern Experimental Investigation', *Report of the Smithsonian Institution* (1912), pp. 421–42.

32 Paul Kammerer, 'Das biologischer Zeitalter: Fortschritte der organischen Technik', *Monistische Bibliothek, kleine Flugschriften, Nr 33* (Hamburger Verlag, 1920). Kammerer also wrote 'Lebensbeherrschung: Grundsteinlegung zur organischen Technik', in *Monistische Bibliothek, kleine Flugschriften, Nr 13* (1st ed. 1919, 2d ed. 1920), a copy of which I have not been able to find.

33 Daniel S. Nadav, *Julius Moses und die Politik der Sozialhygiene in Deutschland,* Schriftenreihe des Instituts für Deutsche Geschichte, Tel Aviv (Geilingen: Belicher Verlag, 1985).

34 William H. Schneider, *Quality and Quantity: The Quest for Biological Regeneration in Twentieth-Century France* (Cambridge University Press, 1990).

35 Quoted and translated in Jacques Donzelot, *The Policing of Families,* trans. Robert Hurley (London: Hutchinson, 1980), p. 186.

36 Mark B. Adams, 'Eugenics as Social Medicine in Revolutionary Russia: Prophets, Patrons and the Dialectics of Discipline Building', in *Health and Society in Revolutionary Russia,* ed. S. Gross Solomons and J.F. Hutchinson (Bloomington: Indiana University Press, 1990), pp. 200–23; for *Salamandr,* see R.R.M. Short and Richard Taylor, 'Soviet Cinema and the International Menace, 1928–1939', *Historical Journal of Film, Radio and Television* 6(1986): 131–59. I am grateful to Mr T. Boon for showing me this article.

37 Raoul H. Francé, *Der Weg zu Mir* (Berlin: Alfred Kröner, 1927).

38 Annie Francé-Harrar, *So war's um Neunzehnhundert: Mein Fin de Siècle* (Munich: Albert Langen, 1962).

39 Adolf Wagner, 'Biotechnik und Plasmatik', in *Der Begründer der Lebenslehre Raoul Francé: Eine Festschrift zu seinem 50 Geburtstag* (Stuttgart: Walter Seifert, 1925), p. 7.

40 R.H. Francé, 'Das biologische Experiment und seine Bedeutung für die Versuchstechnik', *Mitteilungen des K.K. Technischen Versuchsamtes* 7 ii (1918): 15–21.

41 R.H. Francé, *Plants as Inventors* (London: Simpkin & Marshall, 1926), p. 62.
42 R.H. Francé, *Bios: Die Gesetze der Welt* (Munich: Hofstaengli, 1921), vol. 2, p. 81: on p. 126 he cites 1917 as the year of his first work on biotechnics, in the *K. K. Mitteillungen* (though it was actually published in 1918) – 'Daß man das Jahr 1917, in dem diese Wende des Denkens einsetzte, dann auch als den Beginn einer neuen Epoche des Kulturlebens im Gedächtnis behalten wird, daran habe ich nicht den geringsten Zweifel, so unvollkommen auch heute noch die Gedanken sind, welche diese Aera eröffnen.'
43 Francé-Harrar, *So war's um Neunzehnhundert,* 141.
44 Fritz Neumeyer, *Mies Van der Rohe – Das Kunstlose Wort: Gedanken zur Baukunst* (Berlin: Siedler, 1986), pp. 138–40.
45 Stanislaus Van Moos, 'The Visualized Machine Age', in *Lewis Mumford: Public Intellectual,* ed. Thomas P. Hughes and Agatha C. Hughes (Oxford University Press, 1990), p. 407.
46 Alfred Giessler, *Biotechnik* (Leipzig: Quelle & Meyer, 1939). K.E. Rothschuh has traced the history of the concept of *Biotechnik* from Francé to recent years in his article, 'Bionomie/Biotechnik', in *Historisches Wörterbuch der Philosophie,* ed. J. Ritter (Darmstadt: Wissenschaftliche Buchgesellschaft, 1971), vol. 1, pp. 946–47.
47 Alan Seaman, 'The Society for Applied Bacteriology: The First Fifty Years', *Journal of Applied Bacteriology* 50 (1981): 425–31.
48 Robert E. Kohler, *From Medical Chemistry to Biochemistry: The Making of a Biomedical Discipline* (Cambridge University Press, 1982), p. 82.
49 Compare membership figures of 283 for the Society of Economic Biology, compiled from the *Annals of Economic Biology,* with the thousands of chemists belonging to each of that discipline's organizations, listed by Colin A. Russell, Noel G. Coley, and Gerrylyn K. Roberts, *Chemists by Profession* (Milton Keynes: Open University Press, 1977), pp. 330–31.
50 Gary Werskey, *The Visible College* (London: Allen Lane, 1978), p. 165.
51 I am grateful to N.W. Pirie, a distinguished pupil of Gowland Hopkins, for the opportunity to talk about the reading of German material at this time.
52 Stefan Zweig, *The World of Yesterday* (London: Atrium Press, 1987), p. 287.
53 Hogben's reply to Smuts was published in an extended form as *The Nature of Living Matter* (London: Kegan Paul, 1930). For an explanation of how this was a retaliation against Haldane's refusal to call himself a vitalist, see Hogben to Julian Huxley, 3 November 1929, 'Selected Correspondence', Huxley Papers, Rice University, Texas.
54 Hogben to Needham dated '1 Feb.', Needham Archive, Box 4, Cambridge University Library. Although the year is not given this letter seems to be from 1936 since it uses the same phrases as found in a letter dated 10 February 1936. It has been reprinted by Maurice Goldsmith in his *Sage: A Life of J.D. Bernal* (London: Hutchinson, 1980), p. 73.
55 See Patrick Geddes, *Cities in Evolution: An Introduction to the Town Planning Movement and the Study of Civics* (London: Williams & Norgate, 1915), p. 59.
56 J.A. Thomson, 'Biological Philosophy', *Nature* 87 (12 October 1911): 475–76.
57 Benton Mackaye, *From Geography to Geotechnics* (Urbana: University of Illinois Press, 1968), p. 22.

58 V.V. Branford and P. Geddes, *The Coming Polity* (London: Le Play House, 2d ed., 1919), pp. 267–68.
59 Geddes to Mumford, 25 February 1921, Mumford Papers, Special Collections, Van Pelt Library, University of Pennsylvania.
60 Geddes to Marcel Hardy, 2 October 1923, National Library of Scotland, coll. 19995, f. 111.
61 Patrick Geddes, 'Ways of Transition Towards Constructive Peace', *Sociological Review* (January 1930): 2–3. It has been reprinted in Philip Boardman, *The Worlds of Patrick Geddes: Biologist, Town Planner, Re-educator, Peace-Warrior* (London: Routledge, 1978), 480–81.
62 Geddes to Arthur Thompson, 21 February 1931. Ms. 10518, f.228, National Library of Scotland.
63 P. Geddes and J.A. Thomson, *Biology* (London: Home University Library, 1925).
64 J.A. Thomson, 'Biology', *Encyclopaedia Britannica* Supplement to the 11th ed., 1 (1926), pp. 383–85.
65 Moreover, even the social biological ideas of Goldscheid were captured by the vision. Thomson showed how, in a 1926 article entitled 'Biology and Social Hygiene', *Quarterly Review* 246 (1926): 28–48, in future people would be healthier and happier, if policy were examined through the 'biological prism' of folk, work, and place. This included supporting the work of artists, 'the salt of the earth', and buying fish which would sustain fishermen.
66 L. Hogben wrote dismissively about Geddes to Julian Huxley, June 1931. 'Selected Correspondence', Huxley Papers, Rice University.
67 'A Sheaf of Tributes to the Late Sir Patrick Geddes', *Supplement to the Sociological Review,* 24 (October 1932): 349–400.
68 J.B.S. Haldane, *Daedalus or Science and the Future* (London: Kegan Paul, 1925), pp. 1–2.
69 Haldane, *Daedalus,* p. 42.
70 Haldane, *Daedalus,* p. 77.
71 J.S. Huxley, 'Biology and Human Life' 2d Norman Lockyer Lecture, British Science Guild, 23 November 1926.
72 J.S. Huxley, 'The Applied Science of the Next Hundred Years: Biological and Social Engineering', *Life and Letters* 11 (1934): 38–46.
73 Huxley, 'Applied Science of the Next Hundred Years', p. 40. There is no unambiguous proof of Huxley's indebtedness to the recently deceased Goldscheid, but the influence of *Menschenökonomie* is certainly implied by the statement 'biological engineering will begin with the premiss that human beings are the most valuable asset of a nation, and human development the most important process of manufacture, with an elaborate technique to be mastered.' P. 41.
74 Joseph Needham, 'Notes on the Way', *Time and Tide* 12 (10 September 1932): 970–72.
75 'Aldous Huxley's Brave New World', Box 4, Needham papers, Cambridge University Library.
76 This was the aspersion made by Hobhouse. See José Harris, *William Beveridge: A Biography* (Oxford University Press, 1981), p. 287.
77 The accession number of Goldscheid's *Höherentwicklung und Menschenö*

konomie at the LSE is 48,359. Although early registers are now missing, this is compatible with an acquisition date of about 1920. The copy at the LSE is inscribed vV.

78 See Harris, *William Beveridge.* For the establishment of the chair of social biology, see pp. 286–87; see too Lord Beveridge, *Power and Influence* (London: Hodder & Stoughton, 1953), pp. 176–77. The department is discussed in Selma Ahmad, 'Institutions and the Growth of Knowledge: The Rockefeller Foundation's Influence on the Social Sciences Between the Wars', doctoral dissertation, Manchester University, 1987. See Janet Beveridge, 'The Chair of Social Biology 1930–1931', Beveridge Papers, supp. 4/8. Archive Collections, London School of Economics. The department has attracted much attention; see Greta Jones, *Social Hygiene in 20th Century Britain* (London: Croom Helm, 1986).

79 'Copy of Memorandum Submitted in July 1925 by the London School of Economics and Political Science to the Trustees of the Laura Spelman Rockefeller Memorial Together with Appendix on the School of Economics Site', Beveridge Papers, supp. 4/8. Archive Collections, London School of Economics.

80 For a description of Hogben's complex career, see G.P. Wells, 'Lancelot Thomas Hogben', *Biographical Memoirs of the Royal Society* 24 (1978): 183–221.

81 Lancelot Hogben, 'The Foundations of Social Biology', *Economica,* no. 31 (February 1931): 4–24.

82 Richard Soloway, *Demography and Degeneration: Eugenics and the Declining Birthrate in 20th Century Britain* (Chapel Hill: University of North Carolina Press, 1990), pp. 226–58, deals with Kuczynski. See also the biography by his son: Jürgen Kuczynski, *René Kuczynski: Ein fortschriftlischer Wissenschaftler in der ersten Hälfte des 20. Jahrhunderts* (Berlin: Aufbau Verlag, 1957).

83 Lancelot Hogben, 'Prolegomenon to Political Arithmetic', in *Political Arithmetic: A Symposium of Population Studies,* ed. Lancelot Hogben (London: Allen & Unwin, 1938), pp. 13–46.

84 'Biotechnology', *Nature* 131 (29 April 1933): 597–99. In the copy held by *Nature*'s library, the article is initialled R.B. for Rainald Brightman, one of *Nature*'s regular columnists. I am grateful to *Nature* for this information.

85 Hogben to Needham, 10 February 1936, Needham Papers, set 4, Cambridge University Library.

86 Hogben, *The Retreat from Reason,* pp. 43–49.

87 Gary Werskey, *The Visible College* (London: Allen Lane, 1978), p. 203.

88 Lancelot Hogben, 'Look Back with Laughter', ed. G.P. Wells, Box A10., f. 190, Birmingham University Archives.

89 On Boyd-Orr, see his autobiography, *As I Recall* (London: McGibbon & Kee, 1966).

90 See Madeleine Mayhew, '1930s Nutrition Controversy', *Journal of Contemporary History* 23 (1988): 445–64.

91 F.A.E. Crew et al., 'Social Biology and Population Improvement', *Nature* 144 (16 September 1939): 521–22. Other signatories were: C.D. Darlington, J.B.S. Haldane, S.C. Harland, L. Hogben, J.S. Huxley, H.J. Muller, J. Needham, G.P. Child, P.R. David, G. Dahlberg, Th. Dobhzhansky, R.A.

Emerson, C. Gordon, J. Hammond, C.L. Huskins, P.C. Koller, W. Landauer, H.H. Plough, B. Price, J. Schultz, A.G. Steinberg, and C.H. Waddington.

92 Pnina Abir-Am, 'The Biotheoretical Gathering, Transdisciplinary Authority and the Incipient Legitimation of Molecular Biology in the 1930s: New Perspective on the Historical Sociology of Science', *History of Science* 25 (1987): 1–70.

93 Hogben, 'Look Back with Laughter', f. 185.

94 N.W. Pirie, 'Recurrent Luck in Research', *Selected Topics in the History of Biochemistry: Personal Recollections,* ed. G. Semenza, Comprehensive Biochemistry, vol. 36 (Amsterdam: Elsevier, 1986), pp. 491–522.

95 Sir Harold Hartley, 'Commentary', *Process Biochemistry* 2, no. 4 (April 1967): 3.

96 Harold Hartley, 'Agriculture as a Source of Raw Materials for Industry?' *Journal of the Textile Institute* 28 (1937), p. 172.

CHAPTER 4. INSTITUTIONAL REALITY

1 H.J. Sauer Jr. and R.G. Nevins, 'Biotechnology and the Mechanical Engineer', *Mechanical Engineering* 87 (December 1965), p. 36.

2 The images of bacteriology in British and U.S. fiction are compared in Ilana Löwy, 'Immunology and Literature in the Early Twentieth Century: *Arrowsmith* and *The Doctor's Dilemma*', *Medical History* 32 (1988): 314–32. The quotation is from W.C. Sellar and R.J. Yeatman, *1066 and All That* (London: Methuen, 1st ed, 1930), p. 123 of 1984 edition.

3 W.M. Kiplinger, 'Causes of Our Unemployment: An Employment Puzzle', *New York Times,* 17 August 1930. This is discussed by Peter J. Kuznick, *Beyond the Laboratory: Scientists as Political Activists in 1930s America* (Chicago: University of Chicago Press, 1987), p. 18.

4 'Charges Industry with Duty to Idle', *New York Times,* 16 February 1931.

5 See D.J. Rhees, *The Chemists' Crusade: The Rise of an Industrial Science in Modern America, 1907–1922,* Doctoral dissertation, University of Pennsylvania, 8714116 (Ann Arbor, Mich.: UMI, 1987).

6 William J. Hale, *Chemistry Triumphant: The Rise and Reign of Chemistry in a Chemical World* (Baltimore: Williams & Wilkins in cooperation with The Century of Progress Exposition, 1932), p. 127.

7 Quoted in Christy Borth, *Pioneers of Plenty: The Story of Chemurgy* (Indianapolis: Bobbs-Merrill, 1939), p. 74.

8 Borth, *Pioneers of Plenty,* p. 28.

9 Thomson to Geddes, 29 April 1930, Ms. 10555 f. 301, National Library of Scotland.

10 Arthur P. Molella, 'The First Generation: Usher, Mumford and Giedion', in *In Context: History and the History of Technology: Essays in Honor of Melvin Kranzberg,* ed. Stephen H. Cutcliffe and Robert C. Post (Bethlehem, Pa.: Lehigh University Press, 1989), pp. 88–105.

11 Lewis Mumford, 'The Disciple's Rebellion', *Encounter* 27 (1966): 11–21; 'Mumford on Geddes', BBC Radio Three, 22 August 1976, and, on television, BBC Scotland, ' "Eye" for the Future', 28 December 1975. Here Mumford compared Geddes with Leonardo.

12 Lewis Mumford, *Technics and Civilization* (New York: Harper, Brace & World, 1934). See also his 'An Appraisal of Lewis Mumford's "Technics and Civilization" (1934)', *Daedalus* 88 (1959): 527–36.
13 See W.E. Wickenden, 'Technology and Culture', Commencement address, Case School of Applied Science, 29 May 1929, and idem, 'Technology and Culture', *Ohio College Association Bulletin* (1933): 4–9.
14 On Wickenden, see David Noble, *America by Design: Science, Technology and the Rise of Corporate Capitalism* (New York: Knopf, 1977); and Edwin R. Layton, *The Revolt of the Engineers: Social Responsibility and the American Engineering Profession* (Cleveland: Case Western Reserve University Press, 1971), pp. 232–34. For Wickenden's abhorrence of seeing food burnt, see W.E. Wickenden, 'The Engineer in a Changing Society', *Electrical Engineering* 51 (July 1932): 467.
15 W.E. Wickenden, 'Training Engineers for Positions of Responsibility', *Cleveland Engineering* (26 December 1929): 3–5, 14.
16 W.E. Wickenden, 'Final Report of the Director of Investigations, June 1933', in *Report of the Investigation of Engineering Education, 1923–1929,* vol. 2 (Pittsburgh: Society for the Promotion of Engineering Education, 1934), p. 1059.
17 Noble, *America by Design,* pp. 82–83.
18 Robert E. Kohler, *Partners in Science, Foundations and Natural Scientists, 1900–1945* (Chicago: University of Chicago Press, 1991), p. 319.
19 For Roosevelt's letter, see 'Asks "Social Mind" in Engineer Study', *New York Times,* 23 October 1936. Compton's reply was published in 'M.I.T. Head Fears "Relief Palliative" Hampers Science', *New York Times,* 25 October 1936.
20 Karl T. Compton and John W.M. Bunker, 'The Genesis of a Curriculum in Biological Engineering', *Scientific Monthly* 48 (January 1939): 5–15.
21 In the cover letter to the Rockefeller Foundation conveying the proposal for a grant for biological engineering, Compton emphasized the parallels between biological and chemical engineering. Karl Compton to the Rockefeller Foundation, 13 February 1939, MIT Office of the President, 1930–1958. (AC4) Institute Archives and Special Collections, MIT Libraries, Cambridge, Mass. (hereafter Compton Papers).
22 Compton and Bunker, 'Genesis of a Curriculum in Biological Engineering', p. 12.
23 'Biological Engineering at the Massachusetts Institute of Technology and an Application for a Grant in Support of It', 'Exhibit A. Proposed Organization of Staff in Biological Engineering Including Suggested Additions to Personnel at M.I.T.' Compton Papers.
24 Vannevar Bush, 'The Case for Biological Engineering', in *Scientists Face the World of 1942* (New Brunswick: Rutgers University Press, 1942), p. 39.
25 Detlev Bronk, 'Commentary', *Scientists Face the World of 1942,* p. 74.
26 Personal information from Professor Myron Tribus who joined the UCLA department shortly after its formation.
27 Craig L. Taylor and L.M.K. Boelter, 'Biotechnology: A New Fundamental in the Training of Engineers', *Science* 105 (28 February 1947): 217–19.
28 Myron Tribus, interview.
29 'How Hot Can a Man Get', *Life* 24 (9 February 1948): 85–87.
30 J.A.R. Kraft, 'The 1961 Picture of Human Factors Research in Business

and Industry in the United States of America', *Ergonomics* 5i (1962): 293–99.

31 These details are summarized in an internal report, John Lyman, 'Re: Biotechnology (Bioengineering)', Department of Engineering, UCLA, c. 1969. I am grateful to Professor George Sines of UCLA for pointing this out to me.

32 Lawrence J. Fogel, *Biotechnology: Concepts and Applications* (Englewood Cliffs: Prentice-Hall, 1963), p. 798.

33 U.S. Congress. Senate Committee on Government Operations. Subcommittee on Reorganization and International Relations, *Hearings . . . to Create a Department of Science and Technology,* 86th Cong., 1st Sess, pt. 1, 16–17 April 1959, pp. 8–16. For this interpretation and reference, I am grateful to Dr Nathan Reingold who kindly gave me a copy of his lecture 'Physics and Engineering in the United States, 1945–1965, A Study of Pride and Prejudice'.

34 Cited in John Lyman 'Biotechnology'. This was the last report of a Ford Foundation supported study at UCLA entitled 'A Study of a Profession and Professional Education' whose principal investigators were Boelter himself and Professor Allen Rosenstein. I am grateful to Professor Rosenstein for a copy of the report.

35 R.R. Roth, 'The Foundation of Bionics', *Perspectives in Biology and Medicine* 26 (1983): 229–42, portrays Francé as the forgotten founder of bionics.

36 Norbert Wiener's classic exposition of his subject was entitled *Cybernetics or Control and Communication in the Animal and the Machine* (Cambridge, Mass: MIT Press, 1948).

37 John Lyman, 'Biotechnology', p. 5. See J.N. Martin, *Biomedical Engineering Education* (Pittsburgh: Chilton, 1966).

38 A.V. Hill, 'Biology and Electronics', *Journal of the British Institution of Radio Engineers* 19 (1959), p. 86.

39 Heinz Wolff, personal communication.

40 See C.N. Smyth, *Medical Electronics: Proceedings of the Second International Conference on Medical Electronics, Paris, 24–27 June 1959* (London: Illiffe & Sons, 1960), particularly the 'Foreword' by V.K. Zworykin (xv–xvi); and 'International Conference on Medical Electronics', *Journal of the British Institution of Radio Engineers* 18 (1958):505.

41 Hill, 'Biology and Electronics', p. 80.

42 'Biological Engineering Society', *Lancet* 2 (23 July 1960): 218.

43 'Biological Engineering Society', *Lancet* 2 (12 November 1960): 1097. I am grateful to Mr Keith Copeland, a founder member and past secretary of the Biological Engineering Society for his help and comments.

44 Robert M. Kenedi, 'Bio-engineering – Concepts, Trend and Potential', *Nature* 202 (25 April 1964): 334–36.

45 'Protokoll' of the first meeting of section X of IVA, 19 June 1943. I am grateful to IVA for access to these minutes. On the general history of IVA, see Gregory Ljunberg, 'Krig och Fred: IVA under 40-talet', *TVF* 40 (1969): 187–95.

46 See Ron R. Eyerman, 'Rationalising Intellectuals: Sweden in the 1930s and 1940s', *Theory and Society* 14 (1985): 777–808.

47 A. Enström, 'Maschinenkraft als Kulturfaktor' reprinted in Torsthin Al-

thin, *Axel F. Enström* (Stockholm: Ingeniörvetenskapsakademien, 1958), pp. 71–76.

48 On Velander, see Gregory Ljunberg, *Edy Velander och Ingenjörsvetenskapsakademien,* IVA-meddelande, vol. 251 (Stockholm: IVA, 1986). I am grateful to the author for translating and interpreting the relevant passages.

49 'P.M. beträffande bioteknisk (biologisk-teknisk) forskning' ref: EV/Z/Ru, 24/8/42, IVA, Stockholm.

50 'Bioteknik', *IVA* 14 (15 February 1943): 1.

51 [E. Velander], 'Några nya utvecklingslinjer inom biotekniken', *IVA,* 13(1942), p. 236. This piece also attributes the concept of *Bioteknik* to Enström. The translation from the Swedish is by Mrs Patricia Crampton.

52 Torbjorn O. Caspersson, 'The Background for the Development of the Chromosome Banding Techniques', *American Journal of Human Genetics* 44(1989): 441–51.

53 'Betr. Bioteknik', ref LS/AKR, 14/11/58, IVA.

54 C.-G. Hedén to E. Gregory Ljunberg, 27 January 1961, IVA.

55 C.-G. Hedén to E. Gregory Ljunberg, 27 January 1961, IVA.

56 Einar Selandar to Gregory Ljunberg, 15 December 1960, IVA.

57 Carl-Göran Hedén, 'The GIAMS – A Contribution to Technology Transfer', in *From Recent Advances in Biotechnology and Applied Biology,* ed. S.T. Chang, K.Y. Chan, and N.Y.S. Woo (Hong Kong: Chinese University Press, 1988), pp. 63–74.

CHAPTER 5. THE CHEMICAL ENGINEERING FRONT

1 Elmer Gaden, 'Editorial', *Biotechnology and Bioengineering* 4(1962), p. 1.

2 Peter H. Spitz, *Petrochemicals: The Rise of an Industry* (New York: Wiley, 1988), p. 264.

3 W.B. Duncan, 'Lesson from the Past, Challenges and Opportunity', in *The Chemical Industry,* ed. D.H. Sharp and T.F. West (Chichester: Ellis Horwood, 1982), pp. 15–30.

4 Richard Carter, *Breakthrough: The Saga of Jonas Salk* (New York: Pocket Books, 1967), p. 259.

5 'The Case for Biochemical Engineering', *Chemical Engineering* 54 (May 1947): 106.

6 Sidney Kirkpatrick, 'A Case Study in Biochemical Engineering', *Chemical Engineering* 54 (1947): 94–101.

7 Leo Hepner, 'Current State of Industrial Microbiology in the USA' (October 1968), enclosed with Hepner to Harold Hartley, 16 October 1968, Box 303; Hartley Archives, Churchill College, Cambridge (hereafter cited as Hartley Archives). I am grateful to Sir Christopher Hartley for permission to consult his father's papers.

8 H.M. Tsuchiya and K.H. Keller, 'Bioengineering – Is a New Era Beginning?' *Chemical Engineering Progress* 61 (May 1965): 60–62.

9 Tsuchiya and Keller, 'Bioengineering – Is a New Era Beginning?', pp. 60–61.

10 W.E. Ranz and A.G. Fredrickson, 'Minneapolis to Host Chemical Engineers' *Chemical Engineering Progress* 61 (July 1965): 112–19.
11 Elmer Gaden to Allan Wittman 17 July 1981. I am grateful to Professor Gaden for showing this letter to me.
12 John Henahan, 'Elmer Gaden: Father of Biochemcial Engineering', *Chemical and Engineering News* 49 (31 May 1971): 27–30.
13 E.M. Crook, 'How Biotechnology Developed at University College London', *Biochemical Society Symposium* 48(1982): 1–7; also Carl-Gören Hedén, 'The GIAMS — A Contribution to Technology Transfer', in *From Recent Advances in Biotechnology and Applied Biology,* ed. S.T. Chang, K.Y. Chan, and N.Y.S. Woo (Hong Kong: Chinese University Press, 1988), 63–74; I am grateful for the opportunity to correspond with Professors Crook, Gaden, and Hedén about the journal's foundation.
14 Gaden to Wittman, 17 July 1981. I am grateful to Professor Gaden for permission to reproduce this section of his letter.
15 See Gladys L. Hobby, *Penicillin: Meeting the Challenge* (New Haven, Conn.: Yale University Press, 1985), provides an introduction to a vast literature.
16 Detailed production figures are available in Allan J. Greene and Andrew J. Schmitz, 'Meeting the Objective', in *The History of Penicillin Production,* ed. by A. A. Elder, Chemical Engineering Progress *Symposium Series* no. 100, 66 (1970): 86–87.
17 A.W.J. Bufton, *Industrial and Economic Microbiology in North America,* DSIR Overseas Technical Report no. 4 (London: HMSO, 1958), p. 6.
18 J.L. Sturchio, ed., *Values and Visions: A Merck Century* (Rahway, N.J.: Merck, 1991), p. 185.
19 J.C. Sheehan, *The Enchanted Ring – The Untold Story of Penicillin* (Cambridge, Mass.: MIT Press, 1982).
20 Emerson J. Lyons, 'Deep Tank Fermentation', in *The History of Penicillin Production,* pp. 31–36.
21 J.C. Hoogerheide, 'Address by the Symposium Co-Sponsor', *Biotechnology and Bioengineering Symposium* 4i (1973): vii.
22 For the development of antibiotics, see Milton Wainwright, *Miracle Cure: The Story of Penicillin and the Golden Age of Antibiotics* (Oxford: Blackwell Publishers, 1990).
23 Paul. R. Burkholder, 'Trends in Antibiotic Research', in *The Pasteur Fermentation Centennial 1857–1957,* ed. Charles Pfizer and Co., Inc. (New York: Charles Pfizer, 1958), p. 44.
24 H.G. Lazell, *From Pills to Penicillin: The Beecham Story* (London: Heinemann, 1975), pp. 135–50.
25 Chain described his original vision in a letter to Sir Charles Dodds, 23 May 1955, PP/EBC F45, Wellcome Institute London. I am grateful to Lady Chain for permission to consult the Chain papers.
26 Fermentation Industries Section, IUPAC, 'Worldwide Survey of Fermentation Industries', *Pure and Applied Chemistry* 13 (1966): 405–17.
27 Durey H. Peterson, 'Autobiography', *Steroids* 45 (1985): 1–17.
28 Bufton, *Industrial and Economic Microbiology in the United States,* p. 6.
29 Jokichi Takamine, 'Enzymes of Aspergillus Oryzae and the Application of Its Amyloclastic Enzyme to the Fermentation Industry', *Industrial and Engineering Chemistry* 6 (1914): 824–28.
30 Kin-ichiro Sakaguchi, *Outline and Characteristics of Japanese Fermentation In-*

dustries (Tokyo: RIKEN, 1961). See also S. Sogusawa, 'History of Japanese Natural Products Research', *Pure and Applied Chemistry* 9 (1964): 1–19

31 Shukuro Kinoshita, 'Amino Acid Production by Fermentative Processes: Metabolic Regulation of Microorganisms and Its Applications', in *Profiles of Japanese Science and Scientists,* ed. Hideki Yukawa (Tokyo: Kodansha, 1970), pp. 98–105.

32 S. Aiba, A.E. Humphrey, and N.F. Millis, *Biochemical Engineering* (New York: Academic Press, 1965). See R.E. Spier, 'So Who is a Biotechnologist', *Biotech Quarterly* 3, nos. 2–3 (1984): 3–4, 15.

33 Zbigniew Towalski, 'The Integration of Knowledge within Science, Technology and Industry: Enzymes: a Case Study', Doctoral dissertation, Aston University, Birmingham, 1985. I am grateful to Dr Towalski for permission to draw on material presented in his thesis. Production figures drawn from reports by B. Wolnak are presented on pp. 183–84.

34 Towalski, "The Integration of Knowledge", p. 184. Dr Towalski extracted these figures from various publications of Bernard Wolnak and Associates.

35 C. Dambmann, P. Holm, V. Jensen, and M.H. Nielsen, 'How Enzymes Got into Detergents' *Development in Industrial Microbiology* 12(1971): 11–23.

36 Fermentation Industries Section, IUPAC, 'Worldwide Survey of Fermentation Industries'; K. Bernhauer, *Gärungschemisches Praktikum,* 2d ed. (Berlin: Springer, 1939), pp. 1–7.

37 J.W. Foster, 'Speculative Discourse on Where and How Microbiological Science as Such May Advise with Application', in *Global Impacts of Applied Microbiology,* ed. M. P. Starr (Stockholm: Almquist & Wiksel, 1964), p. 70.

38 Towalski, 'The Integration of Knowledge within Science, Technology, and Industry', p. 379.

39 M. Lamb and G.D. Walker, 'Industrial Applications of Continuous Culture Processes', in *Continuous Culture of Microorganisms,* Society of Chemical Industry Monograph 12 (1961): 254–64.

40 J. Monod, 'La technique de culture continue', *Annales Institut Pasteur* 79 (1950): 390–410; A. Novick and Leo Szilard, 'Experiments with the Chemostat on Spontaneous Mutations of Bacteria', *Proceedings of the National Academy of Sciences* 36 (1950): 708–19.

41 Sir Charles Dodds in a letter to Steven Roskill, quoted in Stephen Roskill, *Hankey: Man of Secrets,* vol. 3 (London: Collins, 1974), p. 603.

42 Quoted in Robert Harris and Jeremy Paxman, *A Higher Form of Killing: The Secret Story of Gas and Germ Warfare* (London: Chatto & Windus, 1982), p. 70.

43 Stephen Roskill, *Hankey: Man of Secrets,* pp. 321–25; Harris and Paxman, *A Higher Form of Killing.*

44 Interestingly, from this point Fildes's work disappeared from the mainstream of academic biochemistry. Robert Kohler concludes of Fildes and his collaborator Woods: 'They spent the years of World War II at the Army's experimental station at Porton doing secret bacteriological work and watching others build upon their innovation. When Fildes returned to reconstitute his research team in 1945, the world had passed him by'. Robert Kohler, 'Bacterial Physiology: The Medical Context', *Bulletin of the History of Medicine* 59 (1985): 54–74. However, the postwar development of continuous fermentation at Porton grew out of Woods and Fildes's interest in bacterial metabolism.

45 Mr Gradon Carter, Private communication.
46 George W. Merck, 'Peacetime Implications of Biological Warfare', *Chemical and Engineering News* 24, no. 10 (25 May 1936): 1346–49.
47 C.E. Gordon Smith, 'The Microbiological Research Establishment, Porton – Research Establishments in Europe: 69 Porton Down', *Chemistry and Industry,* 4 March 1967, pp. 338–46.
48 Keith Norris, Interview with the author, 21 September 1989.
49 Postgate to Mr Gradon Carter, 20 August 1980. I am grateful for permission to quote this letter.
50 This comment was made at an after-dinner talk given by Herbert at the July 1975 Continuous Culture Symposium in Oxford. I am grateful to Mr Charles Evans for a tape-recording of this speech.
51 The atmosphere of the time is recalled by A.T. James, 'The Discovery of Gas-Liquid Chromatography: A Personal Recollection', in *Historical Aspects of Gas Separation,* Royal Society of Chemistry Special Publication 62 (London: Royal Society of Chemistry, 1987), pp. 175–200.
52 D. Herbert, 'Stoichemetric Aspects of Microbial Growth', in *Continuous Culture 6: Applications and New Fields,* ed. A.C.R. Dean et al. (Chichester: Ellis Horwood for the Society of Chemical Industry, 1976), p. 3.
53 Charles Evans, Interview with the author, 21 September 1989.
54 I. Málek, 'The Role of Continuous Processes and Their Study in the Present Development of Science and Production', in *Theoretical and Methodological Basis of Continuous Culture of Micro-organisms,* ed. I. Málek and Z. Fencl (Prague: Czech Academy of Science, 1966), pp. 11–30.
55 D.W. Tempest, 'The Place of Continuous Culture in Microbiological Research', *Advances in Microbial Physiology,* 4 (1970): 223–50.
56 The issues are briefly described by John Postgate, in his *Microbes and Man* (London: Penguin, 1969), pp. 33–34. I am grateful to Professor Postgate, who was a member of Butlin's team, for clarifying this development for me.
57 J.S. Hough, 'Production of Beer by Continuous Fermentation', in *Continuous Culture of Microorganisms,* pp. 219–29.
58 R.N. Greenshields, 'Acetic Acid: Vinegar', in *Economic Microbiology,* vol. 2 *Primary Products of Metabolism,* ed. A.H. Rose (New York: Academic Press, 1978), pp. 121–86.
59 J.S. Hough, *The Biotechnology of Malting and Brewing* (Cambridge University Press, 1985), p. 129.
60 R. Seligman, 'Brewing Engineering and the Future', *Journal of the Institute of Brewing* 36 (1930): 288–97.
61 R. Seligman, 'The Research Scheme of the Institute – Its Past and Its Future', *Journal of the Institute of Brewing* 54, no. 3 (1948): 133–44.
62 'Guinness's New Laboratories at Park Royal', *Brewers Guardian* (August 1955): 26; BIRF 'Half-Yearly Report No. 2', 26 October 1953, 57/47, p. 93, Institute of Brewing, London.
63 BIRF Research Board Minutes, 1952–56, 5, Sixth Progress Report, p. 29, Institute of Brewing.
64 A.J.R. Purssell and M.J. Smith, 'Continuous Fermentation', *Proceedings of the 11th European Brewery Convention* (Madrid: European Brewery Convention, 1967), pp. 155–67.

65 J.N.W. Payne, 'Mashing on the Larger Scale', *Brewers Guardian* (November 1962): 75–78.
66 I. Heilbron, 'Brewing in Relation to Natural Science', *Proceedings of the Royal Society*, B, 143 (1955): 178–199; I. Heilbron, 'Reflections on Science in Relation to Brewing', Horace Brown Memorial Lecture, *Journal of the Institute of Brewing* 65 (1959): 144–54; A.H. Cook, 'Unfolding Pattern of Research: Brewing Industry of the Future', *Brewer's Guardian*, June 1960, pp. 17–25.
67 G.A. Dummett, 'Chemical Engineering in the Biochemical Industries', *The Chemical Engineer* (July–August 1969): 306–310.
68 G.A. Dummett, 'The Engineering of Continuous Brewing', *Wallerstein Laboratories Communications* 25 (April 1962), 19–36. S.R. Green, 'Past and Current Aspects of Continuous Beer Fermentation', *Wallerstein Laboratories Communications* 25 (December 1962): 337–48.
69 'Replies to Mr F.E. Lord's Questionnaire dated 30.11.61', in Arthur Guinness Son & Co. Collection, Ms. 2013, Science Museum Library, London (hereafter Guinness Papers).
70 D.T. Shore, 'Chemical Engineering of the Continuous Brewing Process', *Chemical Engineer* (May 1968): 99–109; [Arthur Guinness Son & Co.] 'General Report on Continuous Brewing Research in Production Research Department Park Royal, 1957–1966', Guinness Papers.
71 [Arthur Guinness Son & Co.] 'General Report on Continuous Brewing Research in Production Research Department Park Royal, 1957–1966', Guinness Papers.
72 G.A. Dummett, *From Little Acorns: A History of the A.P.V. Company Limited* (London: Hutchinson Benham, 1981), pp. 196–98. See also R.N. Greenshields and E.L. Smith, 'Tower Fermentation Systems and the Applications', *The Chemical Engineer* (May 1971): 182–90.
73 J.H. Hastings, 'Development of the Fermentation Industries in Great Britain', *Advances in Applied Microbiology* 14(1971), p. 42.
74 Aiba, Humphrey, and Millis, *Biochemical Engineering*, pp. 128–62.
75 R.C. Righelato and R. Elsworth, 'Industrial Applications of Continuous Culture: Pharmaceutical Products and Other Products and Processes', *Advances in Applied Microbiology* 13 (1970):399–465.

CHAPTER 6. BIOTECHNOLOGY – THE GREEN TECHNOLOGY

1 Carl-Göran Hedén, 'Biological Research Directed towards the Needs of Underdeveloped Areas', *TVF* 7 (1961), p. 298.
2 This phrase is used by Richard M. Krause in his 'Is the Biological Revolution a Match for the Trinity of Despair?' *Technology in Society* 4 (1982): 267–82.
3 Leo Szilard, *The Voice of the Dolphins and Other Stories* (London: Victor Gollancz, 1961), pp. 21–101.
4 Richard A. Cellarius and John Platt, 'Councils of Urgent Studies', *Science* 177 (25 August 1972): 670–76.
5 R. Jungk, *The Everyman Project: Resources for a Human Future* (London:

Thames & Hudson, 1976), pp. 21–23. The original is entitled *Jahrtausendmensch.*

6 Janine Clarke and Robin Clarke, 'The Biotechnic Research Community', *Futures* 4 (June 1972): 169–73. A pen portrait of Clark is provided by Jungk, *The Everyman Project,* pp. 119–23.

7 Johan Galtung, 'Foreword' to Russell E. Anderson, *Biological Paths to Self-Reliance: A Guide to Biological Solar Energy Conversion* (New York: Van Nostrand, 1979), p. ix.

8 Raymond Ewell, 'The Rising Giant: The World Food Problem', in *Engineering of Unconventional Protein Production,* ed. Herman Bieber, Chemical Engineering Progress, Symposium series no. 93, 65 (1969): 1–4.

9 R. Barker, 'Socio-economic Impact', in *Agricultural Biotechnology: Opportunities for International Developments,* ed. Gabrielle J. Persley (Wallingford, Oxon: CAB International, 1990), 299–310.

10 This parallel was drawn by D.E. Hughes, 'The Scope of the Symposium', in *Microbiology: Proceedings of a Conference Held in London 19 and 20 September 1967,* ed. P. Hepple (London: Institute of Petroleum, 1968), pp. 1–2.

11 For a more balanced assessment, see Robert Walgate, *Miracle or Menace: Biotechnology and the Third World* (London: Panos Institute, 1990). For an optimistic vision, see Albert Sasson, *Biotechnologies and Development* (Paris: UNESCO, 1988). A sceptical analysis from a Marxist viewpoint is expressed in David Goodman, Bernardo Sorj, and John Wilkinson, *From Farming to Biotechnology: A Theory of Agro-Industrial Development* (Oxford: Blackwell Publishers, 1987).

12 See, e.g., the papers in S. Lackoff, 'Biotechnology and Developing Countries', *Politics and the Life Sciences* 2ii (1984): 151–83.

13 Russell E. Anderson, *Biological Paths to Self-Reliance: A Guide to Biological Solar Energy Conversion* (New York: Van Nostrand, 1979), p. xvii. For Hedén's own account of his work, see Carl-Gören Hedén, 'The GIAMS – A Contribution to Technology Transfer', in *From Recent Advances in Biotechnology and Applied Biology,* ed. S.T. Chang, K.Y. Chan, and N.Y.S. Woo (Hong Kong: Chinese University Press, 1988), pp. 63–74.

14 J.G. Baer, 'Biology and Humanity: The International Biological Programme', *Impact of Science on Society* 17 (1967): 315–28.

15 Paul Forman, 'Behind Quantum Electronics: National Security as Basis for Physical Research in the United States, 1940–1960', *Historical Studies in the Physical and Biological Sciences* 18 (1987): 149–229.

16 Marcel Florkin, 'Ten Years of Science at UNESCO', *Impact of Science on Society* 7 (1956): 121–46.

17 [UNESCO], *Records of the General Conference,* 8th Sess. 1954 (Paris: UNESCO, 1955), pp. 694–96.

18 C.-G. Hedén, 'Microbiology in World Affairs', *Impact of Science on Society* 17 (1967): 187–208.

19 C.-G. Hedén, Interview with the author, 1 March 1989.

20 Hedén, 'Microbiology in World Affairs', p. 190.

21 'The Contribution of Microbiology to World Food Supplies'. A resolution prepared by the Permanent Section on Food Microbiology & Hygiene and adopted unanimously by the plenary session of IAMS, 24 August 1962, in *Recent Progress in Microbiology.* Symposia held at the 8th Inter-

national Congress for Microbiology. Montreal, 1962, ed. N.E. Gibbons (Toronto: University of Toronto Press, 1963), pp. 719–20.

22 Gibbons, *Recent Progress in Microbiology*, p. 717.

23 Mortimer P. Starr and Carl-Göran Hedén, 'Preface', in *Global Impacts of Applied Microbiology*, ed. Mortimer P. Starr (Stockholm: Almquist & Wiksell, 1964), p. i.

24 Starr and Hedén, 'Preface', p. iv.

25 E.L. Gaden, Jr, 'Process and Equipment Design for Less-Developed Areas', in *Global Impacts of Applied Microbiology*, ed. Starr, pp. 338–43.

26 C.-G. Hedén, 'Recommendations Accepted at the Plenary Meeting, August 2 1963: Conference on Global Impacts of Applied Microbiology', *Global Impacts of Applied Microbiology*, ed. Starr, pp. 520–30.

27 J.W.M. la Rivière, 'Biotechnology in Development Cooperation: A Donor Countries' View', in *Biotechnology in Developing Countries*, ed. P.A. Van Hemert et al. (Delft: Delft University Press, 1983), pp. 1–17.

28 E.J. DaSilva and C.-G. Hedén, 'The Role of International Organizations in Biotechnology: Cooperative Efforts', in *Comprehensive Biotechnology*, ed. Murray Moo Young, vol. 4 (London: Pergamon, 1985), pp. 717–49.

29 Instrumental in these developments were the late Roger Pater (USA), J.W.M. la Rivière (now secretary-general of ICSU), Dr A.C.J. Burgers, and Dr E. DaSilva (UNESCO).

30 Raymond A. Zilinskas, 'The International Centre for Genetic Engineering and Biotechnology: A New International Scientific Organisation', *Technology in Society* 9 (1987):47–61.

31 Quoted in K.R. Butlin, 'Survey of Research on the Biological Fixation of Nitrogen'. Shell Research, 1962. I am grateful to Professor John Postgate for showing me his copy of this report.

32 Butlin, 'Survey of Research on the Biological Fixation of Nitrogen'.

33 E.J. DaSilva, A.C.J. Burgers, and R.J. Olembo, 'UNESCO, UNEP and the International Community of Culture Collections', in *Proceedings of the Third International Conference on Culture Collections*, ed. Fabian Fernandes and Raby Costa Pereira (Bombay: University of Bombay 1977), pp. 107–28.

34 D.N. Li, 'Biogas Production in China: An Overview', in *Microbial Technology in the Developing World*, ed. E.J. DaSilva et al. (Oxford University Press, 1987), pp. 196–208.

35 Hal Bernton, William Kovarik, and Scott Sklar, *The Forbidden Fuel: Power Alcohol in the Twentieth Century* (New York: Boyd Griffin, 1982), p. 147.

36 Bernton, Kovarik, and Sklar, *The Forbidden Fuel*, p. 160.

37 Lennart Enebo, 'Growth of Algae for Protein: State of the Art', in *Engineering of Unconventional Protein Production*, pp. 80–86.

38 A.C. Thaysen, 'Food and Fodder Yeast', in *Yeasts*, ed. W. Roman (The Hague: W. Junk, 1957), pp. 155–210.

39 See A.C. Thaysen, 'Food Yeast: Its Nutritive Value and Its Production from Empire Sources', *Journal of the Royal Society of Arts* 93 (8 June 1945): 353–64.

40 H.J. Bunker, 'Microbial Food', in *Global Impacts of Applied Microbiology*, pp. 234–40, cited at 238.

41 A.E. Humphrey, 'A Critical Review of Hydrocarbon Fermentations and Their Industrial Utilization', *Biotechnology and Bioengineering* 9 (1967): 3–24. See Felix Just and Willy Schnabel, 'Submerse Massenzüchtung von

Bakterien auf nicht-kohlenhyrathaltigen Nährstoffen 1. Ein Beitrag zur biotechnischen Fettsynthese', *Die Branntweinwirtschaft* 2(1948):113–15. I am grateful to Professor H. Dellweg for bringing this paper to my attention.

42 Alfred Champagnat and Jean Adrian, *Petrole et Proteines* (Paris: Doin, 1974).

43 Richard I. Mateles and Steven R. Tannenbaum, *Single Cell Protein* (Cambridge, Mass: MIT Press, 1968).

44 G.B. Carter, 'Is Biotechnology Feeding the Russians?' *New Scientist* 90 (23 April 1981): 215–17.

45 [PAG], *The PAG Compendium.* Collected Papers Issued by the Protein-Calorie Advisory Group of the United Nations System, 1956–1973. Vol. C2 (New York: Worldmark, 1979).

46 David H. Sharp, *Bio-Protein Manufacture: A Critical Assessment* (Chichester: Ellis Horwood, 1989).

47 K. Arima, 'The Problems of Public Acceptance of S.C.P.-s.', in *International Symposium on SCP,* ed. J.C. Senez, Proceedings of symposium held 28–31 January 1981 organized by DGRST, IAMS, and APRIA (Paris: Technique et Documentation (Lavoisier), 1983), pp. 145–62.

48 J. Edelman, A. Fewell, and G.L. Solomons. 'Myco-Protein – A New Food', *Reviews in Clinical Nutrition* 53 (1983): 471–80.

49 Arima, 'The Problems of Public Acceptance of S.C.P.-s', p. 158.

50 See Kiyoaki Katoh, 'Statement on Current Status of SCP Production in Japan', in *Single Cell Protein: Proceedings of the International Symposium Held in Rome, Italy, on November 7–9, 1973,* ed. P. Davis (London: Academic Press, 1974), pp. 223–32, including translations of two petitions, one from five major consumer organizations and one from the Tokyo Metropolitan Consumers Council.

51 la Rivière, 'Biotechnology in Developing Countries: A Donor Countries' View', pp. 1–2.

CHAPTER 7. FROM PROFESSIONAL TO POLICY CATEGORY

1 Pearce Wright, 'Time for Bug Valley', *New Scientist* 82 (5 July 1979): 27–29.

2 Jean-Jacques Servan-Schreiber's best selling *Le défi americain* (Paris: Editions Denoel, 1967) was the classic expression of anxiety about American dominance.

3 Kendall Bailes, *Environmental History: Critical Issues in Comparative Perspective* (Lanham, Md.: University Presses of America, 1985); Donald Fleming, 'Roots of the New Conservation Movement', *Perspectives in American History* 6 (1972): 7–91. See Max Nicholson, *The Environmental Age* (Cambridge University Press, 1987), for a British perspective.

4 Hal Bernton, William Kovarik, and Scott Sklar, *The Forbidden Fuel: Power Alcohol in the Twentieth Century* (New York: Boyd Griffin, 1982).

5 For the Carter years, see Frank Press, 'Science and Technology in the White House, 1977 to 1980'. Pt. 1, *Science* 211 (9 January 1981): 139–45; Pt. 2, *Science* 211 (16 January 1981): 249–56; on the shift of the Reagan

years, see Norman Waks, 'Consequences of the Shifts in U.S. Research and Development Policy', *R&D Management* 15 (1985): 191–96.

6 D. Thomas, *Production of Biological Catalysts, Stabilization and Exploitation,* EUR 6079 (Luxembourg: Office for Official Publications of the European Communities, 1978), 66–70.

7 Arnold Thackray, P. Thomas Carroll, J.L. Sturchio, and Robert Bud, *Chemistry in America, 1876–1976: Historical Indicators* (Dordrecht: Reidel, 1984), p. 132.

8 See Stephen P. Strickland, *Politics, Science and Dread Disease: A Short History of United States Medical Research Policy* (Cambridge, Mass: Harvard University Press, 1972).

9 National Science Board, *Science Indicators 1972: Report of the National Science Board 1973* (Washington D.C.: U.S. Government Printing Office, 1973), p. 118.

10 June Goodfield, *Cancer Under Siege* (London: Hutchinson, 1975), p. 168. By 1 April 1985, 670,000 people were engaged.

11 The information here is taken from Dirk Hanson, *The New Alchemists: Silicon Valley and the Microelectronics Revolution* (Boston: Little, Brown, 1982).

12 John W. Bennett and Solomon B. Levine, 'Industrialization and Social Deprivation: Welfare, Environment and the Post-Industrial Society in Japan', in *Japanese Industrialization and Its Social Consequences,* ed. Hugh Patrick (Berkeley: University of California Press 1976), pp. 439–92.

13 Bennett and Levine, 'Industrialization and Social Deprivation', p. 456.

14 Japan, Science and Technology Agency, *Outline of the White Paper on Science and Technology: Aimed at making Technological Innovations in Social Development,* February 1977. Trans. Foreign Press Centre, pp. 176–78.

15 Kin-ichirō Sakaguchi and Yōnosuke Ikeda, 'Applied Microbiology in Japan: An Outline of its Historical Development and Characteristics', in *Profiles of Japanese Science and Scientists,* ed. Hideki Yukawa (Tokyo: Kodansha, 1970), p. 96.

16 Yukawa, *Profiles of Japanese Science and Scientists.*

17 See Keiko Nakamura, 'Studies on Life Sciences. Part II. Role of Life Sciences with Special Reference to Research in the Mitsubishi-Kasei Institute', *Technocrate* 6 (1973): 48–52.

18 *Outline of the White Paper on Science and Technology.* p. 18.

19 A. Wada, 'One step from Chemical Automatons', *Nature* 257 (23 October 1975): 633–34. The subsequent reports were 'Present and Future of Enzyme Technology' and 'Report of Current Advances in Research of Enzyme Technology' whose chapter titles and introductions are reprinted in Appendix 1, 'Report on the Current State of Planning of Life Science Promotion in Japan', in A. Rörsche, *Genetic Manipulations in Applied Biology: A Study of the Necessity, Content and Management Principles of a Possible Community Action* EUR 6078 (Luxembourg: Office for Official Publications of the European Communities, 1979), pp. 59–63.

20 Rörsche, *Genetic Manipulations in Applied Biology,* p. 62.

21 For a reflection upon German technology policy, see Ernst-Jürgen Horn, *Management of Industrial Change in Germany,* Sussex European Papers no. 13 (Brighton: University of Sussex, 1982).

22 Hartmut Bossel, 'Die vergessenen Werte', in *Der Grüne Protest: Heraus-*

forderung durch die Umweltparteien, ed. Rudolf Brun (Frankfurt: Fisher, 1978), pp. 7–17. The entire book is an interesting collection of manifestos for the then new green movement.

23 Shelia Jasanoff, 'Technological Innovation in a Corporatist State: The Case of Biotechnology in the Federal Republic of Germany', *Research Policy* 14 (1985): 23–38.

24 [BMBW], *Erster Ergebnisbericht des ad hoc Ausschusses 'Neue Technologien' des Beratenden Ausschusses für Forschungspolitik,* Schriftenreihe Forschungsplanung 6 (Bonn: BMBW, December 1971). All students of the development of German biotechnology policy owe a great debt to Klaus Buchholz, himself a key participant, for his article 'Die gezielte Förderung und Entwicklung der Biotechnologie', in *Geplante Forschung,* ed. Wolfgang van den Daele, Wolfgang Krohn, and Peter Weingart (Frankfurt: Suhrkamp, 1979), pp. 64–116.

25 R. Brunner and W. Friedrich, 'In Memoriam: Konrad Bernhauer', *Mitteilungen der Versuchsstation für das Gärungsgewerbe in Wien,* no. 2 (1976): 22–28.

26 The change in name is indicated in his 1964 article 'Synthesen auf dem Vitamin B-12 Gebiet', *Biochemische Zeitschrift* 340 (1964), p. 471. I am indebted to Professor Herbert Dellweg for pointing out this article and association to me.

27 H. Dellweg, 'Die Geschichte der Fermentation – Ein Beitrag zur Hundertjahrfeier des Instituts für Gärungsgewerbe und Biotechnologie zu Berlin', in *100 Jahre Institut für Gärungsgewerbe und Biotechnologie zu Berlin, 1874–1974. Festschrift* (Berlin: Institut für Gärungsgewerbe und Biotechnologie, 1974), pp. 40–41.

28 H.-J. Rehm, 'Modern Industrial Microbiological Fermentations and Their Effects on Technical Developments', *Angewandte Chemie,* International ed. 9 (1970): 936–45, p. 945.

29 [Wilhelm Schwartz], 'Biotechnik und Bioengineering', *Nachrichten aus Chemie und Technik* 17 (1969): 330–331. I am grateful to Dr Rudolph of Verlag Chemie for his information that the author was Dr Wilhelm Schwartz who had since 1967 led a group at the Gesellschaft für Molekularbiologische Forschung (GMBF).

30 Jasanoff, 'Technological Innovation in a Corporatist State', p. 29.

31 DECHEMA, *Biotechnologie: Eine Studie über Forschung und Entwicklung – Möglichkeiten, Aufgaben und Schwerpunkte der Förderung* (Frankfurt: DECHEMA, January 1974). BMFT had been represented at the first meeting of the DECHEMA working group which had been established following pressure from Professor Patat. I am grateful to Klaus Buchholz for advice and to Professor Behrens for the loan of his copy of the report.

32 'Die Biotechnologie behandelt den Einsatz biologischer Prozesse im Rahmen technischer Verfahren und industrieller Produktionen. Sie ist also eine anwendungsorientierte Wissenschaft der Mikrobiologie und Biochemie in enger Verbindung mit der technischen Chemie und der Verfahrenstechnik'. DECHEMA, *Biotechnologie.* p. vii.

33 [BMFT], *Biotechnologie,* BMFT-Leistungplan 04, Plan Periode: 1979–1983 (Bonn: BMFT, 1978), p. 15.

34 GMBF/GBF, *Entwicklung eines Forschungsinstituts 1965–1975* (Hannover: Stifung Volkswagenwerke, 1975).

35 GMBF/GBF, *Entwicklung eines Forschungsinstituts,* p. 4.
36 'The European Federation of Biotechnology', *Chemistry and Industry* (21 October 1978): 781. See also Dieter Behrens, 'Europäische Föderation Biotechnologie: Vier Jahre nach der Gründung', *Swiss Biotech* 1 (1983): 11–16.
37 [K. Buchholz], 'Editorial', *EFB Newsletter,* no. 4 (December 1981), p. 2.
38 Harold Hartley, 'Chemical Engineering at the Cross-Roads', *Transactions of the Institution of Chemical Engineers* 30 (1952): 13–19.
39 Harold Hartley, 'Chemical Engineering – The Way Ahead', *Transactions of the Institution of Chemical Engineers* 33 (1955): 20–26.
40 T.K. Walker, 'A History of the Development of a School of Biochemistry in the Faculty of Technology, University of Manchester', *Advances in Applied Microbiology* 12 (1970): 1–10.
41 Ronald W. Clark, *The Life of Ernst Chain: Penicillin and Beyond* (London: Weidenfeld, 1985), pp. 176–97. See Ernst Chain, 'Thirty Years of Penicillin Therapy', *Proceedings of the Royal Society of London* B 179 (1971): 293–319.
42 Gradon Carter, 'The Microbiological Research Establishment and Its Precursors at Porton Down: 1940–1979' (pt. 1), *The ASA Newsletter* 91–6, issue 27, December 1991, and Gradon Carter, 'The Microbiological Research Department and Establishment: 1946–1979' (pt. 2), *The ASA Newsletter* 92–1, issue 28, 1992.
43 Working Party on Industrial Microbiology (P.W. Brian, Chairman), 'Report on the State of Research into Economic Microbiology in the United Kingdom', 1960.
44 Postgate to G.B. Carter 20 August 1980. I am grateful to Professor Postgate for permission to quote from his letter.
45 Ken Sargeant to the author, 24 October 1991.
46 I am grateful for the advice of David Sharp, then secretary of the SCI, in a personal communication.
47 Sir Harold Hartley, 'Commentary', *Process Biochemistry* 2, no. 4 (April 1967), p. 3. I am grateful to Dr Leo Hepner for clarifying the background to this article, which is reflected in correspondence in the Hartley Archive, Churchill College, Cambridge. See Hartley to H. Krebs, 22 January 1969, Box 303.
48 See R.F. Homer to Hartley, 19 January 1967, Hartley Archives, Box 247.
49 A.C.R. Dean to Hartley, 30 December 1968, Hartley Archives, Box 303.
50 A.N. Emery, 'Biochemical Engineering: An Industrial Fellowship Report for the SRC and Institution of Chemical Engineers' (Birmingham, University of Birmingham, 1976).
51 See S. Brenner, B.S. Hartley, and P.J. Rodgers, eds., 'New Horizons in Industrial Microbiology', *Philosophical Transactions of the Royal Society* B 290 (1980): 277–430.
52 Advisory Council for Applied Research and Development, Advisory Board for the Research Councils, and The Royal Society, *Biotechnology: Report of a Joint Working Party* (London: HMSO, 1980). Note that the phrase 'service industries' is omitted in the definition on p. 14 and only appears in the otherwise identical formulation on p. 7.
53 Alan T. Bull, Derek C. Ellwood, and Colin Ratledge, 'The Changing Scene in Microbial Technology', in *Microbial Technology: Current State, Future*

Prospects, ed. A.T. Bull, D.C. Ellwood, and C. Ratledge, Society of General Microbiology, Symposium no. 29 (Cambridge University Press for the Society for General Microbiology, 1979).

54 Alan Bull and John Bu'Lock, 'The Living Micro Revolution', *New Scientist* 82 (7 June 1979): 808–10.

55 On the Netherlands, see A. Rip and A. Nederhof, 'Between Dirigism and Laissez-faire: Effects of Implementing the Science Policy Priorities for Biotechnology in the Netherlands', *Research Policy* 15 (1986): 253–68.

56 These papers were compared and analyzed by Ken Sargeant in a FAST paper, 'Notes on Three French Biotechnology Reports: A Review Paper' FAST XII/1160/81/EN.

57 Michel Godet and Oliver Ruyssen, *The Old World and the New Technologies: Challenges to Europe in a Hostile World* (Luxembourg: Office for Official Publications of the European Communities, 1981). The French original was entitled *L'Europe en mutation.*

58 Commission of the European Communities, 'FAST Subprogramme C: Bio-Society', FAST/ACPM/79/14–3E, 1979, p. 3.

59 Dieter Behrens, K. Buchholz, and H.-J. Rehm, *Biotechnology in Europe – A Community Strategy for European Biotechnology* (Frankfurt: DECHEMA for European Federation of Biotechnology, 1983). The bottlenecks were applied genetics; cell physiology (here defined as the 'key to progress'); animal and plant cell culture, enzymes and immobilized catalysts; analysis, measurement, and control; biochemical engineering; and downstream processing. The six products and processes were energy, chemical feedstocks, feed and food, fine chemicals and pharmaceuticals, environmental process, and medical analytical systems.

60 An edited version of the final report was published as FAST Group, *Eurofutures: The Challenge of Innovation* (London: Butterworth, 1987).

CHAPTER 8. THE WEDDING WITH GENETICS

1 G. Rattray Taylor, *The Biological Time Bomb* (New York: New American Library, 1968), p. 13.

2 Arnold L. Demain, 'The Marriage of Genetics and Industrial Microbiology – After a Long Engagement, A Bright Future', in *Genetics of Industrial Microorganisms*, ed. Z. Vaněk, Z. Hošťálek, and J. Cudlin (Prague: Academia, 1973), p. 19.

3 Elmer Gaden, 'Biochemical Engineering: Where Has it Been and Where is it Going?' *Biotechnology Letters* 2(1980): 336, 'During the last decade, biochemical engineering has been much less productive. An important reason for this has undoubtedly been the reduced pace of development of biotechnology itself'.

4 Max Kennedy, 'The Evolution of the Word "Biotechnology" ', *Trends in Biotechnology* 9(1991): 218–20.

5 Sharon McAuliffe and Kathleen McAuliffe, *Life for Sale* (New York: Coward, McCann & Geoghegan, 1981).

6 For an attempt at 'total history' of public attitudes to nuclear power, see Spencer Weart, *Nuclear Fear: A History of Images* (Cambridge, Mass.: Harvard University Press, 1988).

7 J.R. Ravetz, 'The DNA Controversy and Its History', in *The Social Assess-*

ment of Science: Proceedings, Wissenschaftsforschung 13 Report, Science Studies (Bielefeld: Kleine Verlag, 1982): 79–90. Discussions of the press coverage of biotechnology's costs and benefits are provided by Rae Goodell, 'How to Kill a Controversy: The Case of Recombinant DNA', in *Scientists and Journalists,* ed. S. Fiedman, S. Dunwoody, and C. Rogers (New York: Free Press, 1986), pp. 170–181. I am grateful to Bruce Lewenstein for pointing me to this article. See also Nancy Pfund and Laura Hofstadter, 'Biomedical Innovation and the Press', *Journal of Communication* 31 (1981): 138–54.

8 John Naisbitt, *Megatrends: Ten New Directions Transforming Our Lives* (London: Macdonald, 1984), particularly p. 73. The U.S. edition is two years older. The assertion that biology would replace chemistry and physics as the science of the future repeats the prophecy by Gordon Rattray Taylor quoted at the beginning of this chapter and, of course, echoes, unconsciously, the assertion by Huxley introducing Hogben almost half a century earlier.

9 See Ronald Rainger, Keith R. Benson, and Jane Maienschein, eds., *The American Development of Biology* (Philadelphia: University of Pennsylvania Press, 1988); Charles Rosenberg, *No Other Gods: On Science and American Social Thought* (Baltimore: Johns Hopkins University Press, 1976), pp. 196–209.

10 The distinction between genetics and microbiology, and its erosion, is explored by Joshua Lederberg, 'Genetic Recombination in Bacteria: A Discovery Account', *Annual Review of Genetics* 21(1987): 23–46.

11 Thomas Brock, *The Emergence of Bacterial Genetics* (Cold Spring Harbor, N.Y.: Cold Spring Harbor Laboratory Press, 1990), p. 2.

12 Joshua Lederberg, 'Edward Lawrie Tatum, 1909–1975', *Biographical Memoirs* (National Academy of Sciences) 57 (1990): 357–86. The work of Tatum's father has recently been discussed by John Patrick Swann, 'Arthur Tatum, Parke-Davis, and the Discovery of Maphersen as an Antisyphilitic Agent', *History of Medicine and Allied Sciences* 40 (April 1985): 167–87.

13 Edward L. Tatum, 'A Case History in Biological Research', Nobel Lecture, 11 December 1958.

14 Edward L. Tatum, 'Molecular Biology, Nucleic Acids, and the Future of Medicine', *Perspectives in Biology and Medicine* 10 (1966–67), p. 31.

15 Gordon Wolstenholme, 'Preface', in *Man and His Future,* ed. Gordon Wolstenholme (London: J. A. Churchill, 1963), p. v.

16 This and the subsequent quotations are taken from Joshua Lederberg, 'Biological Future of Man', in Wolstenholme, ed., *Man and His Future,* pp. 263–73.

17 The phrase seems to have been first used by the radical sociologist Bernhard J. Stern, 'Human Heredity and Environment', *Science and Society* 14 (1950): 122–33. Here he refers to the Lysenkoist controversy over the fundamental role of genetics. It is however unclear whether this usage had any connection with later uses of the term (although Lederberg first drew attention to this paper).

18 Edward L. Tatum, 'Perspectives from Physiological Genetics', in *The Control of Human Heredity and Evolution,* ed. T.M. Sonneborn (New York: Macmillan, 1965), p. 22.

19 Rollin D. Hotchkiss, 'Portents for a Genetic Engineering', *Journal of Heredity* 56 (1965), p. 202.
20 H.J. Muller, 'Perspectives for the Life Sciences', *Bulletin of the Atomic Scientists* 20 (January 1964): 3–7.
21 Joshua Lederberg, 'Experimental Genetics and Human Evolution', *The American Naturalist* 100 (1966): 519–31.
22 [Syntex], *A Corporation and a Molecule* (Palo Alto: Syntex Laboratories, 1966), p. 102.
23 B.M. Richards and N.H. Carey, 'Insertion of Beneficial Genetic Information', Searle Research Laboratories, 16 January 1967. I am grateful to Dr Richards, who later founded British Bio-technology Ltd, for giving me a copy of this memorandum and to himself, Dr Carey and G.D. Searle & Co. for permission to quote from it.
24 Kornberg graphically describes the excitement that followed his announcement in *For the Love of Enzymes: The Odyssey of a Biochemist* (Cambridge, Mass.: Harvard University Press, 1989), pp. 200–206.
25 'All the Way with DNA', *New Scientist* 39 (18 July 1968): 142.
26 For the book and the reaction, see J.D. Watson, *The Double Helix,* ed. G.S. Stent (New York: Norton, 1980).
27 Taylor, *The Biological Time Bomb.*
28 D. Paterson, ed., *Genetic Engineering* (London: BBC, 1969).
29 J. Lederberg, interview with the author, 3 July 1991.
30 Ronald Cape interview with Charles Weiner, 19 April 1978, p. 16. MC100, MIT Institute Archives and Special Collections, MIT Libraries, Cambridge, Mass.
31 Ronald Cape interview with Charles Weiner, 19 April 1978, p. 14.
32 U.S. Congress, House of Representatives, Subcommittee of the Committee on Appropriations, *Genetics Research,* 91st Cong., 2d sess. 1971, pp. 914–51.
33 'Genetic Engineering – Certainties and Doubts', *New Scientist* 47 (24 September 1971): 614.
34 See James D. Watson and John Tooze, *The DNA Story: A Documentary History of Gene Cloning* (San Francisco: Freeman, 1981).
35 R. Roblin, interview with Charles Weiner, 21 April and 2 May 1975, pp. 9–10, MC100, MIT.
36 Kass to Berg, 30 October 1970, MC100, MIT. Reprinted in Sheldon Krimsky, *Genetic Alchemy: The Social History of the Recombinant DNA Controversy* (Cambridge, Mass.: MIT Press, 1982), pp. 33–36.
37 See the evocation by June Goodfield, *Playing God: Genetic Engineering and the Manipulation of Life* (London: Hutchinson, 1977), p. 111.
38 Krimsky, *Genetic Alchemy.*
39 Joshua Lederberg, 'A Geneticist on Safeguards', *New York Times,* 11 March 1975. This was a reply to a report by V.K. McElheny, 'World Biologists Tighten Rules on "Genetic Engineering" Work', *New York Times,* 28 February 1975.
40 U.S. Congress, U.S. House of Representatives, Committee on Science and Technology, Subcommittee on Science Research and Technology, *Science Policy Implications of DNA Recombinant Molecule Research,* 95th Cong. 1st sess., 1977, Testimony of Leon Kass, 'Human Genetic Engineering', p. 1102.

41 See, e.g., the resentment in David Perlman, 'Preface', *Advances in Applied Microbiology* 11 (1970), p. viii; also Dunnill to Berg, 19 September 1974, MC 100, MIT Libraries.
42 Goodfield, *Playing God,* p. 101.
43 BBC, 'Certain Types of Genetic Research Should Be Suspended', Broadcast, BBC2, 16 September 1974, transcript, p. 8.
44 See Ronald W. Clark, *The Life of Ernst Chain: Penicillin and Beyond* (London: Weidenfeld, 1985), pp. 176–97.
45 Chakrabarty to Roy Curtiss III, 2 October 1974. MC100, MIT Libraries.
46 Watson Fuller, ed., *The Biological Revolution: Social Good or Social Evil?* (New York: Doubleday, 1972), p. 232.
47 Joshua Lederberg, 'DNA Research: Uncertain Peril and Certain Promise', *PRISM* (AMA Policy Journal), 15 June 1975, p. 2. I have been unable to obtain the published version, and I am grateful to Professor Lederberg for a copy of the typescript and for the information that he circulated this paper at Asilomar, Interview with the author, 3 July 1991.
48 See, e.g., a call made by Walter Gilbert to the shareholders of Biogen, 'Shareholders Meeting', 19 June 1984, p. 2. I am grateful to Dr H. Strimpel for showing me this.
49 This race has been vividly described by Stephen Hall, *Invisible Frontiers: The Race to Synthesise a Human Gene* (London: Sidgwick & Jackson, 1987).
50 Diana B. Dutton and Nancy E. Pfund, 'Genetic Engineering: Science and Social Responsibility', in *Worse than the Disease: Pitfalls of Medical Progress,* ed. Diana B. Dutton (Cambridge University Press, 1988), pp. 174–225.
51 BBC, 'Certain Types of Genetic Research Should be Suspended', p. 2.
52 The evolution of rules and their underlying logic are well described in Krimsky, *Genetic Alchemy,* pp. 181–93.
53 This enumeration is taken from Dutton and Pfund, 'Genetic Engineering', p. 188.
54 J.R. Ravetz, *The Merger of Knowledge with Power: Essays in Critical Science* (London: Mansell, 1990), pp. 72–73.
55 *Science Policy Implications of DNA Recombinant Molecule Research,* p. 484.
56 'Research with Genetic Recombinations Generates Promise and Controversy', *Lilly News* 21 (1977), reprinted in *Science Policy Implications of DNA Recombinant Molecule Research,* pp. 474–79.
57 U.S. Congress, Office of Technology Assessment, *Impacts of Applied Genetics: Micro-organisms, Plants and Animals,* OTA-HR-132 (Washington D.C.: U.S. Government Printing Office, 1981), p. iii.
58 David Dickson, *The New Politics of Science,* 2d ed. (Chicago: University of Chicago Press, 1988), pp. 242–55.
59 Zsolt Harsanyi, Interview with the author, 13 June 1990.
60 'Industry Starts to Do Biology with Its Eyes Open', *The Economist* 269 (2 December 1978): 95.
61 On the interferon story, see Sandra Panem, *The Interferon Crusade* (Washington D.C.: Brookings Institution, 1984).
62 'Statement of Nelson Schneider', in U.S. Congress, House of Representatives, Committee on Science and Technology, Subcommittee on Investigations and Oversight and the Subcommittee on Science, Research and Technology, *Commercialization of Academic Biomedical Research,* 97th Cong. 1st sess. 1981, p. 126.

63 Nelson Schneider, personal communication.
64 Zsolt Harsanyi, Interview with the author, 13 June 1990. The meeting has also been described by Robert Teitelman, *Gene Dreams: Wall Street, Academia and the Rise of Biotechnology* (New York: Basic, 1989), p. 26.
65 Quoted in McAuliffe and McAuliffe, *Life for Sale,* p. 26.
66 U.S. Trademark 1180658. Provisional Registration 3 December 1979, 'for magazine reporting scientific and financial developments in the field of genetics' (US Class 38). First use 9 November 1979. I am grateful to Mr Schneider for his help in understanding this development.
67 Zsolt Harsanyi, 'Biotechnology and the Environment: An Overview', in *Biotechnology and the Environment: Risk and Regulation,* ed. Albert H. Teich, Morris A. Levin, and Jill H. Pace (Washington, D.C.: AAAS, 1985), p. 16.
68 Ronald Cape, 'Statement', at the Annual Meeting of the American Association for the Advancement of Science, Session on Recombinant DNA, Public Health and Research Policy. 15 February 1978, MC100, MIT.
69 Office of Technology Assessment, *Impacts of Applied Genetics,* p. 4.
70 U.S. National Science Board, *Science and Engineering Indicators 1987* (Washington D.C.: U.S. Government Printing Office, 1987), p. 253. Watson and Tooze, *The DNA Story,* entitle their treatment of European regulatory debates as 'The European Side-Show'.
71 Hiuga Saito, 'Biotechnology R&D: Japan and the World', *Science and Technology in Japan* 4 (April–June 1985): 8–11.
72 S. Brenner, B.S. Hartley, and P.J. Rodgers, eds., 'New Horizons in Industrial Microbiology', *Philosophical Transactions of the Royal Society* B 290 (1980), p. 430.
73 E.J. Yoxen, 'Assessing Progress with Biotechnology', in *Science and Technology Policy in the 1980s and Beyond,* ed. M. Gibbons et al. (London: Longman, 1984), p. 210.
74 Ravetz, *The Merger of Knowledge with Power,* pp. 72–73.
75 See the study of GBF by Klaus Amann et al., 'Kommerzialisierung der Grundlagenforschung: Das Beispiel Biotechnologie', *Science Studies Report 28* (Bielefeld: Kleine Verlag, 1985).
76 These figures are taken from Mark F. Cantley, 'Democracy and Biotechnology: Popular Attitudes, Information, Trust and the Public Interest' *Swiss Biotech* 5 (1987):5–15. They are based on Commission of the European Communities, *The European Public's Attitudes to Scientific and Technological Development,* XII/201/79, 1979.
77 Annemieke J.M. Roobeek, *Beyond the Technology Race: An Analysis of Technology Policy in Seven Industrial Countries* (Amsterdam: Elsevier, 1990), p. 120.
78 D. de Nettancourt, 'Applied Molecular and Cellular Biology: Background Note on a Possible Action of the European Communities for the Optimal Exploitation of the Fundamentals of the New Biology', Commission of the European Communities, Directorate General for Research Science and Education, XII/207/77-E, 15 June 1977.
79 A. Rörsch, *Genetic Manipulations in Applied Biology: A Study of the Necessity, Content and Management Principles of a Possible Community Action,* EUR 6078, (Luxembourg: Office for Official Publications of the European Communities, 1978); D. Thomas, *Production of Biological Catalysts, Stabilization and Exploitation,* EUR 6079 (Luxembourg: Office for Official Publications

of the European Communities, 1978); C. de Duve, *Cellular and Molecular Biology of the Pathological State*, EUR 6348 (Luxembourg: Office for Official Publications of the European Communities, 1979).

80 For a discussion of transatlantic differences, see Gerald E. Markle and Stanley S. Robin, 'Biotechnology and the Social Reconstruction of Molecular Biology', *Science, Technology and Human Values* 10 (1985):70–79.

CHAPTER 9. THE 1980s: BETWEEN LIFE AND COMMERCE

1 Advisory Council on Science and Technology [ACOST], *Developments in Biotechnology* (London: HMSO, 1990), p. 23.

2 Max Kennedy, 'The Evolution of the Word "Biotechnology" ', *Trends in Biotechnology* 9 (1991): 218–20.

3 Gregory A. Daneke, 'Bureaucratization of U.S. Biotechnology', *Technology Analysis and Strategic Management* 2 (1990), p. 131.

4 The ways in which these trade associations defined biotechnology are explored by L. Christopher Plein, 'Popularizing Biotechnology: The Influence of Issue Definition', *Science, Technology and Human Values* 16 (1991): 474–90.

5 ACOST, *Developments in Biotechnology*, p. 49.

6 John Elkington, *The Gene Factory: Inside the Biotechnology Business* (London; Century Publishing, 1985), pp. 43–45.

7 Alex Beam, 'Biotech Bubble', *Boston Globe*, 13 June 1990.

8 United Nations, *Transnational Corporations in Biotechnology 1988*, cited in ACOST, *Developments in Biotechnology*, p. 16.

9 One can compare a survey of U.S. biotechnology in 1988 by a scientific attaché at the British Embassy in Washington, D. Yarrow, 'U.S. Biotechnology in 1988 – A Review of Current Trends' (London: Department of Trade and Industry, 1988), with an analogous report four years earlier by his predecessor, M.G. Norton, 'The U.S. Biotechnology Industry' (London: Overseas Technical Information Unit, 1984).

10 'Glaxo Drug Approved', *The Independent*, 14 August 1991.

11 Quoted by Elkington, *Inside the Gene Factory*, p. 70.

12 Yarrow, 'U.S. Biotechnology in 1988', Figure 16.

13 'The Genetic Economy', *The Economist*, Supp. 307 (30 April 1988): 10.

14 'Maturing of Product Sales up to $4b a year by 1993', *Los Angeles Times*, 17 March 1991.

15 'The Genetic Economy', 10.

16 ACOST, *Developments in Biotechnology*, p. 1.

17 'The Unfulfilled Promise of Biotech "Cures"', *Los Angeles Times*, 31 January 1991.

18 U.S. Congress, Office of Technology Assessment, *Commercial Biotechnology: An International Analysis*, OTA-BA-218 (Washington D.C.: U.S. Government Printing Office, 1984), p. 93.

19 G. Steven Burrill with the Arthur Young High Technology Group, *Biotech 89: Commercialization* (New York: Mary Ann Liebert, 1988), p. 22.

20 Mark D. Dibner and Nancy G. Bruce, 'The Greening of Biotechnology: The Growth of the US Biotechnology Industry', *Trends in Biotechnology* 5(1987): 270–72.

21 'Statement of Carl Djerassi', U.S. House of Representatives, Committee on Science and Technology, Subcommittee on Investigations and Oversight and the Subcommittee on Science, Research and Technology, *Commercialization of Academic Biomedical Research,* 97th Cong. 1st sess. 8–9 June 1981, p. 149.
22 See 'The Biotechnology Roundtable on Commercialization', *Bio/technology* 4i (January 1986): 21–26.
23 E.C. Dart, 'Exploiting Biotechnology in a Large Company', *Philosophical Transactions of the Royal Society* B 324 (1989): 599–611, p. 599. Anthony Walker of Arthur D. Little has pointed out that European companies do not appear to regard biotechnology as a business per se. Anthony Walker, 'Europe Steals the Lead in Plant Biotechnology', in *Biotech '89* Proceedings of the conference held in London, May 1989 (London: Blenheim Online, 1989): 121–28.
24 Arthur Kornberg, 'The Two Cultures: Chemistry and Biology', *Biochemistry* 26 (1987), p. 6888.
25 Emily A. Arakaki, 'A Study of the U.S. Competitive Position in Biotechnology', in *Biotechnology – High Technology in Industry: Profiles and Outlooks,* U.S. Department of Commerce (Washington D.C.: U.S. Government Printing Office, 1985), p. 63.
26 David Hounshell and John Kenly Smith, Jr, *Science and Corporate Strategy: Du Pont R&D, 1902–1980* (Cambridge University Press, 1988), p. 589.
27 Anthony Walker, 'Europe Steals the Lead in Plant Biotechnology'.
28 Mark Ratner, 'Bandshift', *Bio/technology* 8 (November 1990): 993.
29 [OECD], *Biotechnology and the Changing Role of Government* (Paris: OECD, 1988), p. 27.
30 Bernard D. Davis, 'Frontiers of Biological Sciences', *Science* 209 (4 July 1980): 78.
31 Erik Baark and Andrew Jamison, 'Biotechnology and Culture: The Impact of Public Debates on Government Regulation in the United States and Denmark', *Technology and Society* 12(1990): 27–44.
32 Steve Olson, *Biotechnology: An Industry Comes of Age* (Washington D.C.: National Academy Press, 1986).
33 R.F. Coleman, 'National Policies and Programmes in Biotechnology', in *Biotechnology and the Changing Role of Government,* pp. 83–94.
34 See Ronald Dore, *A Case Study of Technology Forecasting in the Next Generation Base Technologies Development Programme* (London: Technical Change Centre, 1983).
35 Michael D. Rogers, 'The Japanese Government's Role in Biotechnology R&D', *Chemistry and Industry* (7 August 1982): 533–37.
36 Yano Research Institute, 'Development of Biotechnology in Japan and the Government', 1983.
37 Rogers, 'The Japanese Government's Role in Biotechnology'.
38 Alfred Scheidegger, 'Biotechnology in Japan – Towards the Year 2000 ... and Beyond', *Trends in Biotechnology* 9 (June 1991): 183–90.
39 Malcolm V. Brock, *Biotechnology in Japan* (London: Routledge, 1989), pp. 89–115.
40 Bioindustry Office, Basic Industries Bureau, MITI, 'New Development of Bioindustry Policy', January 1989.
41 Thomas F. Lee, *The Human Genome Project: Cracking the Genetic Code of Life*

(New York: Plenum, 1991), pp. 217–18. See also U.S. Congress, Office of Technology Assessment, *Mapping Our Genes – Genome Projects: How Big How Fast* (Washington D.C.: U.S. Government Printing Office, 1987).

42 Office of Technology Assessment, *Commercial Biotechnology.*

43 Office of Technology Assessment, *Commercial Biotechnology,* p. 308.

44 Zsolt Harsanyi, 'Biotechnology and the Environment: An Overview', in *Biotechnology and the Environment: Risk and Regulation*, ed. Albert A. Teich, Morris A. Levin, and Jill A. Pace (Washington D.C.: AAAS, 1985), p. 15.

45 S. Silver, ed., *Biotechnology: Potentials and Limitations,* Life Science Research Report 35 (New York: Springer, 1986). He explicitly argued with the position expressed in the same volume by B. Mattiesson in 'Technological Processes for Biotechnological Utilization of Micro-Organisms', pp. 113–125.

46 See Henry Etzkowitz, 'The Capitalization of Knowledge', *Theory and Society* 19(1990): 107–21.

47 Mark Dibner, *Directory of State Biotechnology Centers,* February 1988, cited in Yarrow, 'U.S. Biotechnology', pp. 8–9.

48 Beam, 'Biotech Bubble'.

49 Mark Ratner, 'The Analysts Take Aim: Biotech Fires Back', *Bio/technology* 4 (November 1988): 1280. The article cited the comments of Paine Webber analyst Linda Miller to the Biopharmaceuticals Futures Conference in October 1988.

50 Lee, *The Human Genome Project.*

51 The policies of individual countries were laid out by Margaret Sharp, *The New Biotechnology: European Governments in Search of a Strategy,* Sussex European Papers no. 15 (Brighton: Science Policy Research Unit, 1985).

52 'Discours prononcé devant le Parlement européen par M. Gaston E. Thorn, président de la Commission des Communautés européennes' cited by Mark Cantley to Paolo Fasella, 16 February 1983, CUBEDOC.

53 Commission of the European Communities, 'Biotechnology in the Community', Com. (83) 672, 1983.

54 F. Rexen and L. Munck, *Cereal Crops for Industrial Use in Europe,* EUR 9617 (Copenhagen: Carlsberg Research Laboratory, 1984), p. 8.

55 Ken Sargeant, in evidence to the House of Lords Select Committee on the European Communities, Session 1986–1987, 4th Report HL 145 (London: HMSO, 1987), p. 58.

56 Commission of the European Communities, 'Biotechnology in the Community: Stimulating Agro-Industrial Development', Com. 86-221/2, 1986.

57 German Federal Government, 'Applied Biology and Biotechnology, 1985–1988' (Bonn: BMFT, 1985), p. 1.

58 See Jacqueline Senker and Margaret Sharp, 'The Biotechnology Directorate of the SERC: Report and Evaluation of Its Achievements – 1981–1987', Science Policy Research Unit Report, submitted to the Management Committee of the Biotechnology Directorate, 5 May 1988, p. 24.

59 Its development is recounted and analyzed in Mark Dodgson, *Celltech: The First Ten Years* (Brighton: SPRU, 1990).

60 B. Balmer, 'Biotechnology and the UK Research Councils', M.Sc. dissertation, University of Sussex, 1990.

61 D. McCormick, 'Not as Easy as It Looked', *Bio/technology* 7 (July 1989): 629.

62 Joanna Blythman, 'Will Daisy Become a Monster?', *Independent*, 29 June 1991.
63 For this description, I am indebted to Dr Bayrhuber of Kiel. It has resonances with the broader tensions within the German Green Party; however it would be simplistic to reduce the tensions over biotechnology particularly outside Germany to the particular issues within this party. See J.L. Cohen and A. Arato, 'The German Green Party: A Movement between Fundamentalism and Modernism', *Dissent* 3 (1984): 327–32.
64 U.S. Congress, Office of Technology Assessment, *New Developments in Biotechnology – Background Paper: Public Perceptions of Biotechnology*, OTA-BP-BA-45 (Washington, D.C.: U.S. Government Printing Office, 1987).
65 Sheldon Krimsky and Alonzo Plough, *Environmental Hazards: Communicating Risks as a Social Process* (New York: Auburn Homer, 1988).
66 Erik Baark and Andrew Jamison, 'Biotechnology and Culture: The Impact of Public Debates on Government Regulation in the United States and Denmark', *Technology in Society* 12 (1990): 27–44.
67 Jack Doyle, *Altered Harvest: Agriculture, Genetics and the Fate of the World's Food Supply* (New York: Viking-Penguin, 1985).
68 The distinction in German is outlined by Adolf Weber, 'Biotechnologie und Gentechnologie', in *Die Herstellung der Natur: Chancen und Risiken der Gentechnologie*, ed. Ulrich Steger (Bonn: Verlag Neue Gesellschaft, 1985), pp. 13–14. Here '*Biotechnologie*' is defined in its DECHEMA sense of industrial microbiology in general whereas '*Gentechnologie*' entailed the targeted alteration of genetic material.
69 Niels Arnfred, Annegrethe Hansen, and Jørgen Lindgaard Pedersen, 'Bio-technology and Politics: Danish Experiences' in *Technology Policy in Denmark*, ed. J.L. Pedersen (Copenhagen: New Social Science Monographs, 1989), pp. 159–84; Sheila Jasanoff, 'Technological Innovation in a Corporatist State: The Case of Biotechnology in the Federal Republic of Germany', *Research Policy* 14 (1985): 23–38.
70 For an introduction to current German debates, see Wolf-Michael Catenhusen, 'Public Debate on Biotechnology: The Experience of the Bundestag Commission of Inquiry on the Opportunities and Risks of Genetic Engineering', in *Biotechnology in Future Society. Scenarios and Options for Europe*, ed. Edward Yoxen and Vittorio Di Martino (Luxembourg: Office for Official Publications of the European Communities, 1989), pp. 117–23.
71 G. Heinson, R. Knieper, and O. Steiger, *Menschenproduktion: Allgemeine Bevölkerungstheorie der Neuzeit* (Frankfurt: Suhrkamp, 1979), pp. 194–96.
72 ACOST, *Developments in Biotechnology*, p. 28.
73 Quoted in K. Heusler, 'The Commercialisation of Government and University Research, and Public Acceptance of Biotechnology', in [OECD], *Biotechnology and the Changing Role of Government*, p. 110.
74 INRA (Europe) sa/nv for 'CUBE', *Opinions of Europeans on Biotechnology in 1991*, Eurobarometer 35.1.
75 The findings of this poll sponsored by Eli Lilly during 1990 were released in a summary by Prima Europe, 'New Poll Backs Biotechnology in Europe', 27 November 1990. Fears of eugenics are cited by 40% of a sample in France and 35% in Germany, though by only 25% in Britain and Italy.

76 European Parliament, Committee on Energy, Research and Technology, Rapporteur Benedikt Härlin, 'Report... on the Proposal from the Commission to the Council (COM/88/424–C2-119/88) for a Decision Adopting a Specific Research Programme in the Field of Health and Predictive Medicine: Human Genome Analysis (1989–1991)', Document A2-0370/88, 30 January 1989, para. 27, p. 27.
77 Edward Yoxen and Vittorio Di Martino, eds., 'Learning about Participation in Biotechnology', in *Biotechnology in Future Society,* ed. Yoxen and Martino, pp. 128–29.
78 European Parliament, Committee on Agriculture, Fisheries and Food, Rapporteur Mr F.W. Gräfe zu Baringdorf, 'Report... on the Effects of the Use of Biotechnology on the European Farming Industry', Working Document A 2-159/86, 26 November 1986, p. 7.
79 Jennifer Van Brunt, 'Environmental Release: A Portrait of Opinion and Opposition', *Bio/technology* 5 (1987): 559–663.
80 Olaf Diettrich, 'Beauty and the Beast: Particular Aspects of the Socioeconomic Integration of Biotechnology in the Context of Activities of the Commission of the European Communities'. Presented at the International Symposium on Environmental Biotechnology, 22–25 April 1991, Ostend. On the importance of regulators in general, see Sheila Jasanoff, *The Fifth Branch: Science Advisors as Policymakers* (Cambridge, Mass.: Harvard University Press, 1990).
81 U.S. Congress, House of Representatives, Committee on Science and Technology, *The Biotechnology Science Coordination Act of 1986,* 99th Cong., 2d sess. 4–5 June 1986, p. 102.
82 Douglas McCormick, 'A-sitting on a Gate', *Bio/technology* 5 (February 1987): 101.
83 Luther Val Giddings, 'The United States Regulatory Framework', in *Biotech '89.* Proceedings of the conference held in London, May 1989 (London: Blenheim Online, 1989), p. 73.
84 U.S. Congress, Office of Technology Assessment, *New Developments in Biotechnology: U.S. Investment in Biotechnology – Special Report,* OTA-BA-360 (Washington, D.C.: U.S. Government Printing Offices, 1988), pp. 29–30.
85 K. Sargeant, 'Biotechnology, Connectedness and Dematerialisation', in *Biotechnology '84.* Proceedings of a conference held in the Royal Irish Academy, 1–2 May 1984, ed. J.P. Arbuthnott (Dublin: Royal Irish Academy, 1985), p. 96.
86 The experience of Gary Strobel was the subject of a 1987 Senate hearing. See U.S. Congress, Senate, Committee on Environment and Public Works, Subcommittee on Hazardous Wastes and Toxic Substances, *Federal Oversight of Biotechnology,* 100th Cong. 1st sess. 5 November 1987.
87 H.I. Miller, and F.E. Young, 'Isn't It about Time We Dispensed with "Biotechnology" and "Genetic Engineering"?' *Bio/Technology* 5 (1987): 184. David T. Kingsbury, 'Safety and Regulations, R&D and International Cooperation in Biotechnology', in [OECD], *Biotechnology and the Changing Role of Government,* p. 118.
88 [OECD], *Biotechnology and the Changing Role of Government,* p. 72.
89 On the dominance of the probiotechnology forces in the United States,

see Plein, 'Popularizing Biotechnology'. He makes the interesting point that the importance of Rifkin has been played up by the proponents of biotechnology in an attempt to delegitimate less extreme opposition.

90 Jeffrey L. Fox, 'More Changes in U.S. Regulatory Bodies', *Bio/technology* 8 (November 1990): 996.

91 European Parliament, Committee on Energy, Research and Technology, Rapporteur Mrs P. Viehoff, 'Biotechnology in Europe: The Need for an Integrated Policy', Draft Report, 21 July 1986. PE 105.423/B.

92 Wolf-Michael Catenhusen, 'Public Debate on Biotechnology: The Experience of the Bundestag Commission of Inquiry on the Opportunities and Risks of Genetic Engineering', in *Biotechnology in Future Society,* ed. Yoxen and Di Martino, pp. 117–23. See also Wolf-Michael Catenhusen and Hanna Neumeister, eds., *Chancen und Risiken der Gentechnologie – Dokumentation des Berichts an den Deutschen Bundestag: Enquete-Kommission des Deutschen Bundestages* (Munich: Schweitzer, 1987); for a study of technology assessment, see Andrew Jamison and Erik Baark, 'Modes of Biotechnology Assessment in the United States, Japan and Denmark', Institute for Samfundstag, Danmarks Tekniske Højskole, *Skriftsserie,* no. 20, 1989.

93 G. Ruhrmann, 'Genetic Engineering in the Press'. In *Biotechnology in Public: A Review of Current Research,* ed. John Durant (London: Science Museum, 1992), pp. 169–201.

94 Commission of the European Communities, 'Council Directive of 23 April 1990 on the Contained Use of Genetically Modified Micro-organisms', 90/219/EEC, and 'Council Directive of 23 April 1990 on the Deliberate Release into the Environment of Genetically Engineered Micro-organisms'. See Olivier Leroy, *Biotechnology: EEC Policy on the Eve of 1993* (Rixensart, Belgium: European Study Service, 1991), pp. 167–80 and 217–29. The discussions are recorded by Gordon Lake, 'Scientific Uncertainty and Political Regulation: European Legislation on the Contained Use and Deliberate Release of Genetically Modified (Micro)organisms', *Project Appraisal* 6 (March 1991): 7–15.

95 F.K. Beier, R.S. Crespi, and J. Straus, *Biotechnology and Patent Protection: An International Review* (Paris: OECD, 1985), p. 67.

96 Deborah MacKenzie, 'Europe Rethinks Patent on the Harvard Mouse', *New Scientist* 132 (19 October 1991): 11.

97 A recent study of biotechnology and business gives its own definition of biotechnology: 'making money with biology'. See Vivian Moses and Ronald Cape, 'The Science and the Business: An Introduction', in *Biotechnology: The Science and the Business,* ed. V. Moses and R. Cape (London: Harwood Academic, 1991), p. 1. For the clearest statement by 'promoters', see Prima Europe, *The Case for Biotechnology* (London: Prima Europe, 1990).

98 Modern technologies have been examined from a wide variety of perspectives. For information technology, e.g., see James R. Beniger, *The Control Revolution: Technological and Economic Origins of the Information Society* (Cambridge, Mass.: Harvard University Press, 1986), and on nuclear technology, see Spencer R. Weart, *Nuclear Fear: A History of Images* (Cambridge, Mass.: Harvard University Press, 1988).

99 Tecknoligi Naevnet, *Consensus Conference on the Application of Knowledge Gained from Mapping the Human Genome: Final Document,* 1–3 November 1989, p. 16.

Sources

ARCHIVES USED

Cambridge University Library, Cambridge
  Needham Archive
Case Western Reserve University, University Archives, Cleveland
  W.E. Wickenden Papers
CUBEDOC, Brussels
  FAST and CUBE Papers
Ingenjörsvetenskapsakademien, Stockholm
  Papers relating to section 10, Bioteknik
Institute of Brewing, London
  Papers relating to Brewing Industry Research Foundation
London School of Economics, London
  W. Beveridge Papers
Massachusetts Institute of Technology, Institute Archives and Special Collections
  Recombinant DNA History Collection
  MIT Office of the President
National Library of Scotland, Edinburgh
  Patrick Geddes Papers
Rice University, Woodson Research Centre, Fondren Library, Houston
  Julian Huxley Papers
Science Museum Library, London
  Arthur Guinness Son & Co. Papers
  Ereky Correspondence
Siebel Institute of Technology, Chicago
  Papers relating to the early days of the institute
Technical University of Denmark, Lyngby
  University Regulations
University of Birmingham Library, Birmingham
  Lancelot Hogben Papers
University of Pennsylvania, Special Collections, Philadelphia
  Lewis Mumford Papers
University of Strathclyde, Glasgow
  Patrick Geddes Papers
University of Sussex Library, Brighton
  R.A. Gregory Papers

Weizmann Archive, Rehovoth, Israel
Chaim Weizmann Papers
Wellcome Institute, London
Sir Ernst Chain Papers

SELECT BIBLIOGRAPHY

Abir-Am, Pnina. 'The Biotheoretical Gathering, Transdisciplinary Authority and the Incipient Legitimation of Molecular Biology in the 1930s: New Perspective on the Historical Sociology of Science'. *History of Science* 25 (1987): 1–70.

Adams, Mark B. 'Eugenics as Social Medicine in Revolutionary Russia: Prophets, Patrons and the Dialectics of Discipline Building'. In *Health and Society in Revolutionary Russia,* ed. S. Gross Solomons and J.F. Hutchinson, pp. 200–23. Bloomington: Indiana University Press, 1990.

Advisory Council for Applied Research and Development, Advisory Board for the Research Councils and The Royal Society. *Biotechnology: Report of a Joint Working Party.* London: HMSO, 1980.

Advisory Council on Science and Technology [ACOST]. *Developments in Biotechnology.* London: HMSO, 1990.

Ahmad, S. 'Institutions and the Growth of Knowledge: The Rockefeller Foundation's Influence on the Social Sciences Between the Wars.' Doctoral dissertation, University of Manchester, 1987.

Aiba, S.; Humphrey, A.E.; and Millis N.F. *Biochemical Engineering.* New York: Academic Press, 1965.

Ainsworth, G.C. *Introduction to the History of Mycology.* Cambridge University Press, 1976.

'All the Way with DNA'. *New Scientist* 39 (18 July 1968): 142.

Althin, Torsthin. *Axel F. Enström.* Stockholm: Ingeniörvetenskapsakademien, 1958.

Amann, Klaus, et al. 'Kommerzialisierung der Grundlagenforschung: Das Beispiel Biotechnologie'. *Science Studies Report 28.* Bielefeld: Kleine Verlag, 1985.

Anderson, Russell E. *Biological Paths to Self-Reliance: A Guide to Biological Solar Energy Conversion.* New York: Van Nostrand, 1979.

Arakaki, Emily A. 'A Study of the U.S. Competitive Position in Biotechnology'. In *Biotechnology – High Technology in Industry: Profiles and Outlooks,* U.S. Department of Commerce. Washington D.C.: U.S. Government Printing Office, 1985.

Arima, K. 'The Problems of Public Acceptance of S.C.P.-s'. In *International Symposium on SCP,* ed. J.C. Senez, pp. 145–62. Proceedings of symposium held 28–31 January 1981 organized by DGRST, IAMS, and APRIA. Paris: Technique et Documentation: Lavoisier, 1983.

Armstrong, H.E. 'The Production of Rubber: With or Against Nature?' *Times Engineering Supplement,* 17 July 1912, p. 21.

Arnfred, Niels; Hansen, Annegrethe; and Pedersen, Jørgen Lindgaard. 'Biotechnology and Politics: Danish Experiences'. In *Technology Policy in Denmark,* ed. J.L. Pedersen, pp. 158–84. Copenhagen: New Social Science Monographs, 1989.

Arnold, John P., and Penman, Frank. *History of the Brewing Industry and Brewing Science in America: Prepared as a Memorial to the Pioneers of American Brewing Science, Dr John E. Siebel and Anton Schwartz*. Chicago: Privately printed, 1933.

'Asks "Social Mind" in Engineer Study'. *New York Times*, 23 October 1936.

Aubry, L. 'Hofrat Dr Carl Lintner'. *Zeitschrift für das gesamte Brauwesen* 23 (1900): 93–96.

Baark, Erik, and Jamison, Andrew. 'Biotechnology and Culture: The Impact of Public Debates on Government Regulation in the United States and Denmark'. *Technology in Society* 12 (1990): 27–44.

Baer, J.G. 'Biology and Humanity: The International Biological Programme'. *Impact of Science on Society* 17 (1967): 315–28.

Bailes, Kendall. *Environmental History: Critical Issues in Comparative Perspective*. Lanham, Md: University Presses of America, 1985.

Balls, A.K. 'Liebig and the Chemistry of Enzymes and Fermentation'. In *Liebig and After Liebig*, ed. F.R. Moulton, pp. 30–39. AAAS pub. 16. Washington D.C.: AAAS, 1942.

Balmer, B. 'Biotechnology and the UK Research Councils'. M.Sc. dissertation, University of Sussex, 1990.

Barker, R. 'Socio-economic Impact'. In *Agricultural Biotechnology: Opportunities for International Development*, ed. Gabrielle J. Persley, pp. 299–310. Wallingford, Oxon: CAB International, 1990.

BBC. 'Certain Types of Genetic Research Should Be Suspended'. Broadcast, BBC2, 16 September 1974.

Beam, Alex. 'Biotech Bubble'. *Boston Globe*, 13 June 1990.

Behrens, Dieter. 'Europäische Föderation Biotechnologie: Vier Jahre nach der Gründung'. *Swiss Biotech* 1 (1983): 11–16.

Behrens, Dieter; Buchholz, K.; and Rehm, H.-J. *Biotechnology in Europe – A Community Strategy for European Biotechnology*. Frankfurt: Dechema for European Federation of Biotechnology, 1983.

Beier, F.K.; Crespi, R.S.; and Straus, J., eds. *Biotechnology and Patent Protection: An International Review*. Paris: OECD, 1985.

Benedict, Elizabeth T.; Stippler, Heinrich Hermann; and Benedict, Mary Reed. *Theodor Brinkmann's Economics of the Farm Business*. Berkeley: University of California Press, 1935.

Benichou, Claude, and Blanckaert, Claude. *Julien-Joseph Virey: Naturaliste et anthropologue*. Paris: Vrin, 1988.

Beniger, James R. *The Control Revolution: Technological and Economic Origins of the Information Society*. Cambridge, Mass.: Harvard University Press, 1986.

Bennett, John W., and Levine, Solomon B. 'Industrialization and Social Deprivation: Welfare, Environment and the Post-Industrial Society in Japan'. In *Japanese Industrialization and Its Social Consequences*, ed. Hugh Patrick, pp. 439–92. Berkeley: University of California Press, 1976.

Benninga, H. *A History of Lactic Acid Making: A Chapter in the History of Biotechnology*. Dordrecht: Kluwer, 1990.

Bergen, Henry. 'Rudolf Goldscheid, "Höherentwicklung u. Menschenökonomie" '. *Eugenic Review* 3 (1911–12): 236–41.

Bergson, Henri. *Creative Evolution*, trans. Arthur Mitchell. London: Methuen, 1954.

Berman, Alex. 'Romantic Hygeia: J.J. Virey (1775–1846), Pharmacist and Philosopher of Nature'. *Bulletin of the History of Medicine* 39 (1965): 134–42.
Bernhauer, K. *Gärungschemisches Praktikum,* 2d ed. Berlin: Springer, 1939.
Bernton, Hal; Kovarik, William; and Sklar, Scott. *The Forbidden Fuel: Power Alcohol in the Twentieth Century*. New York: Boyd Griffin, 1982.
Beveridge, William. *Power and Influence*. London: Hodder & Stoughton, 1953.
'Bierproduction in den verschiedenen Ländern des Kontinents und in Nord-Amerika'. *Bayerische Bierbraver* 19 (1884): 342.
Bioindustry Office. Basic Industries Bureau. MITI. 'New Development of Bioindustry Policy', January 1989.
'Biological Engineering Society', *Lancet* 2 (23 July 1960): 218.
'Biological Engineering Society', *Lancet* 2 (12 November 1960): 1097.
'The *Bio/technology* Roundtable on Commercialization'. *Bio/technology* 4 (January 1986): 21–26.
'Bioteknik', *IVA* 14 (15 February 1943): 1.
Blom-Björner, J. 'Alfred Jørgensen'. *Journal of the Institute of Brewing* 32 (1926): 198.
Blythman, Joanna. 'Will Daisy Become a Monster?' *The Independent,* 29 June 1991.
[BMBW]. *Erster Ergebnisbericht des ad hoc Ausschusses "Neue Technologien" des Beratenden Ausschusses für Forschungspolitik.* Shriftenreihe Forschungsplanung 6. Bonn: BMBW, December 1971.
[BMFT.] *Biotechnologie.* BMFT-Leistungsplan 04, Plan Periode. 1979–1983. Bonn: BMFT, 1978.
Boardman, Philip. *The Worlds of Patrick Geddes: Biologist, Town Planner, Re-educator, Peace-Warrior.* London: Routledge, 1978.
Borkenhagen, E. 'Gesellschaft für die Geschichte und Bibliographie des Brauwesens'. In *100 Jahre Institut für Gärungsgewerbe und Biotechnologie zu Berlin, 1874–1974. Festschrift,* pp. 245–52. Berlin: Institut für Gärungsgewerbe und Biotechnologie, 1974.
Borth, Christy. *Pioneers of Plenty: The Story of Chemurgy.* Indianapolis, Ind.: Bobbs-Merrill, 1939.
Bossel, Hartmut. 'Die vergessenen Werte'. In *Der grüne Protest. Herausforderung durch die Umweltparteien,* ed. Rudolf Brun, pp. 7–17. Frankfurt: Fisher, 1978.
Bowler, Peter J. 'Theodor Eimer and Orthogenesis: Evolution by "Definitely Directed Variation" '. *Journal of the History of Medicine and Allied Sciences* 34 (1979): 40–73.
The Eclipse of Darwinism: Anti-Darwinian Evolution Theories in the Decades around 1900. Baltimore: Johns Hopkins University Press, 1983.
Boyd-Orr, John. *As I Recall.* London: McGibbon & Kee, 1966.
Branford, V.V. and Geddes, P. *The Coming Polity* 2d ed. London: Le Plais House, 1919.
Braude, R. 'Dried Yeast as Fodder for Livestock'. *Journal of the Institute of Brewing* 48 (October 1942): 206–12.
Brenner, S.; Hartley, B.S.; and Rodgers, P.J., eds. 'New Horizons in Industrial Microbiology'. *Philosophical Transactions of the Royal Society* B 290 (1980): 277–430.
[Brightman, Rainald.] 'Biotechnology'. *Nature* 131 (29 April 1933): 597–99.

Brinkmann, Theodor. 'Die Dänische Landwirtschaft: Die Entwicklung ihrer Produktion seit dem Auftreten der internationalen Konkurrenz und ihre Anpassung an den Weltmarkt Vermittels genossenschaftlicher Organisation'. *Abhandlungen des Staatswissenschaftlichen Seminars zu Jena* 6, pt. 1 (1908).

Brock, Malcolm V. *Biotechnology in Japan*. London: Routledge, 1989.

Brock, Thomas D. *Robert Koch: A Life in Medicine and Bacteriology*. Madison: Science-Tech Publishers, 1988.

The Emergence of Bacterial Genetics. Cold Spring Harbor, N.Y.: Cold Spring Harbor Laboratory Press, 1990.

Brock, W.H. 'Liebigiana: Old and New Perspectives'. *History of Science* 19 (1981): 201–18.

Brooke, J.H. 'Organic Chemistry'. In *Recent Developments in the History of Chemistry*, ed. C.A. Russell, pp. 107–9. London: Royal Society of Chemistry, 1985.

Brunner, R., and Friedrich, W. 'In Memoriam: Konrad Bernhauer', *Mitteilungen der Versuchsstation für das Gärungsgewerbe in Wien*, no. 2 (1976): 22–28.

Buchholz, Klaus, 'Die gezielte Förderung und Entwicklung der Biotechnologie'. In *Geplante Forschung*, ed. Wolfgang van den Daele, Wolfgang Krohn, and Peter Weingart, pp. 64–116. Frankfurt: Suhrkamp, 1979.

[Buchholz, K.] 'Editorial'. *EFB Newsletter*, no. 4, December 1981.

Bud, Robert. 'Biotechnology in the Twentieth Century'. *Social Studies of Science* 21 (1991): 415–57.

Bud, Robert, and Roberts, Gerrylynn K. *Science versus Practice: Chemistry in Victorian Britain*. Manchester: Manchester University Press, 1984.

Bull, Alan, and Bu'Lock, John. 'The Living Micro Revolution'. *New Scientist* 82 (7 June 1979): 808–10.

Bull, Alan T.; Ellwood, Derek C.; and Ratledge, Colin. 'The Changing Scene in Microbial Technology'. In *Microbial Technology: Current State, Future Prospects*, ed. A.T. Bull, D.C. Ellwood, and C. Ratledge, pp. 1–28. Society of General Microbiology, Symposium no. 29. Cambridge University Press for the Society for General Microbiology, 1979.

Bull, Alan T.; Holt, T. Geoffrey; and Lilly, Malcolm D. *Biotechnology: International Trends and Perspectives*. Paris: OECD, 1982.

Burrill, G. Steven, with the Arthur Young High Technology Group, *Biotech 89: Commercialization*. New York: Mary Ann Liebert, 1988.

Bush, Vannevar. 'The Case for Biological Engineering'. In *Scientists Face the World of 1942*, pp. 33–45. New Brunswick. Rutgers University Press, 1942.

Byer, Doris. *Rassenhygiene und Wohlfahrtspflege: Zur Entstehung eines sozialdemokratischen Machtdispositivs in Oesterreich bis 1934*. Frankfurt am Main: Campus Verlag, 1988.

Cantley, Mark F. 'Democracy and Biotechnology: Popular Attitudes, Information, Trust and the Public Interest'. *Swiss Biotech* 5 (1987): 5–15.

Čapek, K. *R.U.R.*, Ed. Henry Shefter. New York: Simon & Schuster, 1973.

Caron, Joseph A. ' "Biology" in the Life Sciences'. *History of Science* 26 (1989): 223–68.

Carter, G.B. 'Is Biotechnology Feeding the Russians?' *New Scientist* 90 (23 April 1981): 216–18.

'The Microbiological Research Establishment and Its Precursors at Porton

Down: 1940–1979' (pt. 1), *The ASA Newsletter,* 91–6, issue 27, December 1991.

'The Microbiological Research Department and Establishment: 1946–1979', (pt. 2), *The ASA Newsletter,* 92–1, Issue 28, 1992.

Carter, Richard. *Breakthrough: The Saga of Jonas Salk.* New York: Pocket Books, 1967.

'The Case for Biochemical Engineering'. *Chemical Engineering* 54 (May 1947): 106.

Catenhusen, Wolf-Michael, and Neumeister, Hanna, eds. *Chancen und Risiken der Gentechnologie – Dokumentation des Berichts an den Deutschen Bundestag: Enquete-Kommission des Deutschen Bundestages.* Munich: Schweitzer, 1987.

Cellarius, Richard A., and Platt, John. 'Councils of Urgent Studies'. *Science* 177 (25 August 1972): 670–76.

Chain, Ernst. 'Thirty Years of Penicillin Therapy', *Proceedings of the Royal Society of London* B 179 (1971): 293–319.

Champagnat, Alfred, and Adrian, Jean. *Petrole et Proteines.* Paris: Doin, 1974.

Channell, David F. *The Vital Machine: A Study of Technology and Organic Life.* Oxford: Oxford University Press, 1991.

Chapman, W. Chaston. 'The Employment of Micro-organisms in the Service of Industrial Chemistry: A Plea for a National Institute of Industrial Microbiology'. *Journal of the Society of Chemical Industry* 38 (1919): 282T–86T.

'Charges Industry with Duty to Idle'. *New York Times,* 16 February 1931.

Cittadino, Eugene. *Nature as the Laboratory: Darwinian Plant Ecology in the German Empire, 1880–1900.* Cambridge University Press, 1990.

Clark, Ronald W. *The Life of Ernst Chain: Penicillin and Beyond.* London: Weidenfeld & Nicolson, 1985.

Clarke, Janine, and Clarke, Robin. 'The Biotechnic Research Community'. *Futures* 4 (June 1972): 168–73.

Coates, Austin. *The Commerce of Rubber: The First 250 Years.* Oxford University Press, 1987.

Cohen, J.L., and Arato, A. 'The German Green Party: A Movement between Fundamentalism and Modernism'. *Dissent* 3 (1984): 327–32.

Coleman, W. 'The Cognitive Basis of the Discipline: Claude Bernard on Physiology'. *ISIS* 76 (1985): 49–70.

'Commercial Solvents Corporation v. Synthetic Products Company Ltd.' In *Reports of Patent Design, Trade Mark and Other Cases,* vol. 43, ed. F.G. Underhay, pp. 185–238. London: HMSO, 1951.

Commission of the European Communities. 'Biotechnology in the Community', Com. (83)–672. Brussels: Commission of the European Communities, 1983.

'Biotechnology in the Community: Stimulating Agro-Industrial Development', Com. 86–221/2. Brussels: Commission of the European Communities, 1986.

FAST Subprogramme C: 'Biosociety'. FAST/ACPM/79/14–3E. Brussels: Commission of the European Communities, 1979.

Compton, Karl T., and Bunker, John W.M. 'The Genesis of a Curriculum in Biological Engineering'. *Scientific Monthly* 48 (January 1939): 5–15.

Connstein, W., and Lüdecke, K. 'Ueber Glycerin-Gewinnung durch Gärung'. *Berichte der Deutschen Chemischen Gesellschaft* 52ii (1919): 1385–91.

Cook, A.H. 'Unfolding Pattern of Research: Brewing Industry of the Future', *Brewer's Guardian* (June 1960): 17–25.

Crew, F.A.E., et al., 'Social Biology and Population Improvement'. *Nature* 144 (16 September 1939): 521–22.

Crook, E.M. 'How Biotechnology Developed at University College London'. *Biochemical Society Symposium* 48 (1982): 1–7.

Curtius, Th. 'Wilhelm Koenig'. *Berichte der Deutschen Chemischen Gesellschaft* 45iii (1912): 3781–92.

Dambmann, C.; Holm, P.; Jensen, V.; and Nielsen, M.H. 'How Enzymes Got into Detergents'. *Development in Industrial Microbiology* 12 (1971): 11–23.

Daneke, Gregory A. 'Bureaucratization of US Biotechnology'. *Technology Analysis and Strategic Management* 2(1990): 129–42.

Dart, E.C. 'Exploitation of Biotechnology in a Large Company'. *Philosophical Transactions of the Royal Society* B 324 (1989): 599–611.

DaSilva, E.J.; Burgers, A.C.J.; and Olembo, R.J. 'UNESCO, UNEP and the International Community of Culture Collections'. In *Proceedings of the Third International Conference on Culture Collections,* ed. Fabian Fernandes and Raby Costa Pereira, pp. 107–28. Bombay: University of Bombay, 1977.

DaSilva, E.J., and Hedén, C.-G. 'The Role of International Organizations in Biotechnology: Cooperative Efforts'. In *Comprehensive Biotechnology,* ed. Murray Moo Young, vol. 4, pp. 717–49. London: Pergamon, 1985.

Davis, Bernard D. 'Frontiers of Biological Sciences', *Science* 209 (4 July 1980): 78–89.

Davis, George E. *Handbook of Chemical Engineering.* Manchester: Davis Bros., 1901–2.

Dean, A.C.R., et al., eds. *Continuous Culture 6: Applications and New Fields.* Chichester: Ellis Horwood for the Society of Chemical Industry, 1976.

DECHEMA. *Biotechnologie: Eine Studie über Forschung und Entwicklung – Möglichkeiten, Aufgaben und Schwerpunkte der Förderung.* Frankfurt: DECHEMA, January 1974.

de Duve, C. *Cellular and Molecular Biology of the Pathological State,* EUR 6348. Luxembourg: Office for Official Publications of the European Communities, 1979.

de Enese, Bela Dorner. 'Pig Breeding in Hungary'. *Pig Breeders Annual* (1934–35): 61–63.

Delbrück, Max. 'Ueber Hefe und Gärung in der Bierbrauerei'. *Bayerische Bierbrauer* 19 (1884): 304–12.

'Hefe ein Edelpilz'. *Wochenschrift für Brauerei* 27 (30 July 1910): 373–76.

Delbrück, Max, and Schrohe, A., eds. *Hefe, Gärung und Fäulnis.* Berlin: Paul Parey, 1904.

Dellweg, H. 'Die Geschichte der Fermentation – Ein Beitrag zur Hundertjahrfeier des Instituts für Gärungsgewerbe und Biotechnologie zu Berlin'. In *100 Jahre Institut für Gärungsgewerbe und Biotechnologie zu Berlin, 1874–1974. Festschrift,* pp. 17–41. Berlin: Institut für Gärungsgewerbe und Biotechnologie, 1974.

Demain, Arnold L. 'The Marriage of Genetics and Industrial Microbiology – After a Long Engagement, A Bright Future'. In *Genetics of Industrial Microorganisms,* ed. Z. Vaňek, Z. Hošťálek, and J. Cudlin, pp. 19–32. Prague: Academia, 1973.

de Nettancourt, D. 'Applied Molecular and Cellular Biology: Background

Note on a Possible Action of the European Communities for the Optimal Exploitation of the Fundamentals of the New Biology'. Commission of the European Communities, Directorate General for Research Science and Education, XII/207/77-E, 15 June 1977.

Dibner, Mark D., and Bruce, Nancy G. 'The Greening of Biotechnology: The Growth of the US Biotechnology Industry'. *Trends in Biotechnology* 5(1987): 270–72.

Dickson, David. *The New Politics of Science.* 2d ed. Chicago: University of Chicago Press, 1988.

Diettrich, Olaf. 'Beauty and the Beast: Particular Aspects of the Socioeconomic Integration of Biotechnology in the Context of Activities of the Commission of the European Communities'. Presented at the International Symposium on Environmental Biotechnology, 22–25 April 1991, Ostend.

Djerassi, Carl. *Steroids Made it Possible.* Washington, D.C.: American Chemical Society, 1990.

Dodgson, Mark. *Celltech: The First Ten Years.* Brighton: SPRU, 1990.

Donzelot, Jacques. *The Policing of Families,* trans. Robert Hurley. London: Hutchinson, 1980.

Dore, Ronald. *A Case Study of Technology Forecasting in the Next Generation Base Technologies Development Programme.* London: Technical Change Centre, 1983.

Doyle, Jack. *Altered Harvest: Agriculture, Genetics and the Fate of the World's Food Supply.* New York: Viking-Penguin, 1985.

Dubos, René. *Louis Pasteur: Freelance of Science.* New York: Scribner, 1976.

Dummett, G.A., 'The Engineering of Continuous Brewing'. *Wallerstein Laboratories Communications* 25 (April 1962): 19–36.

'Chemical Engineering in the Biochemical Industries'. *The Chemical Engineer* (July–August 1969): 306–10.

From Little Acorns: A History of the A.P.V. Company Limited. London: Hutchinson Benham, 1981.

Duncan, W.B. 'Lesson from the Past, Challenges and Opportunity'. In *The Chemical Industry,* ed. D.H. Sharp and T.F. West, pp. 15–30. Chichester: Ellis Horwood, 1982.

Dutton, Diana B., and Pfund, Nancy E. 'Genetic Engineering: Science and Social Responsibility'. In *Worse Than the Disease: Pitfalls of Medical Progress,* ed. Diana B. Dutton, pp. 174–225. Cambridge University Press, 1988.

'E.A. Siebel Co.' *Western Brewer and Journal of the Barley, Malt and Hop Trades* (January 1918): 25.

Edelman, J.; Fewell, A.; and Solomons, G.L. 'Myco-Protein – A New Food'. *Reviews in Clinical Nutrition* 53 (1983): 471–80.

Elkington, John. *The Gene Factory: Inside the Biotechnology Business.* London: Century Publishing, 1985.

Emery, A.N. 'Biochemical Engineering: An Industrial Fellowship Report for the Science Research Council and Institution of Chemical Engineers'. Birmingham: University of Birmingham, 1976.

Enebo, Lennart. 'Growth of Algae for Protein: State of the Art'. In *Engineering of Unconventional Protein Production,* ed. Herman Bieber. Chemical Engineering Progress, Symposium series no. 93, 65 (1969): 80–86.

Engel, Michael. *Chemie im achtzehnten Jahrhundert: Auf dem Weg zu einer inter-*

nationalen Wissenschaft – Georg Ernst Stahl (1659–1734) zum 250 Todestag. Exhibition in the Staatsbibliothek Preußischer Kulturbesitz, 29 May to 7 July 1984, Ausstellungskataloge 23. Berlin: Staatsbibliothek Preußischer Kulturbesitz, 1984.

Ereky, Karl. *Nahrungsmittelproduktion und Landwirtschaft.* Budapest: Friedrich Kilians Nachfolger, 1917.

'Die großbetriebsmäßige Entwicklung der Schweinemast in Ungarn'. *Mitteilungen der Deutschen Landwirtschafts-Gesellschaft* 34 (25 August 1917): 541–50.

Biotechnologie der Fleisch-, Fett- und Milcherzeugung im landwirtschaftlichen Großbetriebe. Berlin: Paul Parey, 1919.

Etzkowitz, Henry. 'The Capitalization of Knowledge'. *Theory and Society* 19(1990): 107–21.

'The European Federation of Biotechnology'. *Chemistry and Industry* (21 October 1978): 781.

European Parliament. Committee on Agriculture, Fisheries and Food. Rapporteur Mr F. W. Gräfe zu Baringdorf. 'Report... on the Effects of the Use of Biotechnology on the European Farming Industry', Working Document A 2-159/86, 26 November 1986.

Committee on Energy, Research, and Technology. Rapporteur Mrs P. Viehoff. 'Biotechnology in Europe: The Need for an Integrated Policy', Draft Report, 21 July 1986. PE 105.423/B.

Rapporteur Benedikt Härlin. 'Report... on the Proposal from the Commission to the Council (COM/88/424–C2-119/88) for a Decision Adopting a Specific Research Programme in the Field of Health and Predictive Medicine: Human Genome Analysis (1989–1991)'. Document A2–0370/88, 30 January 1989.

Ewell, Raymond. 'The Rising Giant: The World Food Problem'. In *Engineering of Unconventional Protein Production,* ed. Herman Bieber. Chemical Engineering Progress, Symposium series no. 93, 65 (1969): 1–4.

Eyerman, Ron R. 'Rationalising Intellectuals: Sweden in the 1930s and 1940s', *Theory and Society* 14 (1985): 777–808.

Eyre, J.V., and Rodd, E.H. 'Raphael Meldola'. In *British Chemists,* ed. Alexander Findlay and W.H. Mills, pp. 96–125. London: Chemical Society, 1947.

FAST Group, *Eurofutures: The Challenge of Innovation.* London: Butterworth, 1987.

Fermentation Industries Section, IUPAC, 'Worldwide Survey of Fermentation Industries'. *Pure and Applied Chemistry* 13 (1966): 405–17.

Finlay, Mark R. 'The German Agricultural Experiment Stations and the Beginnings of American Agricultural Research'. *Agricultural History* 62 (1988): 41–50.

Fischer, Emil. 'Synthetical Chemistry in its Relation to Biology'. *Journal of the Chemical Society* 91ii (1907): 1749–65.

Fleming, Donald. 'Roots of the New Conservation Movement'. *Perspectives in American History* 6 (1972): 7–91.

Florkin, Marcel. 'Ten Years of Science at UNESCO'. *Impact of Science on Society* 7 (1956): 121–46.

Fogel, Lawrence J. *Biotechnology: Concepts and Applications.* Englewood Cliffs, N.J.: Prentice-Hall, 1963.

Forman, Paul. 'Behind Quantum Electronics: National Security as Basis for Physical Research in the United States, 1940–1960'. *Historical Studies in the Physical and Biological Sciences* 18 (1987): 149–229.

Fox, Jeffrey L. 'More Changes in U.S. Regulatory Bodies'. *Bio/technology* 8 (November 1990): 996.

Francé, Raoul H. 'Das biologische Experiment und seine Bedeutung für die Versuchstechnik'. *Mitteilungen des K. K. Technischen Versuchsamtes* 7 ii (1918): 15–21.

Bios: Die Gesetze der Welt. 2 vols. Munich: Hofstaengli, 1921.

Plants as Inventors. London: Simpkin & Marshall, 1926.

Der Weg zu Mir. Berlin: Alfred Kröner, 1927.

Francé-Harrar, Annie. *So war's um Neunzehnhundert: Mein Fin de Siècle.* Munich: Albert Langen, 1962.

Freudenthal, Gad, ed. 'Etudes sur Hélène Metzger'. *Corpus des oevres de philosophie en langue francaise* 8/9, 1988.

Fuller, Watson, ed. *The Biological Revolution: Social Good or Social Evil?* New York: Doubleday, 1972.

Fulmer, Ellis I. 'The Chemical Approach to Problems of Fermentation'. *Industrial and Engineering Chemistry* 22 (November 1930): 1148–50.

Gaden, Elmer. 'Editorial', *Biotechnology and Bioengineering* 4 (1962): 1–3.

'Biochemical Engineering: Where Has it Been and Where is it Going?' *Biotechnology Letters* 2 (1980): 336.

Gaissinovitch, A.E. 'The Origins of Soviet Genetics and the Struggle with Lamarckism, 1922–1929'. *Journal of the History of Biology* 13 (1980): 1–51.

Garrett, J.F. 'Lactic Acid'. *Industrial and Engineering Chemistry* 22 (November 1930): 1153–54.

Geddes, Patrick. *Cities in Evolution: An Introduction to the Town Planning Movement and the Study of Civics.* London: Williams & Norgate, 1915.

Geddes, P., and Thomson, J.A. *Biology.* London: Home University Library, 1925.

Geison, Gerald L. 'Pasteur, Roux and Rabies: Scientific *versus* Clinical Mentalities'. *Journal of the History of Medicine and Allied Sciences* 45 (1990) 341–65.

Gejl, Ib., and Vinding, Povl. 'Gustav Hagemann'. *Dansk Biografisk Lexicon* 5 472–77.

'The Genetic Economy'. *The Economist,* Supp. 307 (30 April 1988):10.

'Genetic Engineering – Certainties and Doubts'. *New Scientist* 47 (24 September 1971): 614.

Gibbons, N.E., ed. *Recent Progress in Microbiology.* Symposia held at the 8th International Congress for Microbiology, Montreal, 1962. Toronto: University of Toronto Press, 1963.

Giddings, Luther Val. 'The United States Regulatory Framework'. In *Biotech '89*. Proceedings of the conference held in London, May 1989, pp. 67–78. London: Blenheim Online, 1989.

Giebelhaus, August W. 'Farming for Fuel: The Alcohol Motor Fuel Movement of the 1930s'. *Agricultural History* 54 (1980): 173–84.

Giessler, Alfred. *Biotechnik.* Leipzig: Quelle & Meyer, 1939.

Glamann, Kristoff. 'The Scientific Brewer: Founders and Successors During the Rise of the Modern Brewing Industry'. In *Enterprise and History: Essays in Honour of Charles Wilson*, ed. D.C. Coleman and Peter Mathias, pp. 186–98. Cambridge University Press, 1984.

Glas, E. *Chemistry and Physiology in Their Historical and Philosophical Relations.* Delft: Delft University Press, 1979.

'Glaxo Drug Approved'. *The Independent,* 14 August 1991.

GMBF/GBF. *Entwicklung eines Forschungsinstituts, 1965–1975.* Hannover: Stiftung Volkswagenwerke, 1975.

Godet, Michel, and Ruyssen, Olivier. *The Old World and the New Technologies: Challenges to Europe in a Hostile World.* Luxembourg: Office for Official Publications of the European Communities, 1981.

Goldscheid, Rudolf. 'Ostwald als Persönlichkeit und Kulturfaktor'. In *Wilhelm Ostwald: Festschrift aus Anlaß seines 60. Geburtstages 2 September 1913,* ed. Monistenbund in Oesterreich, pp. 57–82. Vienna: Suschitzky, 1913.

'A Sociological Approach to Problems of Public Finance'. In *Classics in the Theory of Public Finance,* ed. R.A. Musgrave and A.T. Peacock, pp. 202–13. London: Macmillan Press, 1958.

Goldsmith, Maurice. *Sage: A Life of J.D. Bernal.* London: Hutchinson, 1980.

Goodell, Rae. 'How to Kill a Controversy: The Case of Recombinant DNA'. In *Scientists and Journalists,* ed. S. Friedman, S. Dunwoody, and C. Rogers, pp. 170–81. New York: Free Press, 1986

Goodfield, June. *Cancer Under Siege.* London: Hutchinson, 1975.

Playing God: Genetic Engineering and the Manipulation of Life. London: Hutchinson, 1977, p. 111.

Goodman, David; Sorj, Bernardo; and Wilkinson, John. *From Farming to Biotechnology: A Theory of Agro-industrial Development.* Oxford: Blackwell Publishers, 1987.

Gordon Smith, C.E. 'The Microbiological Research Establishment, Porton – Research Establishments in Europe: 69, Porton Down', *Chemistry and Industry,* 4 March 1967, pp. 338–46.

Gouhier, Henri. *Bergson dans l'histoire de la pensée occidentale.* Paris: Vrin, 1989.

Graber, Vitus. *Die äuseren mechanischen Werkzeuge der Wirbeltiere.* Leipzig: Frentag, 1886.

Green, S.R. 'Past and Current Aspects of Continuous Beer Fermentation'. *Wallerstein Laboratories Communications* 25 (December 1962): 337–48.

Greene, Allan J., and Schmitz, Andrew J. 'Meeting the Objective'. In *The History of Penicillin Production,* ed. A.A. Elder. Chemical Engineering Progress, Symposium Series, no. 100, 66 (1970): 79–88.

Greenshields, R.N. 'Acetic Acid: Vinegar'. In *Economic Microbiology.* vol. 2, *Primary Products of Metabolism,* ed. A.H. Rose, pp. 121–86. New York: Academic Press, 1978.

Greenshields, R.N., and Smith, E.L. 'Tower Fermentation Systems and the Applications', *The Chemical Engineer* (May 1971): 182–90.

Grogin, R.C. *The Bergsonian Controversy in France.* Calgary: University of Calgary Press, 1988.

Grote, L.R. 'Wilhelm Roux in Halle a.S.'. In *Die Medizin der Gegenwart in Selbstdarstellungen,* vol. 1, pp. 141–206. Leipzig: F. Meiner, 1924.

Grotjahn, Alfred. *Soziale Pathologie: Versuch einer Lehre von den sozialen Beziehungen der menschlichen Krankheiten als Grundlage der sozialen Medizin und sozialen Hygiene,* 2d ed. Berlin: Hirschwald, 1915.

'Guinness's New Laboratories at Park Royal'. *Brewers Guardian* (August 1955): 26.

Gunst, Péter, and Gaál, Lázlo. *Animal Husbandry in Hungary in the 19th–20th Centuries.* Budapest: Akadamiai Kiado, 1977.

Haldane, J.B.S. *Daedalus or Science and the Future.* London: Kegan Paul, 1925.

Hale, William J. *Chemistry Triumphant: The Rise and Reign of Chemistry in a Chemical World.* Baltimore: Williams & Wilkins in cooperation with the Century of Progress Exposition, 1932.

Hall, Stephen S. *Invisible Frontiers: The Race to Synthesise a Human Gene.* London: Sidgwick & Jackson, 1988.

Hamlin, Christopher. 'William Dibdin and the Idea of Biological Sewage Treatment'. *Technology and Culture* 29 (1988): 189–218.

Hamstra, Anneke M., and Feenstra, Marijke H. *Consument en Biotechnologie: Kennis en meninvorming van consumenten over biotechnologie.* Report no. 85. The Hague: Instituut voor consumentenonderzoek, 1989.

Hansen, Emil Christian. *Practical Studies in Fermentation,* trans. Alex K. Miller. London: Spon, 1896.

Hanson, Dirk. *The New Alchemists: Silicon Valley and the Microelectronics Revolution.* Boston: Little, Brown, 1982.

Haraway, Donna Jean. *Crystals, Fabrics and Fields: Metaphors of Organicism in Twentieth Century Developmental Biology.* New Haven, Conn.: Yale University Press, 1976.

Simians, Cyborgs and Women: The Reinvention of Women. London: Free Association Books, 1991.

Harris, José. *William Beveridge: A Biography.* Oxford: Oxford University Press, 1981.

Harris, Robert, and Paxman, Jeremy. *A Higher Form of Killing: The Secret Story of Gas and Germ Warfare.* London: Chatto & Windus, 1982.

Harsanyi, Zsolt. 'Biotechnology and the Environment: An Overview'. In *Biotechnology and the Environment: Risk and Regulation,* ed. Albert H. Teich, Morris A. Levin, and Jill H. Pace, pp. 15–27. Washington, D.C.: AAAS, 1985.

Hartley, Harold. 'Agriculture as a Source of Raw Materials for Industry?' *Journal of the Textile Institute* 28 (1937): 151–72.

'Chemical Engineering at the Cross-Roads'. *Transactions of the Institution of Chemical Engineers* 30 (1952): 13–19.

'Chemical Engineering – The Way Ahead'. *Transactions of the Institution of Chemical Engineers* 33 (1955): 20–26.

'Commentary'. *Process Biochemistry* 2, no. 4 (April 1967): 3.

Hase, Albrecht. 'Ueber technische Biologie: Ihre Aufgaben und Ziele, ihre prinzipielle und wirtschaftliche Bedeutung'. *Zeitschrift für Technische Biologie* 8 (1920): 23–45.

Hastings, J.H. 'Development of the Fermentation Industries in Great Britain', *Advances in Applied Microbiology* 14 (1971): 1–45.

Haushofer, Heinz. *Die Agrarreformen der Oesterreich-ungarischen Nachfolgestaaten.* Munich: Dresler, 1929.

Hawksworth, David L. 'The Commonwealth Mycological Institute (CMI)'. *Biologist* 32 (1985): 7–12.

Hayduck, F. 'Max Delbrück'. *Berichte der Deutschen Chemischen Gesellschaft* 53i (1920): 48A–62A.

'Das Institut für Gärungsgewerbe in Vergangenheit und Zukunft'. In *Das Institut für Gärungsgewerbe und Stärkefabrikation zu Berlin,* pp. 1–15. Berlin: Paul Parey, 1925.

Hedén, Carl-Göran. 'Biological Research Directed towards the Needs of Underdeveloped Areas'. *TVF* 7 (1961): 297–304.

'Microbiology in World Affairs'. *Impact of Science on Society* 17 (1967): 187–208.

'The GIAMS – A Contribution to Technology Transfer'. *In From Recent Advances in Biotechnology and Applied Biology,* ed. S.T. Chang, K.Y. Chan, and N.Y.S. Woo, pp. 63–74. Hong Kong: Chinese University Press, 1988.

Heertje, Arnold. 'Schumpeter and Technical Change'. In *Evolutionary Economics: Application of Schumpeter's Ideas,* ed. Horst Hanusch, pp. 71–89. Cambridge University Press, 1988.

Heilbron, I. 'Brewing in Relation to Natural Science'. *Proceedings of the Royal Society,* B, 143 (1955): 178–99.

'Reflections on Science in Relation to Brewing', Horace Brown Memorial Lecture, *Journal of the Institute of Brewing* 65 (1959): 144–54.

Heinson, G.; Knieper, R.; and Steiger, O. *Menschenproduktion: Allgemeine Bevölkerungstheorie der Neuzeit.* Frankfurt: Suhrkamp, 1979.

Held, Joseph, ed. *The Modernization of Agriculture: Rural Transformation in Hungary, 1848–1975.* Boulder, Colo.: East European Quarterly Press, 1975.

Henahan, John. 'Elmer Gaden: Father of Biochemical Engineering'. *Chemical and Engineering News* 49 (31 May 1971): 27–30.

Hepner, Leo. 'Training for the Biochemical Industries'. *Advances in Applied Microbiology* 11 (1969): 283–88.

Hepple, P., ed. *Microbiology: Proceedings of a Conference Held in London 19 and 20 September 1967.* London: Institute of Petroleum, 1968.

Herf, J. *Reactionary Modernism: Technology, Culture and Politics in Weimar and the Third Reich.* Cambridge University Press, 1984.

Herter, M., and Wilsdorf G. *Die Bedeutung des Schweines für die Fleischversorgung,* Arbeiten der Deutschen Landwirtschafts-Gesellschaft, vol. 270. Berlin. Paul Parey, 1914.

Hill, A.V. 'Biology and Electronics'. *Journal of the British Institution of Radio Engineers* 19 (1959): 80–86.

Hobby, Gladys L. *Penicillin: Meeting the Challenge.* New Haven, Conn.: Yale University Press, 1985.

Hogben, Lancelot. *The Nature of Living Matter.* London: Kegan Paul, 1930.

'The Foundations of Social Biology'. *Economica,* no. 31 (February 1931): 4–24.

The Retreat from Reason: Conway Memorial Lecture Delivered at Conway Hall... May 20, 1936. London: Watts, 1936.

'Prolegomenon to Political Arithmetic'. In *Political Arithmetic: A Symposium of Population Studies,* ed. Lancelot Hogben, pp. 13–46. London: Allen & Unwin, 1938.

Holmes, F.L. 'Introduction'. Reprint edition of Liebig's *Animal Chemistry,* pp. vii–cxvi. New York: Johnson Reprint, 1964.

Hoogerheide, J.C. 'Address by the Symposium Co-Sponsor'. *Biotechnology and Bioengineering Symposium* 4i (1973): vii.

Hoppe, Brigitte. 'Biologische und technische Bewegungslehre in 19. Jahr-

hundert'. In *Geschichte der Naturwissenschaften und der Technik im 19. Jahrhundert,* ed. B. Hoppe et al., pp. 9–35. Dusseldorff: VDI, 1969.
Horn, Ernst-Jürgen. *Management of Industrial Change in Germany.* Sussex European Papers no. 13. Brighton: University of Sussex, 1982.
Hotchkiss, Rollin D. 'Portents for a Genetic Engineering'. *Journal of Heredity* 56 (1965): 197–202.
Hough, J.S. *The Biotechnology of Malting and Brewing.* Cambridge University Press, 1985.
Hounshell, David, and Smith, John Kenly, Jr. *Science and Corporate Strategy: Du Pont R&D, 1902–1980.* Cambridge University Press, 1988, p. 589.
'How Hot Can a Man Get'. *Life* 24 244 (9 February 1948): 85–87.
Hufbauer, Carl. *The Formation of the German Chemical Community (1720–1795).* Berkeley: University of California Press, 1982.
Hughes, Thomas P., and Agatha, C., eds. *Lewis Mumford: Public Intellectual.* Oxford: Oxford University Press, 1990.
Hulme, E.W. *Statistical Bibliography in Relation to the Growth of Modern Civilization.* London: Butler & Tanner, 1923.
Humphrey, A.E. 'A Critical Review of Hydrocarbon Fermentations and Their Industrial Utilization'. *Biotechnology and Bioengineering* 9 (1967): 3–24.
Huxley, J.S. 'Biology and Human Life'. 2d Norman Lockyer Lecture, British Science Guild, 23 November 1926.
'The Applied Science of the Next Hundred Years: Biological and Social Engineering'. *Life and Letters* 11 (1934): 38–46.
'Industry Starts to Do Biology with Its Eyes Open'. *The Economist* 269 (2 December 1978): 95.
INRA (Europe) sa/nv for 'CUBE'. *Opinions of Europeans on Biotechnology in 1991,* Eurobarometer 35.1.
'International Conference on Medical Electronics'. *Journal of the British Institution of Radio Engineers* 18 (1958): 505.
James, A.T. 'The Discovery of Gas-Liquid Chromatography: A Personal Recollection'. In *Historical Aspects of Gas Separation,* Royal Society of Chemistry Special Publication 62, pp. 175–200. London: Royal Society of Chemistry, 1987.
Jamison, Andrew, and Baark, Jamison. 'Modes of Biotechnology Assessment in the United States, Japan and Denmark'. Institut for Samfundstag, Danmarks Tekniske Højskole, *Skriftsserie,* no. 20, 1989.
Japan. Science and Technology Agency. *Outline of the White Paper on Science and Technology: Aimed at Making Technological Innovations in Social Development.* February 1977. Trans. Foreign Press Centre.
Jasanoff, Sheila. 'Technological Innovation in a Corporatist State: The Case of Biotechnology in the Federal Republic of Germany'. *Research Policy* 14 (1985): 23–38.
The Fifth Branch: Science Advisors as Policymakers. Cambridge, Mass.: Harvard University Press, 1990.
Jensen, Einar. *Danish Agriculture – Its Economic Development: A Description and Economic Analysis Centering on the Free Trade Epoch, 1870–1930.* Copenhagen: Schultz, 1937.
Jewson, N.D. 'The Disappearance of the Sick-Man from Medical Cosmology, 1770–1870'. *Sociology* 10 (1976): 225–44.

Jones, Greta. *Social Hygiene in 20th Century Britain.* London: Croom Helm, 1986.

Jørgensen, Alfred. *Micro-organisms and Fermentation.* New ed., trans. A.K. Miller and E.A. Lemdden. London: F.W. Lyons, 1893.

[Jørgensen, Alfred.] 'Alfred Jørgensens Gjaeringsfysiologiske Laboratorium'. *Zymotechnisk Tidsskrift* 19 (1903): 80.

[Jørgensen, Alfred.] 'Ansættelser fra Laboratoriet i sidste Semester'. *Zymotechnisk Tidsskrift* 23 (1907): 90.

Jungk, R. *The Everyman Project: Resources for a Human Future.* London: Thames & Hudson, 1976.

Just, Felix, and Schnabel, Willy. 'Submerse Massenzüchtung von Bakterien auf nicht-kohlenhyrathaltigen Nährstoffen 1. Ein Beitrag zur biotechnischen Fettsynthese'. *Die Branntweinwirtschaft* 2 (1948): 113–15.

Just, John. 'The Commercial Utilization of Milk Waste and the More Recent Products of Milk in a Dry Form'. In 5th International Congress for Applied Chemistry, Berlin, 2–8 June 1903, *Bericht,* vol 3, pp. 870–91. Berlin: Deutsche Verlag, 1904.

Kammerer, Paul. 'Adaptation and Inheritance in the Light of Modern. Experimental Investigation'. *Report of the Smithsonian Institution* (1912): 421–42.

'Höherentwicklung und Biologie'. *Archiv für Rassen und Gesellschaftsbiologie* 11 (1914): 222–33.

'Das biologischer Zeitalter: Fortschritte der organischen Technik'. *Monistische Bibliothek, kleine Flugschriften, Nr 33.* Vienna: Hamburger Verlag, 1920.

Kapp, Ernst. *Grundlinien einer Philosophie der Technik.* Braunschweig: Westermann, 1877.

Katoh, Kiyoaki. 'Statement on Current Status of SCP Production in Japan'. In *Single Cell Protein: Proceedings of the International Symposium Held in Rome, Italy, on November 7–9, 1973,* ed. P. Davis, pp. 223–32. London: Academic Press, 1974.

Kemp, A.F.; la Rivière, J.W.M.; and Verhoeven, W. *Albert Jan Kluyver: His Life and Work.* Amsterdam: North Holland, 1959.

Kenedi, Robert M. 'Bio-engineering – Concepts, Trend and Potential'. *Nature* 202 (25 April 1964): 334–36.

Kennedy, Max. 'The Evolution of the Word "Biotechnology" '. *Trends in Biotechnology* 9 (1991): 218–20.

Kenney, Martin. *Biotechnology: The University/Industrial Complex.* New Haven, Conn.: Yale University Press, 1986.

Kiplinger, W.M. 'Causes of Our Unemployment: An Employment Puzzle'. *New York Times,* 17 August 1930.

Kirkpatrick, Sidney. 'A Case Study in Biochemical Engineering'. *Chemical Engineering* 54 (1947): 94–101.

Kloppenburg J.R., Jr. *First The Seed: The Political Economy of Plant Biotechnology, 1492–2000.* Cambridge University Press, 1988.

Koestler, Arthur. *The Case of the Midwife Toad.* London: Hutchinson, 1971.

Kohler, Robert E. 'The Reception of Eduard Buchner's Discovery of Cell-Free Fermentation'. *Journal of the History of Biology* 5 (1972): 327–53.

From Medical Chemistry to Biochemistry: The Making of a Biomedical Discipline. Cambridge University Press, 1982.

'Bacterial Physiology: The Medical Context'. *Bulletin of the History of Medicine* 59 (1985): 54–74.

Partners in Science, Foundations and Natural Scientists, 1900–1945. Chicago: University of Chicago Press, 1991.

Kornberg, Arthur. 'The Two Cultures: Chemistry and Biology'. *Biochemistry* 26 (1987): 6588–91.

For the Love of Enzymes: The Odyssey of a Biochemist. Cambridge, Mass.: Harvard University Press, 1989.

Kraft, J.A.R. 'The 1961 Picture of Human Factors Research in Business and Industry in the United States of America'. *Ergonomics* 5i (1962): 293–99.

[Král, F.]. *Králs Bakteriologisches Laboratorium, Der gegenwärtige Bestand der Králschen Sammlung von Mikroorganismen. März 1904.* Prague: Privately printed, 1904.

Krause, Richard M. 'Is the Biological Revolution a Match for the Trinity of Despair?' *Technology in Society* 4 (1982): 267–82.

Krimsky, Sheldon. *Genetic Alchemy: The Social History of the Recombinant DNA Controversy.* Cambridge, Mass.: MIT Press, 1982.

Krimsky, Sheldon, and Plough, Alonzo. *Environmental Hazards: Communicating Risks as a Social Process.* Dover, Mass.: Auburn House, 1988.

Kroeger, Gertrud. *The Concept of Social Medicine: As Presented by Physicians and Other Writers in Germany, 1779–1932.* Chicago: Julius Rosenwald Fund, 1937.

Kuczynski, Jürgen. *René Kuczynski: Ein fortschriftlischer Wissenschaftler in der ersten Hälfte des 20 Jahrhunderts.* Berlin: Aufbau Verlag, 1957.

Kuznick, Peter J. *Beyond the Laboratory: Scientists as Political Activists in 1930s America.* Chicago: University of Chicago Press, 1987.

Lackoff, S. 'Biotechnology and Developing Countries'. *Politics and the Life Sciences* 2ii (1984): 151–83.

Lafar, Franz. *Technical Mycology: The Utilization of Micro-organisms in the Arts and Manufactures,* trans. T.C. Salter. London: Griffin, 1898.

Lake, Gordon. 'Scientific Uncertainty and Political Regulation: European Legislation on the Contained Use and Deliberate Release of Genetically Modified (Micro) organisms'. *Project Appraisal* 6 (March 1991): 7–15.

Lamb, M., and Walker, G.D. 'Industrial Applications of Continuous Culture Processes'. In *Continuous Culture of Microorganisms,* Society of Chemical Industry Monograph 12 (1961): 254–64.

la Rivière, J.W.M. 'Biotechnology in Development Cooperation: A Donor Countries' View,' In *Biotechnology in Developing Countries,* ed. P.A. Van Hemert et al., pp. 1–17. Delft: Delft University Press, 1983.

Latour, Bruno. *Microbes: Guerre et paix; suivie de irreduction.* Paris: A.-M. Métilié, 1984.

Layton, Edwin R. *The Revolt of the Engineers: Social Responsibility and the American Engineering Profession.* Cleveland: Case Western Reserve University Press, 1971.

Lazell, H.G. *From Pills to Penicillin: The Beecham Story.* London: Heinemann, 1975.

Lederberg, Joshua. 'Experimental Genetics and Human Evolution'. *The American Naturalist* 100 (1966): 519–31.

'A Geneticist on Safeguards'. *New York Times,* 11 March 1975.

'DNA Research: Uncertain Peril and Certain Promise', *PRISM* (AMA Policy Journal) 15 June 1975.

'Genetic Recombination in Bacteria: A Discovery Account'. *Annual Review of Genetics* 21 (1987): 23–46.

'Edward Lawrie Tatum, 1909–1975'. *Biographical Memoirs* (National Academy of Sciences) 50 (1990): 357–86.

Lee, Thomas F. *The Human Genome Project: Cracking the Genetic Code of Life.* New York: Plenum, 1991.

Lenoir, T. 'Science for the Clinic: Science Policy and the Formation of Carl Ludwig's Institute in Leipzig'. In *The Investigative Enterprise: Experimental Physiology in Nineteenth-Century Medicine,* ed. W. Coleman and F.L. Holmes, pp. 139–78. Berkeley: University of California Press, 1988.

Leroy, Olivier. *Biotechnology: EEC Policy on the Eve of 1993.* Rixensart, Belgium: European Study Service, 1991.

Li, D.N. 'Biogas Production in China: An Overview'. In *Microbial Technology in the Developing World,* ed. E.J. DaSilva et al., pp. 196–208. Oxford University Press, 1987.

Liebig, Justus. *Familiar Letters on Chemistry in Its Relations to Physiology, Dietetics, Agriculture, Commerce and Political Economy,* 3d ed. London: Taylor, Walton & Mabberly, 1851.

Lindner, Paul. 'Die botanische und chemische Charackerisierung der Gärungsmikoben und die Notwendigkeit der Errichtung einer biologischen Zentrale'. *Seventh International Congress of Applied Chemistry,* Section Vib, 'Fermentation', pp. 169–72. London: Partridge & Cooper, 1910.

'Allgemeines aus dem Bereich der Biotechnologie'. *Zeitschrift fur Technische Biologie* 8 (1920): 54–56.

[Lindner, Paul.] 'Förderung eines Institutes für Erforschung technischwichtiger Mikroben in England.' *Zeitschrift für technische Biologie* 8 (1920): 64–67.

Lintner, Carl. 'C.J.H. Balling's Leben und Wirken'. *Bayerische Bierbrauer* 1 (1866): 29–34, 47–49, 62–66.

Lipman, Timothy O. 'Vitalism and Reductionism in Liebig's Physiological Thought'. *Isis* 58 (1967): 167–84.

Ljunberg, Gregory. 'Krig och Fred: IVA under 40-talet', *TVF* 40 (1969): 187–95.

Edy Velander och Ingenjörsvetenskapsakademien, IVA-meddelande, vol. 251. Stockholm: IVA, 1986.

Löwy, Illana. 'Immunology and Literature in the Early Twentieth Century: *Arrowsmith* and *The Doctor's Dilemma*'. *Medical History* 32 (1988): 314–32.

Lukacs, John. *Budapest 1900: A Historical Portrait of a City and Its Culture.* New York: Weidenfeld & Nicolson, 1988.

Lyons, Emerson J. 'Deep Tank Fermentation'. In *The History of Penicillin Production,* ed. A.A. Elder. Chemical Engineering Progress Symposium Series no. 100, 66 (1970): 31–36.

McAuliffe, Sharon, and McAuliffe, Kathleen. *Life for Sale.* New York: Coward, McCann & Geoghegan, 1981.

McCormick, D. 'Not as Easy as It Looked', *Bio/technology* 7 (July 1989): 629.

'A-sitting on a Gate'. *Bio/Technology* 5 (February 1987): 101.

McElheny, V.K. 'World Biologists Tighten Rules on "Genetic Engineering" Work'. *New York Times,* 28 February 1975.

Mackaye, Benton. *From Geography to Geotechnics.* Urbana: University of Illinois Press, 1968.

MacKenzie, Deborah. 'Europe Rethinks Patent on the Harvard Mouse'. *New Scientist* 132 (19 October 1991): 11.

McMillen, Wheeler. *New Riches from the Soil: The Progress of Chemurgy.* New York: Van Nostrand, 1946.

Málek, I. 'The Role of Continuous Processes and Their Study in the Present Development of Science and Production'. In *Theoretical and Methodological Basis of Continuous Culture of Micro-organisms,* ed. I. Málek and Z. Fencl, pp. 11–30. Prague: Czech Academy of Science, 1966.

Markle, Gerald E., and Robin, Stanley S. 'Biotechnology and the Social Reconstruction of Molecular Biology'. *Science, Technology and Human Values* 10 (1985): 70–79.

Martin, J.N. *Biomedical Engineering Education.* Pittsburgh: Chilton, 1966.

März, Edouard. *Joseph Alois Schumpeter: Forscher, Lehrer und Politiker.* Vienna: Verlag für Geschichte und Politik, 1983.

Mason, F.A. 'Microscopy and Biology in Industry'. *Bulletin of the Bureau of Biotechnology* 1 (1920): 3–15.

Mateles, Richard I., and Tannenbaum, Steven R. *Single Cell Protein.* Cambridge, Mass.: MIT Press, 1968.

Mathias, Peter. *The Brewing Industry in England, 1700–1830.* Cambridge University Press, 1959.

'Maturing of Product Sales up to $4b a year by 1993'. *Los Angeles Times,* 17 March 1991.

Max Henius Memoir Committee. *Max Henius: A Biography.* Chicago: Privately printed, 1936.

May, O.E., and Herrick, H.T. 'Some Minor Industrial Fermentations'. *Industrial and Engineering Chemistry* 22 (November 1930): 1172–76.

Meadows, Donella, et al. *The Limits to Growth.* A Report for the Club of Rome. New York: Universe, 1972.

Meldola, Raphael. *The Chemical Synthesis of Vital Products and the Inter-relations Between Organic Compounds.* London: Arnold, 1904.

Merck, George W. 'Peacetime Implications of Biological Warfare'. *Chemical and Engineering News* 24, no. 10 (25 May 1936): 1346–49.

Metzger, Hélène. *Newton, Stahl, Boerhaave et la doctrine chimique.* Paris: Alcan, 1930.

Miller, H.I., and Young, F.E. 'Isn't It about Time We Dispensed with "Biotechnology" and "Genetic Engineering"?' *Bio/Technology* 5 (1987): 184.

'M.I.T. Head Fears "Relief Palliative" Hampers Science'. *New York Times,* 25 October 1936.

Molella, Arthur P. 'The First Generation: Usher, Mumford and Giedion'. In *In Context – History and the History of Technology: Essays in Honor of Melvin Kranzberg,* ed. Stephen H. Cutcliffe and Robert C. Post, pp. 88–105. Bethlehem, Pa.: Lehigh University Press, 1989.

Monod, J. 'La technique de culture continue'. *Annales Institut Pasteur* 79 (1950): 390–410.

Montet, Pierre. *Scènes de la vie privée dans les tombeaux égyptiens.* Publications de la Faculté des Lettres de l'Université de Strasbourg. Strasbourg: University of Strasbourg, 1925.

Moses, Vivian, and Cape, Ronald E., eds. *Biotechnology: The Science and the Business.* London: Harwood Academic, 1991.
Muller, H.J. 'Perspectives for the Life Sciences'. *Bulletin of the Atomic Scientists* 20 (January 1964): 3–7.
Multhauf, Robert P. *The Origins of Chemistry.* New York: Franklin Watts, 1967.
Mumford, Lewis. *Technics and Civilization.* New York: Harper, Brace & World, 1934.
'An Appraisal of Lewis Mumford's "Technics and Civilization" (1934)'. *Daedalus* 88 (1959): 527–36.
'The Disciple's Rebellion'. *Encounter* 27 (1966): 11–21.
Munch-Petersen, H., ed. *Aarbog for Københavns Universitet, Kommuniteter og den polytekniske Læreanstalt, indeholdende Meddelelser for det akademiske Aar, 1907–1908.* Copenhagen: Universitetebogytrykkeriat, 1912.
Muspratt, J.S., and Hoffman, A.W. 'On Toluidine'. *Memoirs and Proceedings of the Chemical Society of London* 2 (1843–45): 367–83.
Nadav, Daniel S. *Julius Moses und die Politik der Sozialhygiene in Deutschland.* Schriftenreihe des Instituts für Deutsche Geschichte, Tel Aviv. Geilingen: Belicher Verlag, 1985.
Naisbitt, John. *Megatrends: Ten New Directions Transforming Our Lives.* London: Macdonald, 1984.
Nakamura, Keiko. 'Studies on Life Sciences. Part II. Role of Life Sciences with Special Reference to Research in the Mitsubishi-Kasei Institute'. *Technocrate* 6 (1973): 48–52.
'Nature and Art'. *Times Engineering Supplement,* 17 July 1912, p. 22.
Needham, Joseph. 'Notes on the Way'. *Time and Tide* 12 (10 September 1932): 970–72.
Neumeyer, Fritz. *Mies van der Rohe – Das Kunstlose Wort: Gedanken zur Baukunst.* Berlin: Siedler, 1986.
Nicholson, Max. *The Environmental Age.* Cambridge University Press, 1987.
Noble, David. *America by Design: Science, Technology and the Rise of Corporate Capitalism.* New York: Knopf, 1977.
Norton, M.G. 'The US Biotechnology Industry'. London: Overseas Technical Information Unit, 1984.
Nou, Joseph. *Studies in the Development of Agricultural Economics in Europe.* Uppsala, Sweden: Lantbrukshögskolan, 1967.
Novick A., and Szilard, Leo. 'Experiments with the Chemostat on Spontaneous Mutations of Bacteria'. *Proceedings of the National Academy of Sciences* 36 (1950): 708–19.
Oermann-Seeste, 'Schweinemastgroßbetriebe, ihre Technik und wirtschaftliche Bedeutung'. *Jahrbuch der Deutschen Landwirtschafts-Gesellschaft* 25 (1911): 956–68.
[OECD.] *Biotechnology and the Changing Role of Government.* Paris: OECD, 1988.
Olson, Steve. *Biotechnology: An Industry Comes of Age.* Washington D.C.: National Academy Press, 1986.
Orla-Jensen, S. *Lidt anvendt Filosofi.* Copenhagen: Det Schønbergske Forlag, 1934.
[PAG.] *The PAG Compendium.* Collected Papers Issued by the Protein-Calorie Advisory Group of the United Nations System, 1956–1973, C2. New York: Worldmark, 1979.

Panem, Sandra. *The Interferon Crusade.* Washington D.C.: Brookings Institution, 1984.

Paterson, D., ed. *Genetic Engineering.* London: BBC, 1969.

Pauly, A. *Darwinismus und Lamarckismus.* Munich: Reinhardt, 1905.

Pauly, Philip. *Controlling Life: Jacques Loeb and the Engineering Ideal in Biology.* (Oxford University Press, 1987.

Payne, J.N.W. 'Mashing on the Larger Scale'. *Brewers Guardian* (November 1962): 75–78.

Perkin, W.H. Jr., 'The Production and Polymerisation of Butadiene, Isoprene and Their Homologues'. *Journal of the Society of Chemical Industry* 31 (1912): 616–25.

Perlman, David. 'Preface'. *Advances in Applied Microbiology* 11 (1970): vii–viii.

Peterson, Durey H. 'Autobiography'. *Steroids* 45 (1985): 1–17.

Pfisterer, Herbert. *Der Polytechnische Verein und sein Wirken im vorindustriellen Bayern (1815–1830).* Miscellanea Bavarica Monacensia, vol. 45. Munich: Stadtarchivs München, 1973.

Pfizer, Charles, & Co. Inc. ed. *The Pasteur Fermentation Centennial, 1857–1957.* New York: Charles Pfizer & Co. Inc, 1958.

Pfund, Nancy, and Hofstadter, Laura. 'Biomedical Innovation and the Press'. *Journal of Communication* 31 (1981): 138–54.

Pirie, N.W. 'Recurrent Luck in Research'. *Selected Topics in the History of Biochemistry: Personal Recollections,* ed. G. Semenza, Comprehensive Biochemistry, vol. 36, pp. 491–522. Amsterdam: Elsevier, 1986.

Plein, L. Christopher. 'Popularizing Biotechnology: The Influence of Issue Definition'. *Science, Technology and Human Values* 16 (1991): 474–90.

Pope, Sir William. 'Address by the President'. *Journal of the Society of Chemical Industry* 40 (1921): 179T–82T.

Porter, Dorothy, and Porter, Roy. 'What was Social Medicine? An Historiographical Essay'. *Journal of Historical Sociology* 1 (1988): 90–106.

Postgate, John. *Microbes and Man.* London: Penguin, 1969.

Press, Frank. 'Science and Technology in the White House, 1977 to 1980'. Pt. 1. *Science* 211 (9 January 1981): 139–45; Pt. 2. *Science* 211 (16 January 1981): 249–56;

Prima Europe. *The Case for Biotechnology.* London: Prima Europe, 1990.

Pursell, Carroll. 'The Farm Chemurgic Council and the United States Department of Agriculture, 1935–1939'. *Isis* 60 (1969): 307–17.

Purssell, A.J.R., and Smith, M.J. 'Continuous Fermentation'. In *Proceedings of the 11th European Brewery Convention,* pp. 155–67. Madrid: European Brewery Convention, 1967.

Querner, Hans. 'Probleme der Biologie um 1900 auf den Versammlungen der Deutschen Naturforscher und Aertzte'. In *Wege der Naturforschung, 1822–1972 im Spiegel der Versammlungen Deutscher Naturforscher und Aertzte,* ed. Hans Querner and Heinrich Shipperges, pp. 186–201. Berlin: Springer, 1972.

Rahn, Otto. 'Theoretische Bakteriologie'. *Naturwissenschaften* 9(1921): 374–76.

Rainger, Ronald; Benson, Keith R.; and Maienschein, Jane, eds. *The American Development of Biology.* Philadelphia: University of Pennsylvania Press, 1988.

Randall, J.H. *Our Changing Civilization: How Science and the Machine are Reconstructing Modern Life.* London: Allen & Unwin, 1929.
Ranz, W.E., and Fredrickson, A.G. 'Minneapolis to Host Chemical Engineers'. *Chemical Engineering Progress* 61 (July 1965): 112–19.
Rapp, Friedrich. 'Philosophy of Technology: A Review'. *Interdisciplinary Science Reviews* 10 (1985): 126–39.
Ratner, Mark. 'The Analysts Take Aim: Biotech Fires Back'. *Bio/technology* 4 (November 1988): 1280.
'Bandshift'. *Bio/technology* 8 (November 1990): 993.
Ravetz, J.R. 'The DNA Controversy and Its History'. In *The Social Assessment of Science: Proceedings,* pp. 79–90. Wissenschaftsforschung 13 Report. Science Studies. Bielefeld: B. Kleine Verlag, 1982.
The Merger of Knowledge with Power: Essays in Critical Science. London: Mansell, 1990.
Rehm, H.-J. 'Modern Industrial Microbiological Fermentations and Their Effects on Technical Developments'. *Angewandte Chemie,* International ed. 9 (1970): 936–45.
Reinharz, Jehuda. *Chaim Weizmann: The Making of a Zionist Leader.* Oxford University Press, 1985.
Rexen, F., and Munck, L. *Cereal Crops for Industrial Use in Europe.* EUR 9617. Copenhagen: Carlsberg Research Laboratory, 1984.
Rhees, D.J. *The Chemists' Crusade: The Rise of an Industrial Science in Modern America, 1907–1922,* Doctoral dissertation, University of Pennsylvania, 8714116. Ann Arbor, Mich.: UMI, 1987.
Righelato, R.C., and Elsworth, R. 'Industrial Applications of Continuous Culture: Pharmaceutical Products and Other Products and Processes'. *Advances in Applied Microbiology* 13 (1970): 399–465.
Rinard, R.G. 'Neo-Lamarckism and Technique: Hans Spemann and the Development of Experimental Embryology'. *Journal of the History of Biology* 21 (1988): 95–118.
Rip, A., and Nederhof, A. 'Between Dirigism and Laissez-faire: Effects of Implementing the Science Policy Priorities for Biotechnology in the Netherlands'. *Research Policy* 15 (1986): 253–68.
Rode, O. *Optegnleser efter Prof. Dr. Phil Orla-Jensens: Efteraars Forelæsninger over Bioteknisk Kemi.* Copenhagen: Det Private Ingeniørsfonds forlag, 1915.
Rogers, Michael D. 'The Japanese Government's Role in Biotechnology R&D'. *Chemistry and Industry* (7 August 1982): 533–37.
Rommel, George. 'The Hog Industry: Selection, Feeding and Management – Recent American Experimental Work, Statistics of Production and Trade'. Bureau of Animal Husbandry, *Bulletin 47.* Washington D.C.: Government Printing Office, 1904.
Roobeek, Annemieke J.M. *Beyond the Technology Race: An Analysis of Technology Policy in Seven Industrial Countries.* Amsterdam: Elsevier, 1990.
Rörsch, A. *Genetic Manipulations in Applied Biology: A Study of the Necessity, Content and Management Principles of a Possible Community Action* EUR 6078. Luxembourg: Office for Official Publications of the European Communities, 1979.
Rosenberg, Charles. 'Science, Technology and Economic Growth: The Case of the Agricultural Experiment Station Scientist, 1875–1914'. In

Nineteenth-Century American Science: A Reappraisal, ed. George F. Daniels, pp. 181–209. Evanston, Ill.: Northwestern University Press, 1974.

No Other Gods: On Science and American Social Thought. Baltimore: Johns Hopkins University Press, 1976.

Roskill, Stephen. *Hankey: Man of Secrets,* vol. 3. London: Collins, 1974.

Rossiter, Margaret. *The Emergence of Agricultural Science in America: Justus Liebig and the Americans, 1840–1880.* New Haven, Conn.: Yale University Press, 1975.

Roth, R.R. 'The Foundation of Bionics'. *Perspectives in Biology and Medicine* 26 (1983): 229–42.

Rothschuh, K.E. 'Bionomie/Biotechnik'. In *Historisches Wörterbuch der Philosophie,* ed. J. Ritter, vol. 1, pp. 946–47. Darmstadt: Wissenschaftliche Buchgesellschaft, 1971.

Roux, Wilhelm, et al., eds., *Terminologie der Entwicklungsmechanik der Tiere und Pflanzen.* Leipzig: Wilhelm Engelmann, 1912.

Ruhrmann, G. 'Genetic Engineering in the Press.' In *Biotechnology in Public: A Review of Current Research,* ed. John Durant, pp. 169–201. London: Science Museum, 1992.

Russell, Colin A.; Coley, Noel G.; and Roberts, Gerrylyn K. *Chemists by Profession.* Milton Keynes: Open University Press, 1977.

Russell, Sir John E. *A History of Agricultural Science in Great Britain.* London: Allen & Unwin, 1966.

Saito, Hiuga. 'Biotechnology R&D: Japan and the World', *Science and Technology in Japan* 4 (April–June 1985): 8–11.

Sakaguchi, Kin-ichiro. *Outline and Characteristics of Japanese Fermentation Industries.* Tokyo: RIKEN, 1961.

Sargeant, K. 'Biotechnology, Connectedness and Dematerialisation'. In *Biotechnology '84.* Proceedings of a conference held in the Royal Irish Academy, 1–2 May 1984, ed. J.P. Arbuthnott, pp. 96–103. Dublin: Royal Irish Academy, 1985.

'Notes on Three French Biotechnology Reports: A Review Paper'. FAST XII/1160/81/EN.

Sasson, Albert. *Biotechnologies and Development.* Paris: UNESCO, 1988.

Sauer, H.J. Jr., and Nevins, R.G. 'Biotechnology and the Mechanical Engineer'. *Mechanical Engineering* 87 (December 1965): 36–39.

Schallmayer, W. 'Antwort auf P. Kammerers Plaidoyer für R. Goldscheid'. *Archiv für Rassen und Gesellschaftsbiologie* 11 (1914): 233–40.

Scheidegger, Alfred. 'Biotechnology in Japan – Towards the Year 2000... and Beyond'. *Trends in Biotechnology* 9 (June 1991): 183–90.

Schling-Brodersen, Ursula. *Entwicklung und Institutionalisierung der Agrikulturchemie im 19. Jahrhundert: Liebig und die landwirtschaftlichen Versuchsstationen.* Braunschweiger Veröffentlichungen zur Geschichte der Pharmazie und der Naturwissenschaften, vol. 31. Braunschweig: University of Braunschweig, 1989.

Schneider, William H. *Quality and Quantity: The Quest for Biological Regeneration in Twentieth-Century France.* Cambridge University Press, 1990.

Schrohe, A. *Aus der Vergangenheit der Gärungstechnik und verwandter Gebiete.* Berlin: Paul Parey, 1917.

[Schwartz, Wilhelm.] 'Biotechnik und Bioengineering'. *Nachrichten aus Chemie and Technik* 17 (1969):330–31.

Seaman, Alan. 'The Society for Applied Bacteriology: The First Fifty Years'. *Journal of Applied Bacteriology* 50 (1981): 425–31.

Seligman, R. 'Brewing Engineering and the Future'. *Journal of the Institute of Brewing* 36 (1930): 288–97.

'The Research Scheme of the Institute – Its Past and Its Future'. *Journal of the Institute of Brewing* 54, no. 3 (1948):133–44.

Senker, Jaqueline, and Sharp, Margaret. 'The Biotechnology Directorate of the SERC: Report and Evaluation of Its Achievements – 1981–1987'. Science Policy Research Unit Report, submitted to the Management Committee of the Biotechnology Directorate, 5, May 1988.

Séris, Jean-Pierre. 'Bergson et la technique.' In *Bergson: Naissance d'une philosophie. Actes du Colloque de Clermond-Ferrand 17 et 18 novembre 1987*, pp. 121–38. Paris: Presses Universitaires de France, 1990.

Servan-Schreiber, Jean-Jacques. *Le défi americain*. Paris: Editions Denoel, 1967.

Sharp, David H. *Bio-Protein Manufacture: A Critical Assessment*. Chichester: Ellis Horwood, 1989.

Sharp, Margaret. *The New Biotechnology: European Governments in Search of a Strategy*. Sussex European Papers no. 15. Brighton: Science Policy Research Unit, 1985.

'A Sheaf of Tributes to the Late Sir Patrick Geddes'. *Supplement to the Sociological Review* 24 (October 1932): 349–400.

Sheehan, J.C. *The Enchanted Ring – The Untold Story of Penicillin*. Cambridge, Mass.: MIT Press, 1982.

Shelley, Mary. *Frankenstein (Or, the Modern Prometheus)*. First published 1818; Clinton, Mass.: Airmont, 1963.

Shore, D.T. 'Chemical Engineering of the Continuous Brewing Process'. *The Chemical Engineer* (May 1968): 99–109.

Short, R.R.M., and Taylor, Richard. 'Soviet Cinema and the International Menace, 1928–1939'. *Historical Journal of Film, Radio and Television* 6 (1986): 131–59.

Siebel, E.A., and Company and Siebel Laboratories, Inc., 'Achievement: Yesterday, Today, Tomorrow'. n.d. Chicago.

[Siebel, John E.] 'The Zymotechnic College. A School for Brewers, Distillers, Maltsters, Wine and Vinegar-Makers.' *American Chemical Review and Zymotechnic Magazine* 4 (1884): 193.

Silver, S., ed. *Biotechnology: Potentials and Limitations*. Life Science Research Report 35. New York: Springer, 1986.

Slosson, Edwin E. *Creative Chemistry*. New York: Century, 1921.

Smyth, C.N. *Medical Electronics: Proceedings of the Second International Conference on Medical Electronics, Paris, 24–27 June 1959*. London: Illiffe & Sons, 1960.

Smyth, Henry Field, and Obold, Walter Lord. *Industrial Microbiology: The Utilization of Bacteria, Yeasts and Molds in Industrial Processes*. Baltimore: Williams & Wilkins, 1930.

Sogusawa, S. 'History of Japanese Natural Products Research'. *Pure and Applied Chemistry* 9 (1964):1–9.

Soloway, Richard. *Demography and Degeneration: Eugenics and the Declining Birthrate in 20th Century Britain*. Chapel Hill: University of North Carolina Press, 1990.

'Some Press Comments'. *Bulletin of the Bureau of Bio-Technology* 1 (1920–3): 82–4.

Sonntag, Otto. 'Religion and Science in the Thought of Liebig'. *Ambix* 24 (1977): 159–69.
Spemann, Hans. *Forschung und Leben.* Stuttgart: J. Engelhorns, Nachfolger, 1943.
Spier, R.E. 'So Who Is a Biotechnologist?' *Biotech Quarterly* 3, nos. 2–3 (1984):3–4, 15.
Spitz, Peter H. *Petrochemicals: The Rise of an Industry.* New York: Wiley, 1988.
Stanbridge, H.H. *History of Sewage Treatment in Britain.* P. 7. *Activated Sludge.* Maidstone: IWPC, 1977.
Starr, M.P., ed. *Global Impacts of Applied Microbiology.* Stockholm: Almquist & Wiksel, 1964.
Stern, Bernhard J. 'Human Heredity and Environment'. *Science and Society* 14 (1950): 122–33.
Strbánová, Soňa. 'On the Beginnings of Biochemistry in Bohemia'. *Acta Historiae rerum naturalium nec non technicarum,* Special issue 9 (1977): 149–221.
Strickland, Stephen P. *Politics, Science and Dread Disease: A Short History of United States Medical Research Policy.* Cambridge, Mass.: Harvard University Press, 1972.
Strube, Wilhelm. *Die Chemie und ihre Geschichte.* Berlin: Akademie Verlag, 1974.
Sturchio, J.L., ed. *Values and Visions: A Merck Century.* Rahway, N.J.: Merck, 1991.
Sulloway, Frank J. *Freud, Biologist of the Mind: Beyond the Psychoanalytic Legend.* New York: Basic, 1979.
Swann, John Patrick. 'Arthur Tatum, Parke-Davis, and the Discovery of Mapharsen as an Antisyphilitic Agent.' *History of Medicine and Allied Sciences* 40 (1985): 167–87.
Szilard, Leo. *The Voice of the Dolphins and Other Stories.* London: Victor Gollancz, 1961.
Takamine, Jokichi. 'Enzymes of Aspergillus Oryzae and the Application of Its Amyloclastic Enzyme to the Fermentation Industry'. *Industrial and Engineering Chemistry* 6 (1914): 824–28.
Tatum, Edward L. 'A Case History in Biological Research'. Nobel Lecture, 11 December 1958.
[Syntex.] *A Corporation and a Molecule.* Palo Alto: Syntex Laboratories, 1966.
Tatum, Edward L. 'Perspectives from Physiological Genetics'. In *The Control of Human Heredity and Evolution,* ed. T.M. Sonneborn, pp. 20–34. New York: Macmillan, 1965.
'Molecular Biology, Nucleic Acids, and the Future of Medicine'. *Perspectives in Biology and Medicine* 10 (1966–67): 19–31.
Taylor, Craig L., and Boelter, L.M.K. 'Biotechnology: A New Fundamental in the Training of Engineers'. *Science* 105 (28 February 1947): 217–19.
Taylor, G. Rattray. *The Biological Time Bomb.* New York: New American Library, 1968.
Tecknoligi Naevnet. *Consensus Conference on the Application of Knowledge Gained from Mapping the Human Genome: Final Document,* 1–3 November 1989.
Teich, Mikuláš. 'Science and the Industrialisation of Brewing: The German Case'. Presented to the conference, Biotechnology: Long Term Development. The Science Museum, London, February 1984.

Teitelman, Robert. *Gene Dreams: Wall Street, Academia and the Rise of Biotechnology.* New York: Basic, 1989.
Temkin, Owsei. 'Materialism in French and German Physiology of the Early Nineteenth Century'. *Bulletin of the History of Medicine* 20 (1946): 322–27.
Tempest D.W. 'The Place of Continuous Culture in Microbiological Research'. *Advances in Microbial Physiology* 4 (1970): 223–50.
Thackray, Arnold; Carroll, P. Thomas; Sturchio, J.L.; and Bud, Robert. *Chemistry in America, 1876–1976: Historical Indicators.* Dordrecht: Reidel, 1984.
Thaysen, A.C. 'Food Yeast: Its Nutritive Value and Its Production from Empire Sources'. *Journal of the Royal Society of Arts* 93 (8 June 1945): 353–64.
'Food and Fodder Yeast'. In *Yeasts,* ed. W. Roman, pp. 155–210. The Hague. W. Junk, 1957.
Thissen, Rudolf. *Die Entwicklung der Terminologie auf dem Gebiet der Sozialhygiene und Sozialmedizin im deutschen Sprachgebiet bis 1930.* Cologne: Westdeutscherverlag, 1969.
Thomas, D. *Production of Biological Catalysts, Stabilization and Exploitation.* EUR 6079. Luxembourg: Office for Official Publications of the European Communities, 1978.
Thomson, J.A. 'Biological Philosophy'. *Nature* 87 (12 October 1911): 475–76.
'Biology'. *Encyclopaedia Britannica.* Supp. to the 11th ed., 1 (1926), pp. 383–5.
'Biology and Social Hygiene'. *Quarterly Review* 246 (1926): 28–48.
Thorndike, Lynn. *History of Magic and Experimental Science.* New York: Columbia University Press, 1934.
Tornier, Gustav. 'Ueberzählige Bildungen und die Bedeutung der Pathologie für die Biontotechnik (mit Demonstrationen)'. In *Verhandlungen des V. Internationalen Zoologen-Congresses zu Berlin, 12–16 August 1901,* ed. Paul Matschie, pp. 467–500. Jena: Gustav Fischer, 1902.
Towalski, Zbigniew. 'The Integration of Knowledge within Science, Technology and Industry: Enzymes: a Case Study'. Doctoral dissertation, Aston University, Birmingham, 1985.
Tracy, Michael. *Agriculture in Western Europe: Crisis and Adaptation since 1880.* London: Jonathan Cape, 1964.
Treviranus, Gottfried Reinhold. *Biologie oder Philosophie der lebenden Natur für Naturforscher und Aertzte.* Göttingen: Rower, 1802.
Tsuchiya, H.M., and Keller, K.H. 'Bioengineering – Is A New Era Beginning?' *Chemical Engineering Progress* 61 (May 1965): 60–62.
[UNESCO.] *Records of the General Conference,* 8th Sess. 1954. Paris: UNESCO, 1955.
'The Unfulfilled Promise of Biotech "Cures" '. *Los Angeles Times,* 31 January 1991.
U.S. Congress. House of Representatives. Committee on Appropriations. *Genetics Research.* 91st Cong. 2d sess. 1971.
Committee on Science and Technology. *The Biotechnology Science Coordination Act of 1986.* 99th Cong. 2d sess. 1986.
Committee on Science and Technology. Subcommittee on Investigations and Oversight and the Subcommittee on Science, Research and Technology. *Commercialization of Academic Biomedical Research.* 97th Cong. 1st sess. 1981.

Subcommittee on Science, Research and Technology. *Science Policy Implications of DNA Recombinant Molecule Research.* 95th Cong. 1st sess. 1977.
U.S. Congress. Office of Technology Assessment. *Commercial Biotechnology: An International Analysis.* OTA-BA-218. Washington, D.C.: U.S. Government Printing Office, 1984.
Impacts of Applied Genetics: Microorganisms, Plants and Animals. OTA-HR-132. Washington, D.C.: U.S. Government Printing Office, 1981.
Mapping Our Genes – Genome Projects: How Big How Fast. Washington, D.C.: U.S. Government Printing Office, 1987.
New Developments in Biotechnology – Background Paper: Public Perceptions of Biotechnology. OTA-BP-BA-45. Washington, D.C.: U.S. Government Printing Office, 1987.
New Developments in Biotechnology. U.S. Investment in Biotechnology – Special Report. OTA-BA-360. Washington, D.C.: U.S. Government Printing Office, 1988.
U.S. Congress. Senate. Committee on Environment and Public Works. Subcommittee on Hazardous Wastes and Toxic Substances. *Federal Oversight of Biotechnology.* 100th Cong. 1st sess. 5 November 1987.
Committee on Government Operations. Subcommittee on Reorganization and International Relations. *Hearings . . . to Create a Department of Science and Technology.* 86th Cong. 1st sess. pt. 1, 16–17 April 1959.
U.S. National Science Board. *Science Indicators 1972: Report of the National Science Board, 1973.* Washington, D.C.: U.S. Government Printing Office, 1973.
Vallery-Radot, René. *The Life of Pasteur,* trans. R.L. Devonshire. London: Constable, 1919.
Van Brunt, Jennifer. 'Environmental Release: A Portrait of Opinion and Opposition'. *Bio/technology* 5 (1987): 559–663.
Vaspinder, S.H. *The Scientific Attitudes in Mary Shelley's Frankenstein.* Ann Arbor, Mich: UMI, 1976.
[Velander, E.]. 'Några nya utvecklingslinjer inom biotekniken'. *IVA* 13 (1942): 236.
Vernon, Keith. 'Pus, Sewage, Beer and Milk: Microbiology in Britain, 1870–1940'. *History of Science* 28 (1990): 289–325.
Virey, J.J. *Hygiène philosophique ou de la santé dans le régime physique, moral et politique de la Civilisation moderne.* Paris: Crochard, 1828.
Wada, A. 'One Step from Chemical Automatons'. *Nature* 257 (23 October 1975): 633–34.
Wagner, Adolf. *Geschichte des Lamarckismus: Als Einführung in die psychobiologische Bewegung der Gegenwart.* Stuttgart: Franck, 1908.
'Biotechnik und Plasmatik'. In *Der Begründer der Lebenslehre Raoul Francé: Eine Festschrift zu seinem 50 Geburtstag,* pp. 5–12. Stuttgart. Walter Seifert, 1925.
Wainwright, Milton. *Miracle Cure: The Story of Penicillin and the Golden Age of Antibiotics.* Oxford: Blackwell Publisher, 1990.
Waks, Norman. 'Consequences of the Shifts in U.S. Research and Development Policy'. *R&D Management* 15 (1985): 191–96.
Walgate, Robert. *Miracle or Menace? Biotechnology and the Third World.* London: Panos Institute, 1990.

Walker, Anthony. 'Europe Steals the Lead in Plant Biotechnology'. In *Biotech '89*. Proceedings of the conference held in London, May 1989, pp. 121–28. London: Blenheim Online, 1989.
Walker, T.K. 'A History of the Development of a School of Biochemistry in the Faculty of Technology, University of Manchester'. *Advances in Applied Microbiology* 12 (1970): 1–10.
Watson, J.D. *The Double Helix,* ed. G.S. Stent. New York: Norton, 1980.
Watson, James D., and Tooze, John. *The DNA Story: A Documentary History of Gene Cloning*. San Francisco: Freeman, 1981.
Weart, Spencer R. *Nuclear Fear: A History of Images*. Cambridge, Mass.: Harvard University Press, 1988.
Weber, Adolf. 'Biotechnologie und Gentechnologie'. In *Die Herstellung der Natur: Chancen und Risiken der Gen-technologie,* ed. Ulrich Steger, pp. 13–14. Bonn.: Verlag Neue Gesellschaft, 1985.
Weindling, Paul. *Health, Politics and German Politics between National Unification and Nazism, 1870–1945*. Cambridge University Press, 1989.
'Degeneration und öffentliches Gesundheitswesen, 1900–1930: Wohnverhältnisse'. In *Stadt und Gesundheit: Zum Wandel von "Volksgesundheit" und kommunaler Gesundheitspolitik im 19. und frühen 20. Jahrhundert,* ed. Jürgen Reulecke and Adelheid Gräfin zu Castell Rüdenhausen, pp. 105–13. Stuttgart: Frans Steiner Verlag, 1991.
Weingart, Peter; Kroll, Jürgen; and Bayertz, Kurt. *Rasse, Blut und Gene: Geschichte der Eugenik und Rassenhygiene in Deutschland*. Frankfurt: Suhrkamp, 1988.
Wells, G.P. 'Lancelot Thomas Hogben'. *Biographical Memoirs of the Royal Society* 24 (1978): 183–221.
Werskey, Gary. *The Visible College*. London: Allen Lane, 1978.
Wickenden, W.E. 'Technology and Culture'. Commencement address, Case School of Applied Science, 29 May 1929.
'Training Engineers for Positions of Responsibility'. *Cleveland Engineering* (26 December 1929): 3–5, 14.
'The Engineer in a Changing Society'. *Electrical Engineering* 51 (July 1932): 467.
'Technology and Culture'. *Ohio College Association Bulletin* (1933): 4–9.
'Final Report of the Director of Investigations, June 1933'. In *Report of the Investigation of Engineering Education, 1923–1929,* vol. 2, pp. 1041–1114. Pittsburgh: Society for the Promotion of Engineering Education, 1934.
Wiener, Norbert. *Cybernetics or Control and Communication in the Animal and the Machine*. Cambridge, Mass.: MIT Press, 1948.
Wiesner, Julius. 'Bedeutung der technischen Rohstofflehre (techn. Waarenkunde) als selbständige Disciplin und über deren Behandlung als Lehrgegenstand an techn. Hochschulen'. *Dinglers Polytechnisches Journal* 237 (1880): 319–26, 400–409, 468–73.
Die Rohstoffe des Pflanzenreiches, 2d ed., 2 vols. Leipzig: Engelmann, 1900.
Wolstenholme, Gordon, ed. *Man and His Future*. London: J.A. Churchill, 1963.
Working Party on Industrial Microbiology (P.W. Brian, Chairman), 'Report on the State of Research into Economic Microbiology in the United Kingdom'. 1960.

Wright, Pearce. 'Time for Bug Valley', *New Scientist* 82 (5 July 1979): 27–29.
Yano Research Institute, 'Development of Biotechnology in Japan and the Government', 1983.
Yarrow, D. 'U.S. Biotechnology in 1988 – A Review of Current Trends'. London: Department of Trade and Industry, 1988.
Yoxen, E.J. *The Gene Business: Who Should Control Biotechnology?* London: Pan 1983.
'Assessing Progress with Biotechnology'. In *Science and Technology Policy in the 1980s and Beyond,* ed. M. Gibbons, P. Gummett, and B.M. Udgaonkar, pp. 207–24. London: Longman, 1984.
Yoxen, Edward, and Di Martino, Vittorio, eds. *Biotechnology in Future Society: Scenarios and Options for Europe.* Luxembourg: Office for Official Publications of the European Communities, 1989.
Yukawa, Hideki, ed. *Profiles of Japanese Science and Scientists.* Tokyo: Kodansha, 1970.
Zilinskas, Raymond A. 'The International Centre for Genetic Engineering and Biotechnology: A New International Scientific Organization'. *Technology in Society* 9 (1987): 47–61.
Zimmerman, Horst. 'Fiscal Pressure on the "Tax State" '. In *Evolutionary Economics: Application of Schumpeter's Ideas,* ed. Horst Hanusch, pp. 255–73. Cambridge University Press, 1988.
Zweig, Stefan. *The World of Yesterday: An Autobiography.* London: Cassell, 1987.
The Zymotechnic Institute. Chicago: Zymotechnic Institute, 1891.

Index

9 780521 476997